Solid Lubrication
Fundamentals and Applications

MATERIALS ENGINEERING

1. Modern Ceramic Engineering: Properties, Processing, and Use in Design: Second Edition, Revised and Expanded, *David W. Richerson*
2. Introduction to Engineering Materials: Behavior, Properties, and Selection, *G. T. Murray*
3. Rapidly Solidified Alloys: Processes • Structures • Applications, *edited by Howard H. Liebermann*
4. Fiber and Whisker Reinforced Ceramics for Structural Applications, *David Belitskus*
5. Thermal Analysis of Materials, *Robert F. Speyer*
6. Friction and Wear of Ceramics, *edited by Said Jahanmir*
7. Mechanical Properties of Metallic Composites, *edited by Shojiro Ochiai*
8. Chemical Processing of Ceramics, *edited by Burtrand I. Lee and Edward J. A. Pope*
9. Handbook of Advanced Materials Testing, *edited by Nicholas P. Cheremisinoff and Paul N. Cheremisinoff*
10. Ceramic Processing and Sintering, *M. N. Rahaman*
11. Composites Engineering Handbook, *edited by P. K. Mallick*
12. Porosity of Ceramics, *Roy W. Rice*
13. Intermetallic and Ceramic Coatings, *edited by Narendra B. Dahotre and T. S. Sudarshan*
14. Adhesion Promotion Techniques: Technological Applications, *edited by K. L. Mittal and A. Pizzi*
15. Impurities in Engineering Materials: Impact, Reliability, and Control, *edited by Clyde L. Briant*
16. Ferroelectric Devices, *Kenji Uchino*
17. Mechanical Properties of Ceramics and Composites: Grain and Particle Effects, *Roy W. Rice*
18. Solid Lubrication Fundamentals and Applications, *Kazuhisa Miyoshi*
19. Modeling for Casting and Solidification Processing, *edited by Kuang-O (Oscar) Yu*

Additional Volumes in Preparation

Solid Lubrication
Fundamentals and Applications

Kazuhisa Miyoshi

NASA Glenn Research Center
Cleveland, Ohio

CRC Press
Taylor & Francis Group
Boca Raton   London   New York

CRC Press is an imprint of the
Taylor & Francis Group, an **informa** business

First published 2001 by Marcel Dekker, Inc.

Published 2019 by CRC Press
Taylor & Francis Group
6000 Broken Sound Parkway NW, Suite 300
Boca Raton, FL 33487-2742

© 2001 by Taylor & Francis Group, LLC
CRC Press is an imprint of Taylor & Francis Group, an Informa business

First issued in paperback 2019

No claim to original U.S. Government works

ISBN 13: 978-0-367-45511-8 (pbk)
ISBN 13: 978-0-8247-8905-3 (hbk)

Visit the Taylor & Francis Web site at
http://www.taylorandfrancis.com

and the CRC Press Web site at
http://www.crcpress.com

Preface

This book reviews studies and observations on the adhesion, friction, wear, and lubrication behavior of dry solid film lubricants and materials, including diamond and related solid films—emphasizing environmental effects and basic material properties. It also gives a reasonably concise treatment of solid lubricant applications to dry lubrication. Many examples are given throughout the book on the nature and character of solid surfaces and their significance in lubrication, friction, and wear. The ultra-high-vacuum environment and surface cleaning techniques used highlight the basic material properties that influence tribological characteristics.

Chapters 1 through 7 should be useful for those who desire a fundamental knowledge of the basic properties of dry solid film lubricants and materials, the surface effects in and characteristics of tribological phenomena, and the basic mechanisms of friction and solid lubrication. An attempt is made to review the comprehensive surface analysis and tribological characterization in this field, to identify new trends applied to controlling the adhesion, friction, wear, and durability of materials, to describe some recent developments and their industrial applications, and to give an appreciation of case studies.

Chapters 8 through 10 concern the specific subject of diamond and diamondlike solid film lubricants. A revolution in diamond technology is in progress, as the low-pressure process becomes an industrial reality. The production of large diamond films or sheets at low cost, a distinct possibility in the not-too-distant future, may drastically change tribology technology. It will soon be possible to take advantage of diamond's demanding properties to develop a myriad of new applications, particularly for dry self-lubricating, wear, protective, and superhard coatings. Diamond and diamondlike solid films may create their own revolution in the development of an entirely new branch of dry solid film lubrication.

Chapter 1 introduces the field of tribology, especially solid lubrication and lubricants, and gives the historical background, setting solid lubricants in the perspective of lubrication techniques and materials through the ages. It also defines solid and liquid lubricant films and describes their applications.

Chapter 2 describes powerful analytical techniques capable of sampling tribological surfaces and solid film lubricants. Some of these techniques can also determine the locus of failure in a bonded structure or coated substrate—information that can strengthen adhesion between a solid film lubricant and a substrate, thus improving performance and life expectancy (durability).

Chapter 3 presents the adhesion, friction, and wear behavior of smooth, atomically clean surfaces of solid-solid couples, such as metal-ceramic couples, in a clean vacuum environment. It also relates surface and bulk material properties to that behavior, emphasizing the nature and character of the metal, especially its surface energy and ductility. The friction and wear mechanisms for clean smooth surfaces are stated.

Chapter 4 presents the adhesion, friction, and wear behavior of smooth but contaminated surfaces of solid-solid couples, such as metal-ceramic couples. It describes the effects of surface contamination formed by interaction with the environment and by the diffusion of bulk compounds as well as surface chemical changes resulting from selective thermal evaporation. The primary emphasis is to relate adhesion, friction, and wear to reactions at the interface and to its composition. The friction and wear mechanisms of contaminated smooth surfaces are stated throughout the chapter.

Chapter 5 presents abrasion, a common wear phenomenon of great economic importance. It has been estimated that 50% of the wear encountered in industry is due to abrasion. Also, it is the mechanism involved in the finishing of many surfaces. Experiments are described to help in understanding the complex abrasion process and in predicting friction and wear behavior in plowing and/or cutting. These experimental modelings and measurements used a single spherical pin (asperity) and a single wedge pin (asperity). Other two-body and three-body abrasion studies used hard abrasive particles.

Chapter 6 focuses attention on the friction and wear properties of selected solid lubricating films to aid users in choosing the best lubricants, deposition conditions, and operational variables. For simplicity, the tribological properties of concern are separated into two parts. The first part of the chapter discusses several commercially developed solid film lubricants: bonded molybdenum disulfide (MoS_2), magnetron-sputtered MoS_2, ion-plated silver, ion-plated lead, magnetron-sputtered diamondlike carbon (MS DLC), and plasma-assisted, chemical-vapor-deposited diamondlike carbon (PACVD DLC) films. Marked differences in friction and wear properties resulted from the environmental conditions (ultrahigh vacuum, humid air, and dry nitrogen) and the solid film lubricant materials themselves. The second part of the chapter concentrates on magnetron-sputtered MoS_2 films, describing their physical and chemical characteristics, friction behavior, and endurance life. The roles of interface species and the effects of applied load, film thickness, oxygen pressure, environment, and temperature on the friction and wear properties are considered.

Chapter 7 relates the practice of tribology to vacuum and space technology. The conclusions may also be useful for general industrial applications. Two case studies describe aspects of real problems in sufficient detail for the reader to understand the tribological situations and wear failures. The nature of the problems is analyzed and the range of potential solutions is evaluated. Courses of action are recommended. The case studies are intended to provoke discussion rather than to reflect good or bad material selection, lubrication engineering, and tribology practice.

Chapter 8 reviews the structures and properties of natural and synthetic diamond to gain a better understanding of the tribological properties of diamond and the related materials described in Chapters 9 and 10. Atomic and crystal structure, impurities, mechanical properties, and indentation hardness of diamond are described.

Chapter 9 describes the nature of clean and contaminated diamond surfaces and the deposition technology, analytical techniques, research results, and general properties for CVD diamond films. Further, it explains the friction and wear properties of these films in the atmosphere, in a controlled nitrogen environment, and in an ultra-high-vacuum environment.

Chapter 10 looks at how the surface design and engineering of CVD diamond can lead to wear-resistant, self-lubricating films and coatings. Three studies examined the effects on the adhesion, friction, and wear behavior in ultrahigh vacuum of three materials couples: an amorphous, hydrogenated carbon film on CVD diamond; an amorphous, nondiamond carbon surface layer formed on CVD diamond by carbon and nitrogen ion implantation; and a combination of cubic boron nitride and CVD diamond. How surface modification and the selected materials couple improved the tribological functionality of coatings, giving low coefficient of friction and good wear resistance, is explained. The friction and solid lubrication mechanisms for CVD diamond are stated.

Kazuhisa Miyoshi

Acknowledgments

A number of people have through the years supported and encouraged the author with his work in the field of tribology and tribological coatings. First of all, the author would like to thank Emeritus Professor Tadasu Tsukizoe of Osaka University for originally inspiring him to enter this field, not least by Professor Tsukizoe's stimulating lectures on his statistical and mechanical explanation of the shape of surfaces; Emeritus Professor Kyuichiro Tanaka of Kanazawa University for his physical explanation of tribological phenomena; and Dr. Donald H. Buckley for his chemical explanation of surface properties and the surface science approach to tribology and his encouragement to conduct fundamental tribology.

The support of his supervisors, colleagues, and technicians at NASA Glenn Research Center through the years is gratefully acknowledged.

The author expresses his thanks to Ms. Carol Vidoli for skillful editing and text processing, and to people in the Publishing Services and Library of NASA Glenn Research Center for meticulous preparation of the text, figures, tables, and references.

Finally, the author wants to warmly thank his wife and children for their support.

Contents

Chapter 1
Introduction and Background

1.1 Definition and Scope of Tribology

Tribology is defined as "the science and technology of interacting surfaces in relative motion, and of associated subjects and practices" [1.1]. This term was introduced and defined in a report by a group set up by the British Department of Education and Science [1.2]. Tribology, having its origin in the Greek word "$\tau \rho \iota \beta o \varsigma$," meaning rubbing or attrition, deals with force transference between surfaces moving relative to each other. This discipline includes such subjects as lubrication, adhesion, friction, and wear of engineering surfaces with a view to understanding surface interactions in detail (Fig. 1.1) and then prescribing improvements in given applications.

The technical function of numerous engineering systems, such as machines, instruments, and vehicles, depends on processes of motion. According to its basic physical definition the term "motion" denotes the change in the position of an object with time. Many processes in nature and technology depend on the motion and the dynamic behavior of solids, liquids, and gases (Table 1.1, [1.3, 1.4]). For example, bearings and gears permit smooth, low-friction rotary or linear movement between two surfaces (Fig. 1.2). Bearings employ either sliding or rolling action and gears have both sliding and rolling action. In these cases a strong attempt is made to provide enough lubrication to keep the bearing and gear teeth surfaces separated by a film of solid lubricant, oil, or other lubricant such as grease. The absence of physical contact provides most bearings and gears with long service lives.

Tribology is a discipline that traditionally belongs to mechanical engineering. However, with the recent push toward higher speeds, loads, and operating temperatures, longer term life, lighter weight and smaller size, and harsh environments in mechanical, mechatronic, and biomechanical systems, the field of tribology is becoming more and more interdisciplinary, embracing physics, chemistry, metallurgy, biology, and engineering.

2

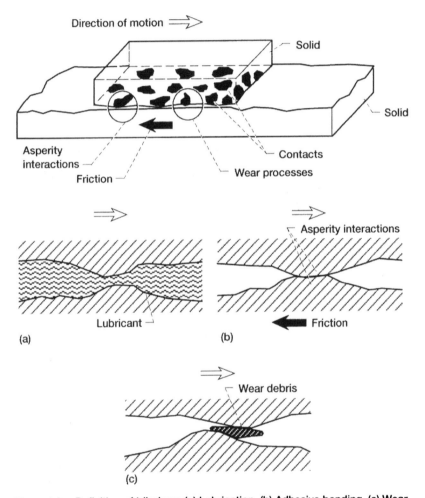

Figure 1.1.—Definition of tribology. (a) Lubrication. (b) Adhesive bonding. (c) Wear.

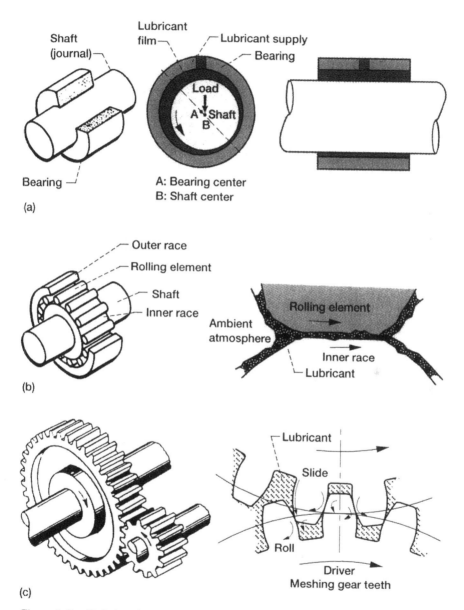

Figure 1.2.—Plain bearings, rolling-element bearings, and gears. (a) Plain bearings based on sliding action. (b) Rolling-element bearings based on rolling action. (c) Gears based on rolling and sliding action.

TABLE 1.1.—TYPES OF SURFACE MOTION AND RELATED SUBJECTS

Related subject	Schematic representation of contact and motion		
	Gas / Solid	Liquid / Solid	Solid / Solid
Lubrication	Gas or air film	Fluid film	Solid film
Resistance to motion	Gas or air friction	Viscous friction	Solid friction
Damage mechanism	Erosion; corrosion	Cavitation erosion; corrosion	Wear; erosion
Technical systems	Gas turbine; wind turbine	Hydraulic turbine	Seals; clutches; brakes; wheel and rail; tires

1.2 Social and Economic Impact of Tribology

The subject of tribology is identified as one of great importance; yet largely because of its multidisciplinary nature, it has received insufficient attention. As a direct result mechanical engineering design is retarded, and many tens of billions of dollars have unnecessarily been lost each year through friction, wear, related breakdowns, wasted energy, etc. These costs are the direct costs of friction and wear. Consideration must also be given to the indirect costs, such as loss of production, product liabilities, failure to accomplish a significant mission, or standby maintenance costs.

Table 1.2 presents some classical estimates of economic losses due to tribology and savings through tribology. The final report of the National Commission on Materials Policy to the Congress of the United States (as reported by Ballard in 1974) stated that "material losses due to tribology (friction and wear) cost the U.S. economy $100 billion per year, with a material component of this loss of about $20 billion." At that time Rabinowicz also estimated the total U.S. cost of wear to be $100 billion per year [1.5]. In 1978 the National Bureau of Standards (presently, the National Institute of Standards and Technology) estimated $70 billion for corrosion and $20 billion for wear [1.6]. Peterson estimated that the wear cost for naval aircraft and ships per year is approximately two-thirds the fuel cost [1.7]. These figures indicate that the cost of tribological losses is large and should be reduced by national efforts.

Jost, of the United Kingdom, suggested that with research efforts it is comparatively easy and inexpensive to save up to 20% of tribological losses. For example, for the United States the calculated savings would have been $12 billion to $16 billion per year in 1974 [1.8]. It is not surprising that estimates of financial savings for the United States in 1977 are significantly larger, and range from $16

TABLE 1.2.—ECONOMIC IMPACTS OF TRIBOLOGY
[Data from Jost [1.2].]

(a) Economic losses in
United States due to
inadequate tribology

Loss	Cost, $billion
Material[a]	100
Total wear[b]	100
Corrosion[c]	70
Wear[c]	20
Wear per year[d]	(e)

(b) Estimated savings reasonably obtainable by U.S.
industry through tribology [1.8]

Reduction in–	Total savings, $billion[f]
Energy consumption through lower friction Manpower Lubricant costs Maintenance and replacement costs Breakdown losses Investment costs due to higher utilization, greater mechanical efficiency, and longer machinery life	12 to 16

[a]National Commission on Materials Policy to U.S.
Congress as reported by Ballard, 1974.
[b][1.5].
[c][1.6].
[d][1.7].
[e]2/3 of fuel cost.
[f]1974 dollars.

billion to more than $40 billion per year [1.9, 1.10]. It is now believed that proper attention to tribology, especially in education, research, and application, could lead to economic savings of between 1.3 and 1.6% of the gross national product (GNP) [1.8]. Thus, tribology impacts strongly on the national economy and on the lifestyles of most people. Wear contributes to short product lives and friction contributes to energy consumption. As material and energy shortages develop, there will be greater demand for longer product lives, increased wear resistance, and reduction in energy consumption through lubrication and accordingly lower friction.

The most effective way to reduce friction and wear is to separate the two sliding surfaces by means of a lubricating film (third body), such as a film of solid lubricant, oil, grease, or gas. Elements of machines (such as plain or rolling-element bearings, slides, guides, ways, gears, cylinders, flexible couplings, chains, cams and cam followers, and wire ropes) have fitted or formed surfaces that move with respect to each other by sliding, rolling, approaching and receding, or combinations of these motions. Therefore, these elements are lubricated to prevent or reduce the actual contact between surfaces. Moving surfaces of machine elements are lubricated by interposing and maintaining films that minimize actual contact between the surfaces and that shear easily so that the frictional force opposing surface motion is low. If actual contact between surfaces occurs, high frictional forces leading to high temperatures and wear will result.

Without lubrication most machines would run for only a short time. With inadequate lubrication excessive wear is usually the most serious consequence, since a point will be reached, usually after a short period of operation, when the machine elements cannot function and the machine must be taken out of service and repaired. Repair costs (material and labor) may be high, but lost production or lost machine availability may be by far the greatest cost. With inadequate lubrication, even before elements fail, frictional forces between surfaces may be so great that drive motors will be overloaded or frictional power losses will be excessive. Finally, with inadequate lubrication machines will not run smoothly and quietly.

1.3 Historical Perspective of Tribology and Solid Lubricants

Historical factors have influenced the development of tribology, in particular solid lubricants. This brief perspective will help the reader understand the present state of the science and technology in this field. A detailed history of tribology, including lubricants and lubrication, can be found in the literature (e.g., [1.11, 1.12]).

We live in a solid world. The earth itself is solid; the stones and sands on its surface are solid; people and their tools and their machines are solid. These solids are in contact with each other. Whenever two solids touch each other so that forces of action and reaction are brought into play, the solids may be said to undergo a surface interaction [1.13]. Naturally, the history of tribology spans a period similar to that of recorded history. Important tribological developments occurred in prehistoric and early historic times.

The first civilization recorded in the history of humanity developed in the fourth millennium B.C., probably about 3500 B.C., in a territory known as Sumer adjacent to the Persian Gulf at the southern end of Mesopotamia (see the earlier chapters of [1.11]). Somewhat later, Egyptian civilization flourished. Five recorded accomplishments of the Sumerian and Egyptian civilizations (3500 B.C. to 30 B.C.) are of great tribological significance:

1. Drills employing alternating rotary motion and simple bearings were developed for making fire and drilling holes.
2. The potter's wheel, employing simple pivot bearings made from wood or stone, was produced to facilitate the throwing of clay at relatively high rotational speeds.
3. The wheeled vehicle appeared.
4. Heavy stone statues and building blocks were transported on sledges.
5. Lubricants were used in a number of applications involving rotation and translation.

Lubricants were mainly of vegetable or animal origin. A most interesting story related to the early use of lubricants comes from the building of the pyramids in the third millennium B.C. Hydrated calcium sulphate (gypsum) was used to form the thin bed of viscid mortar. The mortar acted as a lubricant to facilitate accurate setting of the huge blocks of stone. Clearly, tribology and the use of lubricants date back to the first recorded civilization. Since this early beginning lubrication and the production of lubricating media have grown to be one of the largest industries in the world, yet from one-third to one-half of all the energy produced still is lost through friction.

Although the slippery feel and appearance of graphite has been known for centuries, its use as a solid lubricant probably dates back to the Middle Ages. Graphite, also known as black lead and plumbago, was long confused with similar-appearing minerals, particularly molybdenite, and was not classified as a separate mineral until 1556. In 1779 it was proved to be carbon when it was oxidized to carbon dioxide. About 1564 the Borrowdale graphite mines in England began producing graphite for pencils. These early pencils were made by encasing slabs of cut graphite in slotted wooden dowels. The name "graphite" did not come into being until 1789, when Werner drew it from the Greek word "graphein," which means to write. The ore molybdenite was known to the early Greeks. It has often been confused with graphite and with lead. The name is derived from the Greek word meaning lead.

Traditional animal and vegetable sources satisfied the ever-increasing demand for lubricants throughout the Industrial Revolution. Table 1.3 lists a selection of the lubricants most commonly employed during the Industrial Revolution.

It is commonly thought that solid lubricants are a relatively recent phenomenon, but their use in the lubrication of heavy, slow-moving machinery was well established during the Industrial Revolution. One of the first patents, issued in 1812, describes the use of graphite, pork lard, beef suet, mutton suet, tallow, oil, goose grease, or any kind of grease or greasy substance as lubricants. Also, instructions were given on the methods of application and amounts of the composition to be used in bearings, steam-engine piston rods, and the stone spindles of

TABLE 1.3.—LUBRICANTS OF INDUSTRIAL REVOLUTION
AND THEIR APPLICATIONS

(a) Liquid lubricants

Lubricant	Application
Animal:	
Sperm oil	Lightly loaded spindles and general machinery as an excellent lubricant
Whale oil	Rare use as a lubricant
Fish oils	Little or occasional use in machinery
Lard oil	Wide use as an excellent lubricant
Neat's-foot oil	Low-temperature applications
Tallow oil	Some use as a good lubricant
Vegetable:	
Olive oil	By far the most common vegetable oil; heavy duty; fully equal to sperm oil
Rapeseed oil	Wide application
Palm oil	Limited use as a lubricant; a constitutent of a special lubricant formulation
Coconut oil	Limited application
Ground-nut oil	Similar application to olive oil
Castor oil	Use in severely loaded machinery
Mineral oil[a]	Quite small; insignificant

(b) Solid (dry) lubricants

Graphite, or plumbago	Use in heavy, slow-moving machinery
Soapstone, or talc	Lightly loaded machinery; silk looms
Molybdenum disulfide	Rotating axles

[a]By distillation or other forms or refining from crude oil, shale, coal, or wells.

mills. The first extensive technical investigation was by Rennie, who, in 1829, measured coefficients of friction with various solid materials as lubricants. In the 1800's a variety of solid lubricants were used in metalworking applications.

Molybdenum is widely distributed over the Earth's crust in the form of molybdenite. The largest commercial source of the mineral is in Climax, Colorado, where it is mined from granite containing the ore in a finely divided state. Molybdenum disulfide (MoS_2) has a metallic luster and is blue-gray to black. Early pioneers traveling through the Climax area used pulverized rock to lubricate the wheels on their Conestoga wagons. This probably was one of the first uses of MoS_2 as a solid lubricant in the United States.

The selection of liquid lubricants changed dramatically in the mid-nineteenth century when the first oil well was drilled in Titusville, Pennsylvania, in 1859. Mineral oils, which had previously been available only in relatively small volumes by distillation from shale, emerged in large quantities from flowing and pumping wells to form the major source of lubricants. The 1850's witnessed a far-reaching transition in the origin of lubricants and the start of a petroleum industry that was to support and be vital to industrial expansion in the nineteenth and twentieth centuries. The technology of lubrication advanced rapidly.

During the mid-1930's petroleum oils were improved through the use of additives, which increased their load-carrying ability, lubricating properties, corrosive protection, and thermal oxidation stability. A trend also developed that required moving parts to operate at higher and higher temperatures. Because petroleum oils could not adequately do the job at these high temperatures, synthetic lubricant materials were introduced. Temperatures now encountered in supersonic aircraft, spacecraft, and certain industrial applications are beyond the useful range of even the synthetic lubricants. This trend to the operation of bearing surfaces at higher temperatures and low pressures has led to the development and use of solid lubricants to attain the necessary lubrication. Solid lubricants have at least one very desirable feature—they do not evaporate under the aforementioned conditions.

A solid lubricant can generally be defined as a material that provides lubrication, under essentially dry conditions, to two surfaces moving relative to each other. The most common dry solid lubricants are graphite, MoS_2, tungsten disulfide (WS_2), and polytetrafluorethylene. The use of bonded, solid lubricant materials is relatively new. The first U.S. patent for a bonded material (phosphoric-acid-bonded graphite film) was issued in the mid-1940's. Several hundred patents for solid lubricating materials and binders have been issued so far. The use of molybdenum disulfide as a lubricating solid also began in the 1940's, and MoS_2 is now used in more applications than any other lubricating solid.

On April 6, 1938, Plunkett was investigating the results of a failed experiment involving refrigeration gases when he found a white, waxy substance. The material, polytetrafluoroethylene (PTFE), commonly known as Teflon, has proved inert to virtually all chemicals. It also is one of the most slippery dry materials known.

The study and application of solid lubricants, as they are now known, is a relatively new field. No systematic study of these materials was begun until long after they were introduced in the aircraft industry. In the 1950's, with the development of the jet engine, a number of research laboratories began a systematic study of solid lubrication for high temperatures. Most of the work was directed toward defining the required characteristics of solid lubricants. In the 1960's space lubrication needs prompted increased research into solid lubrication with emphasis on the role of atmosphere. Ways of using solid lubricants were explored. By the early 1970's, when many of the problems had been resolved and their limitations defined, most of the research stopped.

Recently, however, a number of new applications have arisen that have prompted renewed interest. These applications are piston rings for low-heat-rejection engines, lubricating cages for advanced gas turbines, gears and bearings for long-term service in space mechanisms, cages for turbopump bearings operating in liquid hydrogen and oxygen, lightweight gear and bearing systems, and low-cost bearing systems for automobiles and industrial machinery. The new requirements are primarily long-term life and broad-temperature-range capability. New solid lubricants are needed that meet these requirements.

Lastly, although the importance of friction and resistance to motion has no doubt been recognized throughout the ages, a full appreciation of the significance of tribology in a technological society is a recent phenomenon.

1.4 Description of Solid and Liquid Lubrication

Lubricating films are classified as three types: solid films, fluid films, and thin films (Table 1.4). They are described briefly here, but more details can be found in the literature (e.g., [1.14–1.16]).

1.4.1 Solid Films

A solid lubricant is any material used as a thin film or a powder on a surface to provide protection from damage during relative movement and to reduce friction and wear. Solid lubrication is achieved by self-lubricating solids or by imposing a solid material having low shear strength and high wear resistance between the interacting surfaces in relative motion. The solid material may be a dispersion in oils and greases, a loose powder, or a coating.

Solid lubricants are used when liquid lubricants do not meet the advanced requirements of modern technology. They are less expensive than oil and grease lubrication systems for many applications. Solid lubricants also reduce weight, simplify lubrication, and improve materials and processes. Figure 1.3 and Table 1.5 [1.17, 1.18] list applications needed to meet critical operating conditions for which fluid lubricants are ineffective or undesirable. Changes in critical evironmental conditions, such as pressure, temperature, and radiation, affect lubricant efficiency. Further, in the cost-conscious automotive industry, solid lubricants are replacing oils and greases in many applications and helping to make highly efficient automobiles possible.

Oils or greases cannot be used in many applications because of the difficulty in applying them, sealing problems, weight, or other factors, such as environmental conditions. Solid lubricants may be preferred to liquid or gas films for several reasons. In high-vacuum environments, in space-vacuum environments, or in food-processing machines, a liquid lubricant would evaporate and contaminate the product, such as optical and electronic equipment or food. At high temperatures liquid lubricants decompose or oxidize; suitable solid lubricants can extend the operating temperatures of sliding systems beyond 250 or 300 °C while maintaining relatively low coefficients of friction. At cryogenic temperatures liquid lubricants are highly viscous and are not effective. Under radiation or corrosive environments liquid lubricants decompose or will be contaminated.

TABLE 1.4.—LUBRICATING FILMS

(a) Types

Type	Lubricating films
Solid films	Lamellar film (MoS_2 and graphite) Nonmetallic film (titanium dioxide, calcium fluoride, glasses, lead oxide, zinc oxide, and tin oxide) Soft metallic film (lead, gold, silver, indium, and zinc) Lamellar carbon compound film (graphite and graphite fluoride) Diamond and diamondlike carbon (diamond, i-carbon, a-carbon, hydrogen, carbon nitride, and boron nitride) Fats, soap, wax (stearic acid) Polymers (PTFE, nylon, and polyethylene)
Fluid films	Hydrodynamic film: Thick hydrodynamic film Elastohydrodynamic film Hydrostatic film Squeeze film
Thin films	Mixed lubricating film Boundary lubricating film

(b) Terms and definitions

Term	Definition
Solid film	Solid films are more or less permanently bonded onto the moving surfaces.
Fluid film	Fluid films are thick enough that during normal operation they completely separate surfaces moving relative to each other.
•Hydrodynamic film	Hydrodynamic films are formed by motion of lubricated surfaces through a convergent zone such that sufficient pressure is developed in the film to maintain separation of the surfaces.
•Hydrostatic film	Hydrostatic films are formed by pumping fluid under pressure between surfaces that may or may not be moving with respect to each other.
•Squeeze film	Squeeze films are formed by movement of lubricated surfaces toward each other.
Thin film	Thin films are not thick enough to maintain complete separation of the surfaces all the time.

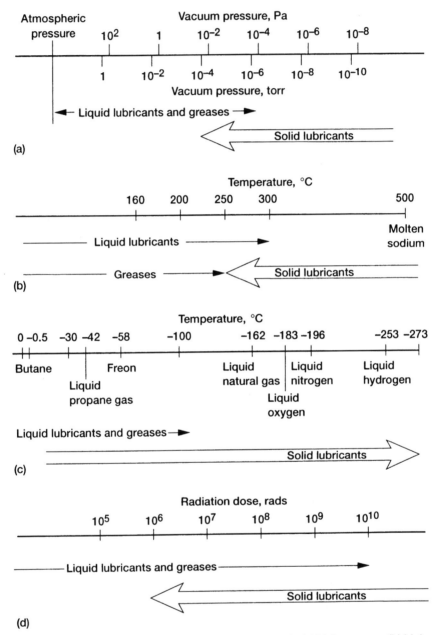

Figure 1.3.—Ranges of application of solid lubricants in (a) high vacuum, (b) high temperature, (c) cryogenic temperature, and (d) radiation environments.

TABLE 1.5.—APPLICATION OF SOLID LUBRICANTS

(a) Areas where fluid lubricants are undesirable

Requirement	Applications
Avoid contaminating product or environment	Food-processing machines Optical equipment Space telescopes Metalworking equipment Surface-mount equipment Tape recorders Microscopes and cameras Textile equipment Paper-processing machines Business machines Automobiles Medical and dental equipment Spectroscopes
Maintain servicing or lubrication in inaccessible or unlikely areas	Aircraft Space vehicles Satellites Aerospace mechanisms Nuclear reactors Consumer durables
Resist abrasion in dirt-laden environments	Aircraft Space vehicles (rovers) Automobiles Agricultural and mining equipment Off-road vehicles and equipment Construction equipment Textile equipment
Provide prolonged storage or stationary service	Aircraft equipment Railway equipment Missile components Nuclear reactors Telescope mounts Heavy plants, buildings, and bridges Furnaces

Further, in the weight-conscious aerospace industry, solid lubricants lead to substantial weight savings relative to the use of liquid lubricants. The elimination (or limited use) of liquid lubricants and their replacement by solid lubricants lessen aircraft or spacecraft weight and therefore have a dramatic impact on mission extent and craft maneuverability. Under high vacuums, high temperatures, cryogenic temperatures, radiation, or space or corrosive environments, solid lubrication may be the only feasible system.

TABLE 1.5.—Continued

(b) Areas where fluid lubricants are ineffective

Environment		Applications
High vacuum	Room temperature or cryogenic temperatures	Vacuum products Space mechanisms Satellites Space telescope mounts Space platforms Space antennae
	Clean room	Semiconductor manu- facturing equipment
	High temperature	X-ray tubes X-ray equipment Furnaces
High temperatures	Air atmosphere	Furnaces Metalworking equipment Compressors
	Molten metals (sodium, zinc, etc.)	Nuclear reactors Molten metal plating equipment
Cryogenic temperatures		Space mechanisms Satellites Space vehicles Space propulsion systems Space telescope mounts Space platforms Space antennae Turbopumps Liquid nitrogen pumps Butane pumps Freon pumps Liquid natural gas pumps Liquid propane pumps Refrigeration plants

Numerous solid lubricants, such as permanently bonded lubricating films, have been developed to reduce friction and wear in applications of this type where fluid lubricants are ineffective and undesirable. The simplest kind of solid lubricating film is formed when a low-friction solid lubricant, such as MoS_2, is suspended in a carrier and applied to the surface like a normal lubricant. The carrier may be a volatile solvent, a grease, or any of several other types of material. After the carrier is squeezed out or evaporates from the surfaces, a layer of MoS_2 provides lubrication.

TABLE 1.5.—Concluded.

(b) Concluded.

Radiation (gamma rays, fast neutrons, x rays, beta rays, etc.)	Nuclear reactors Space mechanisms Satellites Space vehicles Space platforms Space antennae
Corrosive gases (chlorine, etc.)	Semiconductor manufacturing equipment
High pressures or loads	Metalworking equipment Bridge supports Plant supports Building supports
Fretting corrosion (general)	Aircraft engines Turbines Landing gear Automobiles

Solid lubricants are also bonded to rubbing surfaces with various types of resin, which cure to form strongly adhering coatings with good frictional properties. In some plastic bearings the solid lubricant is sometimes incorporated into the plastic. During operation some of the solid lubricant may be transferred to form a lubricating coating on the mating surface.

In addition to MoS_2, PTFE, polyethylene, and a number of other materials are used to form solid films. Sometimes, combinations of several materials, each contributing specific properties to the film, are used.

Because of recent innovations in the physical and chemical vapor deposition processes, solid lubricating materials, such as MoS_2, WS_2, diamond, and PTFE films, are grown economically on ceramics, polymers, and metals and used as solid lubricating films.

1.4.2 Fluid Films

Fluid film lubrication is the most desirable form of lubrication, since during normal operation the films are thick enough to completely separate the load-carrying surfaces. Thus, friction is at a practical minimum, being due only to shearing of the liquid lubricant films; and wear does not occur, since there is essentially no mechanical contact. Fluid films are formed in three ways: hydrodynamically, hydrostatically, and by squeezing (Table 1.4(b)).

Hydrodynamic films.—The most effective way to separate two sliding surfaces by means of a fluid and to reduce friction and wear is known as hydrodynamic lubrication. It provides coefficients of friction on the order of 0.003, or less, depending on the sliding velocity, load, and fluid viscosity. It eliminates wear entirely, since the solids do not touch or collide with each other. Gyroscope bearings are one example where the ideal conditions of hydrodynamic lubrication are substantially achieved. Two types of hydrodynamic film lubrication are now recognized: thick hydrodynamic films and elastohydrodynamic films (Table 1.4(a)).

Plain journal bearings and tilting-pad or tapered-land thrust bearings (Figs. 1.2(a) and 1.4) have thick hydrodynamic films, usually more than 25 μm thick. In applications the loads are low enough, and the areas over which the loads are distributed are large enough, that the load-carrying area does not deform enough to significantly alter that area. Load-carrying surfaces of this type are often referred to as "conforming," although it is obvious that in tapered-land thrust bearings, for example, the surfaces do not conform in the normal concept of the word. However, the term is a convenient opposite for the term "nonconforming," which quite accurately describes the types of surface where elastohydrodynamic films are formed.

The surfaces of the balls in a ball bearing theoretically make contact with the raceways at points; the rollers in a roller bearing make contact with the raceways along lines; and meshing gear teeth also make contact along lines (e.g., Figs. 1.2(b) and (c)). These types of surface are nonconforming. Under the pressures applied to these elements by the lubricating film, however, the metals deform elastically,

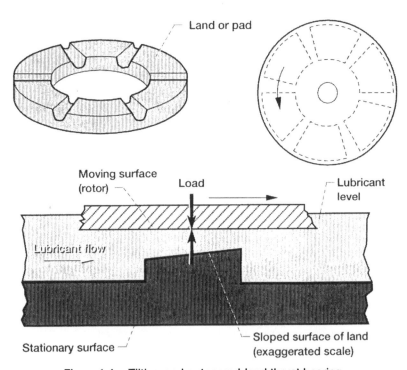

Figure 1.4.—Tilting-pad or tapered-land thrust bearing.

expanding the theoretical points or lines of contact into discrete areas. Since a convergent zone exists immediately before these areas of contact, a lubricant will be drawn into the contact area and can form a hydrodynamic film. This type of film is referred to as an "elastohydrodynamic film." The "elasto" part of the term refers to the fact that elastic surface deformation must occur before the film can be formed. This type of lubrication is elastohydrodynamic lubrication (EHL), a condition of lubrication in which the friction and film thickness between two surfaces in relative motion are determined by the elastic properties of the surfaces in combination with the viscous properties of the lubricant. The viscous properties include variation of viscosity with pressure, temperature, and shear rate. EHL films are very thin, on the order of 0.25 to 1.25 μm thick. However, even with these thin films, complete separation of the contacting surfaces can be obtained.

Any material that will flow at the shear stresses available in the system may be used for fluid film lubrication. In most applications petroleum-derived lubricating oils are used. There are some applications for greases. Some materials not usually considered to be lubricants, such as liquid metals, water, and gases, are also used. For example, magnetic recording is accomplished by relative motion between a magnetic medium and a magnetic head (Fig. 1.5); under steady operating conditions a load-carrying hydrodynamic gas (air) film is formed [1.19].

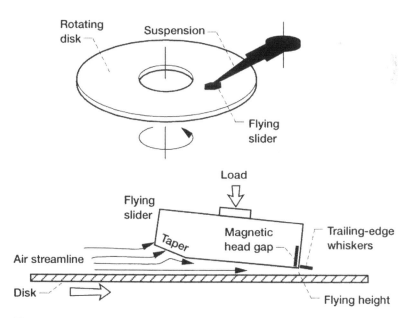

Figure 1.5.—Hydrodynamic gas (air) film lubrication in magnetic hard-disk drive system.

Ideal conditions of hydrodynamic lubrication can rarely be maintained in practice. Starting, stopping, misalignment, heavy loads, and other service conditions can cause the fluid film to be squeezed out or allow the surface asperities to break through the film, so that the two solids are pressed into contact with one another. Even in modern high-end computer tape and disk drives, there is physical contact between the medium and the head during starting and stopping. Ideal hydrodynamic lubrication then ends, and boundary lubrication, or lubrication by solids, begins.

Hydrostatic films.—In hydrostatic film lubrication the pressure in the fluid film that lifts and supports the load is provided from an external source. Thus, relative motion between opposing surfaces is not required to create and maintain the fluid film. The principle is used in plain and flat bearings of various types, where it offers low friction (at very low speeds or when there is no relative motion), more accurate centering of a journal in its bearing, and freedom from stick-slip effects.

Figure 1.6 illustrates the simplest type of hydrostatic bearing. Oil under pressure is supplied to the recess or pocket. If the supply pressure is sufficient, the load will be lifted and floated on a fluid film. The total force developed by the pressure in the pocket and across the lands will be such that the total upward force is equal to the applied load. The clearance space and the oil film thickness will be such that all the oil supplied to the bearing can flow through the clearance spaces under the pressure conditions prevailing.

Squeeze films.—As the applied pressure on an oil increases, its viscosity increases. This fact contributes to the formation of what are called squeeze films [1.14]. Figure 1.7(a) shows the principle of the squeeze film, where application of

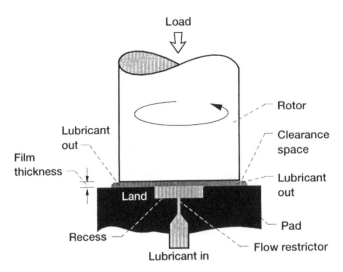

Figure 1.6.—Simple hydrostatic thrust bearing.

a load causes plate A to move toward stationary plate B. As pressure develops in the oil layer, the oil starts to flow away from the area. However, the increase in pressure also increases the oil viscosity, so that the oil cannot escape as rapidly and a heavy load can be supported for a short time. Sooner or later, if load continues to be applied, all the oil will flow or be forced from between the surfaces and metal-to-metal contact will occur, but for short periods such a lubricating film can support very heavy loads.

One application where squeeze films are formed is in piston pin bushings (Fig. 1.7(b)). At the left the load is downward on the pin and the squeeze film develops at the bottom. Before the film is squeezed so thin that contact can occur, the load reverses (right view) and the squeeze film develops at the top. The

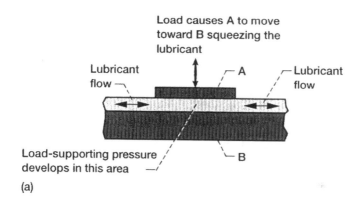

(a)

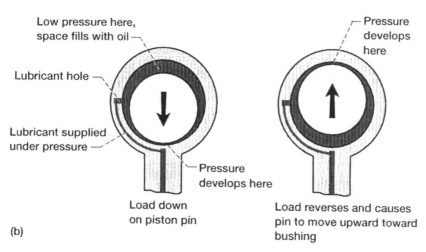

(b)

Figure 1.7.—Squeeze film bearing. (a) Squeeze film principle. (b) Squeeze film in piston pin bushing.

bearing oscillates with respect to the pin, but this motion probably does not contribute much to film formation by hydrodynamic action. Nevertheless, bearings of this type have high load-carrying capacity.

Effect of viscosity, speed, and load.—The oil film (wedge) formed in a hydrodynamic bearing is a function of speed, load, and oil viscosity. Under fluid film conditions an increase in viscosity or speed increases the oil film thickness and the coefficient of friction; an increase in load decreases them. It is now generally accepted that the coefficient of friction μ can be shown by a curve such as that in Fig. 1.8. The coefficient of friction is plotted as a function of a single dimensionless factor, viscosity times velocity divided by load [1.3, 1.14]. A similar type of curve could be developed experimentally for any fluid film bearing. The curve is also called the Stribeck curve. The accurate experimental measurements of Stribeck from 1900 to 1920 served as a basis for the theoretical work of many researchers in establishing the theory of hydrodynamically lubricated bearings.

In Fig. 1.8 three main lubrication regimes may be distinguished:

1. Fluid film lubrication (thick hydrodynamic lubrication, elastohydrodynamic lubrication, etc.) exists in the zone to the right of c.
2. Mixed film lubrication or partial elastohydrodynamic lubrication (EHL) exists in the portion of the curve between a and c, including the minimum value of μ to the dimensionless value indicated by b.

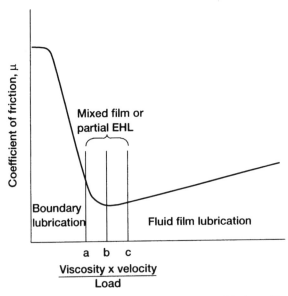

Figure 1.8.—Variation in coefficient of friction with viscosity, velocity, and load for fluid-lubricated sliding bearing (the Stribeck curve).

3. Boundary lubrication exists to the left of a, where conditions are such that a full fluid film cannot be formed, some friction and wear commonly occur, and very high coefficients of friction may be reached.

Mixed film lubrication and boundary lubrication are discussed next.

1.4.3 Thin Films

A copious, continuous supply of liquid lubricant is necessary to maintain fluid films. In many cases it is not practical or possible to provide such an amount of lubricant to machine elements. In other cases, as for example during starting of a hydrodynamic film bearing, loads and speeds are such that fluid films cannot be maintained. Under these conditions lubrication is by what are called thin films.

When surfaces run together under thin film conditions, enough oil is often present so that part of the load is carried by fluid films and part is carried by contact between the surfaces. This condition is often called mixed film lubrication. With less oil present, or with higher loads, a point is reached where fluid oil plays little or no part. This condition is often called boundary lubrication, the condition of lubrication in which the friction and wear between two surfaces in relative motion are determined by the properties of the surfaces and by the properties of the lubricant other than bulk viscosity. Many circumstances that are referred to as "boundary lubrication" are in fact elastohydrodynamic.

When rubbing (sliding) contact is made between the surface peaks, known as asperities, a number of actions take place, as shown in Fig. 1.9, which represents a highly magnified contact area of two surfaces:

1. At locations a in Fig. 1.9, sliding surfaces completely separate and thick films are formed. Friction there is due only to shearing of the liquid lubricant films.
2. There is heavy rubbing at location b, as surface films are sheared, and elastic or plastic deformation occurs. Real (in contrast to apparent) areas of contact are extremely small and unit stresses are very high.

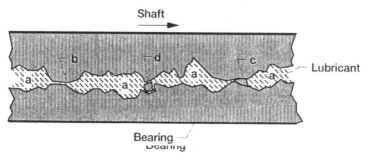

Figure 1.9.—Actions involved in boundary lubrication.

3. As some areas at location c are rubbed or sheared, the clean surfaces weld together. The minute welds break as motion continues. But depending on their strength, the welds may break at another section so that metal is transferred from one member to the other.
4. The harder shaft material plows through the softer bearing material at location d, breaking off wear particles and creating new roughnesses.

These actions account for friction and wear in boundary lubrication.

References

1.1 *Glossary of Terms in the Field of Friction, Wear, and Lubrication*, Research Group on Wear of Engineering Materials, Organization for Economic Cooperation and Development (OECD), Paris, 1969.

1.2 H.P. Jost, *Lubrication (Tribology) Education and Research—A Report on the Present Position and Industry's Needs*, Her Majesty's Stationery Office, London, 1966.

1.3 H. Czichos, *Tribology—A Systems Approach to the Science and Technology of Friction, Lubrication, and Wear*, Elsevier Scientific Publishing Co., New York, Vol. 1, 1978.

1.4 M.B. Peterson, *Wear Control Handbook* (M.B. Peterson and W.O. Winer, eds.), American Society of Mechanical Engineers, New York, 1980.

1.5 M.J. Devine, *Proceedings of a Workshop on Wear Control To Achieve Product Durability*, AD–A055712, Naval Air Development Center, Warminster, PA, 1976.

1.6 L.H. Bennet, et al., *Economic Effects of Metallic Corrosion in the United States*, National Bureau of Standards SP–511–1–PT–1, 1978.

1.7 M.B. Peterson, *Technical Options for Conservation of Metals:Case Studies of Selected Metals and Products*, OTA–M–97, 1979.

1.8 H.P. Jost, Tribology—origin and future, *Wear, 136*: 1–17 (1990).

1.9 O. Pinkus and D.F. Wilcock, *Strategy for Energy Conservation Through Tribology*, American Society of Mechanical Engineers, New York, 1977.

1.10 P.M. Ku, Energy and materials conservation through tribology, *Lubr. Eng.: 34*, 2: 131–134 (1978).

1.11 D. Dowson, *History of Tribology*, Longman, London and New York, 1979.

1.12 M.E. Campbell, *Solid Lubricants—A Survey*, NASA SP–5059(01), 1972.

1.13 E. Rabinowicz, *Friction and Wear of Materials*, John Wiley & Sons, New York, 1965.

1.14 J.G. Wills, *Lubrication Fundamentals*, Marcel Dekker, New York, 1980.

1.15 E.R. Booser, ed., *CRC Handbook of Lubrication—Theory and Practice of Tribology*, CRC Press, Boca Raton, FL, Vols. I and II, 1984.

1.16 J.J. O'Connor, J. Boyd, and E.A. Avallone, eds., *Standard Handbook of Lubrication Engineering*, McGraw-Hill, New York, 1968.

1.17 K. Kakuda, ed., NSK Technical Journal 648, Nippon Seiko Co., Tokyo, 1988.

1.18 J.K. Lancaster, Solid lubricants, *CRC Handbook of Lubrication—Theory and Practice of Tribology* (E.R. Booser, ed.), CRC Press, Boca Raton, FL, Vol. II, 1984, pp. 269–290.

1.19 B. Bhushnan, ed., *Handbook of Micro/Nano Tribology*, CRC Press, Boca Raton, FL, 1995.

Chapter 2
Characterization of Solid Surfaces

2.1 Introduction

Materials and surface analysis has been practised for a relatively long time and has been a fixture of the field of tribology from the beginning [2.1–2.11]. Fifty or so material and surface analysis techniques exist, yet analysis equipment and techniques are undergoing rapid and constant development in response to user needs. Each technique provides specific measurement results in its own unique way. Interpreting those results to extract information that can help tribologists and lubrication engineers solve technical problems takes years of training. Reaching the level of senior analyst usually takes three to five years.

Materials and surface analysis has made considerable contributions, ranging from fundamental tribology to the design, development, evaluation, and quality control of materials and solid film lubricants. The principal desiderata of a solid film lubricant for technical applications, especially aerospace, automotive, and other demanding applications, are low shear strength yet high cohesion and adhesion to the substrate, long-term stability under given operating environmental conditions, and long-term compatibility with the substrate. To understand the benefits provided by solid lubricants, and ultimately to provide better solid lubrication, it is necessary to characterize the physical, chemical, topographical, mechanical, and tribological properties of solid lubricants and substrates. These properties must not be seriously affected by environmental conditions, in particular elevated temperatures, vacuum, ions, radiation, gases, particulates, and water.

Although it is now not always the case, characterization problems must be approached from the end of the tribologists, materials users, and lubrication engineers rather than from that of an analytical technique expert. Currently, materials and surface analytical techniques are used mostly either in quality assurance and control of parts and components used in tribology or in investigation of the mechanisms of friction, lubrication, wear, and tribological failures.

This chapter describes powerful analytical techniques capable of sampling tribological surfaces and solid film lubricants. Some of these techniques may also be used to determine the locus of failure in a bonded structure or coated substrate; such information is important when seeking improved adhesion between a solid film lubricant and a substrate and when seeking improved performance and long life expectancy of solid lubricants. Many examples are given here and throughout the book on the nature and character of solid surfaces and their significance in lubrication, friction, and wear. The analytical techniques used include the latest spectroscopic methods.

2.2 Nature and Structure of Real Surfaces

A surface, by definition, is an interface, a marked discontinuity from one material to another. Because no change in nature is ever instantaneous, any real surface has a finite depth, and in characterizing a surface one must at some point consider just what this depth is.

Almost every industrial process involves—even depends on—the behavior of a surface. The surface region, which affects a broad spectrum of properties such as surface energy and tension, adhesion, bonding, friction, lubrication, wear, contamination, oxidation, corrosion, chemical activity, deformation, and fracture, encompasses the first few hundred atomic layers. All surfaces obey the laws of physics and chemistry in their formation, reactions, and combinations. Because any major discontinuity in the solid affects the electronic energy states, effects also arise from surface energy and tension.

Figure 2.1 depicts the surface structure and chemistry of a solid film lubricant exposed to a surrounding environment and to a bulk substrate material. The solid-film-lubricated surface region includes (1) the single-crystal bulk substrate with defects or the polycrystalline bulk substrate with grain boundaries, (2) a worked layer and an oxide layer, (3) the solid film lubricant, (4) another oxide layer covered by adsorbed contaminants, and (5) the surrounding environmental species. To understand this surface region, it is first necessary to know how the individual parts (1) to (4) perform in the tribological process. Thus, the nature and properties of parts (1), (2), and (4) are discussed, using examples, in this section.

2.2.1 Bulk Materials

Almost all surfaces of bulk substrate materials that are prepared by mechanical preparation techniques contain defects resulting from plastic deformation, fracture, heating, and contamination. Even cleavage faces are rarely defect free.

Metals and polymers readily deform plastically. Ceramics, while having high strength and high elastic modulus, are normally brittle and fracture with little or no evidence of plastic flow. However, plastic flow has been observed in the surface

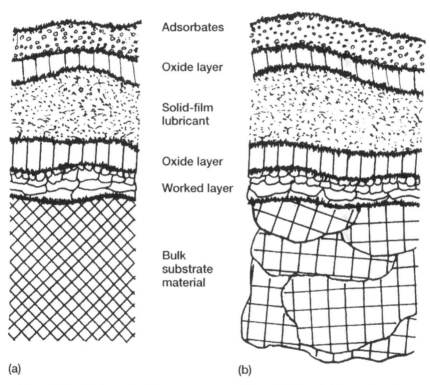

Environment:
gas, vacuum,
liquid, and/or
solid

Adsorbates

Oxide layer

Solid-film
lubricant

Oxide layer

Worked layer

Bulk
substrate
material

(a)

(b)

Figure 2.1.—Schematic diagram representing solid-film-lubricated surface
region. (a) Single-crystal substrate. (b) Polycrystalline substrate.

layers of several ceramics in solid-state contact under load and relative motion. These include magnesium oxide [2.12], aluminum oxide [2.13], manganese-zinc ferrites [2.14], and silicon carbide [2.15] under relatively modest conditions of rubbing contact.

Single-Crystal Materials

Properties of a single crystal, such as atomic density, spacing of atomic planes, surface energy, modulus of elasticity, slip systems, influence of imperfections, deformation, fracture, hardness, coefficient of friction, and wear rate, have all been related to crystal orientation. Some examples of the tribological behavior and properties of single crystals are discussed here.

The simplest type of solid surface is a cleavage face of a single crystal. This surface, a regular array of atoms, is flat and smooth. As an example, Fig. 2.2 presents the bulk microstructure and crystallinity of cleaved single-crystal molybdenite. The transmission electron micrograph and transmission electron diffraction pattern reveal homogeneous crystalline and regularly arrayed diffraction spots, indicating a single-crystal structure.

Indentation deformation.—Figure 2.3 shows the distribution of dislocation etch pits on a well-defined single-crystal magnesium oxide (MgO) surface [2.16–2.18]. The MgO bulk crystals were first cleaved along the {001} surface in air and then subjected to hardness indentation or cavitation erosion, which introduced a certain amount of plastic deformation into the {001} surface. Next, the MgO surfaces were chemically etched in a solution of five parts saturated ammonium chloride, one part sulfuric acid, and one part distilled water at room temperature. Then scanning electron micrographs were taken of the etched surfaces. The dislocation-etch-pit pattern on the surface exposed to cavitation (Fig. 2.3(b)) is definitely different from that on the indented surface (Fig. 2.3(a)).

In Fig. 2.3(a) the dislocation-etch-pit pattern around the hardness indentation on the {001} surface contains screw dislocations in the $\langle 001 \rangle$ directions ([100] and [010]) and edge dislocations in the $\langle 011 \rangle$ directions ([110] and [1̄10]). The screw and edge dislocation arrays are 4.9 and 7.7 times wider, respectively, than the average length of the two diagonals of hardness indentation. On the other hand, a cross-shaped deformation, which contains no edge dislocations, can be seen in Fig. 2.3(b). Similar observations were reported by J. Narayan [2.19, 2.20], who used optical and transmission electron microscopy to investigate the plastic damage introduced into MgO single crystals by bombardment with aluminum oxide (Al_2O_3) particles. He indicated that most of the damage was dislocation dipoles and that, because the resolved stress on $\{110\}_{90°}$ was small, the dislocation-etch-pit pattern was elongated along the $\langle 110 \rangle$ directions. His results are consistent with the results presented in Fig. 2.3(b).

After an MgO {001} surface was exposed to cavitation erosion, the MgO bulk crystal was cleaved along a {001} surface perpendicular to the {001} surface exposed

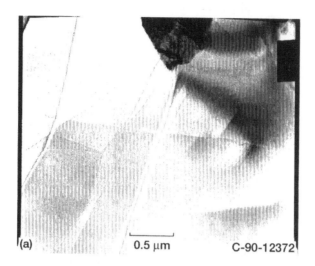

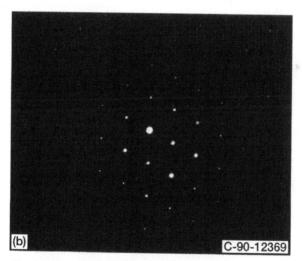

Figure 2.2.—Single-crystal molybdenite. (a) Transmission electron micrograph. (b) Transmission electron diffraction pattern.

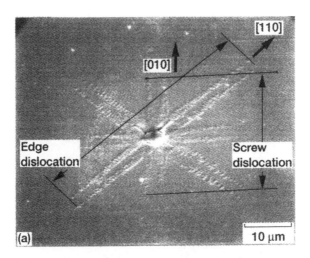

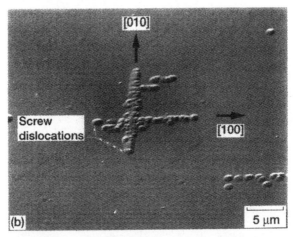

Figure 2.3.—Scanning electron micrographs showing distribution of dislocation etch pits on {001}
MgO surface. (a) Around Vickers indentation
made by diamond indenter at load of 0.1 N.
(b) After exposure to cavitation for 10 s at
20 kHz (double amplitude, 50 mm) with 2 mm
between vibrating disk and specimen in distilled
water.

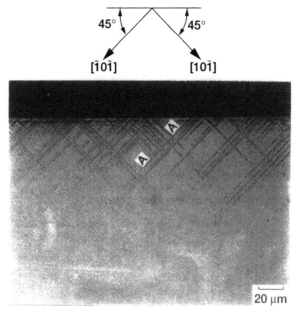

Figure 2.4.—Optical micrograph of dislocation-
etch-pit patterns in cross section of MgO {001}
surface after exposure to cavitation.

to cavitation. And finally all surfaces of the MgO crystal were chemically etched. In Fig. 2.4, which shows a cross section of the cavitation-damaged area, the dislocation rows intersect the specimen surface at 45° and cross the dislocation rows of the other set of $\{110\}_{45°}$ planes.

Explanations for the slip behavior of MgO are known [2.18–2.20]. MgO is highly ionic and slip occurs not on the close-packed {001} cubic planes but on the {011} cubic diagonals. MgO has six easy-slip planes, as shown in Fig. 2.5. In Fig. 2.5(a) two sets of {110} planes intersect the top surface at 45° (to be referred to as $\{110\}_{45°}$ planes). The dislocation lines on these planes that emerge on the top surface are screw dislocations because they lie parallel to their glide directions. In Fig. 2.5(b) the {110} planes intersect the top surface at 90° (to be referred to as $\{110\}_{90°}$ planes). The dislocation lines on these planes lie perpendicular to their glide directions and therefore are edge dislocations.

Figure 2.6 shows the length of the dislocation rows and the length of the diagonal of indentation as functions of load on a log-log scale. As expected, the gradient of the diagonal length is approximately 0.5 because the Vickers hardness is independent of indentation load. Almost the same gradient is shown for the length of edge dislocations. However, the gradient for the screw dislocations is slightly smaller, possibly because cross slips occur easily at higher loads. The row of edge dislocations is always longer than that of screw dislocations for the hardness indentations.

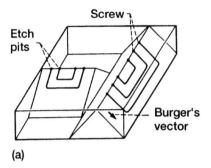

(a)

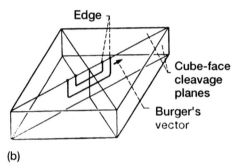

(b)

Figure 2.5.—Slip systems of MgO single crystal.
(a) $\{110\}_{45°}$ slip planes. (b) $\{110\}_{90°}$ slip planes.

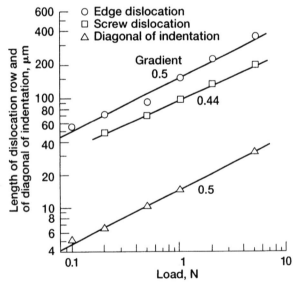

Figure 2.6.—Lengths of dislocation row and
diagonal of indentation as function of load.

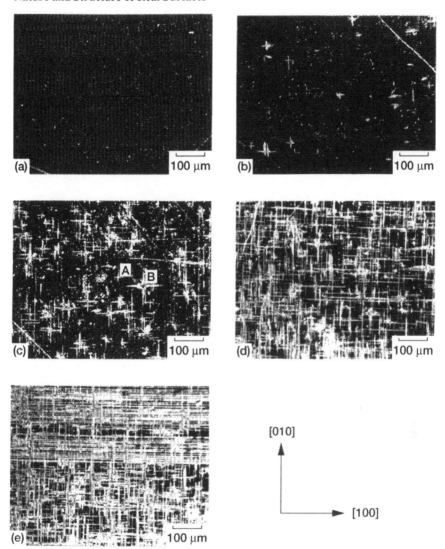

Figure 2.7.—Dark-field micrographs of dislocation-etch-pit patterns on
M$_g$O {001} surface after exposure to cavitation. (a) New, cleaved surface.
(b) After 10 s. (c) After 30 s. (d) After 60 s. (e) After 180 s.

Figure 2.7 shows dark-field optical micrographs of the MgO {001} surface after
exposure to cavitation for 10 to 180 s. The specimen surface exposed for 10 s
(Fig. 2.7(b)) has a number of cross-shaped damage areas with independent
nuclei. The specimen surface exposed for 30 s (Fig. 2.7(c)) has both more and
larger cross-shaped deformations (see a newly formed deformation at point A and

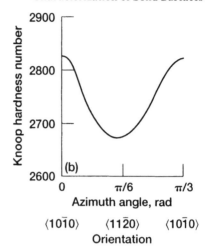

Figure 2.8.—Indentation hardness anisotropy of SiC {0001} basal plane.
Measuring load, 2.9 N. (a) Back-reflection Laue photograph. (b) Knoop
hardness as function of angle between long axis of Knoop indenter and
⟨10$\bar{1}$0⟩ direction axis.

a large deformation at point B). The specimen surfaces exposed to cavitation for
60 and 180 s (Figs. 2.7(d) and (e), respectively) are fully covered with large cross-
shaped deformations. The deformation pattern for 180 s is especially netlike, with
wide dislocation rows. Thus, deformation damage along a preferred orientation
accumulates with increased exposure time.

Indentation hardness anisotropy.—As hardness is a conventionally used
parameter for indicating the abrasion resistance of tribomaterials, it is useful and
important to consider the hardness anisotropy. Figure 2.8, as an example, shows
a back-reflection Laue photograph and the results of Knoop hardness experiments
made on the {0001} plane of single-crystal silicon carbide (SiC). The specimen is
within ±2° of the low index {0001} plane. The Knoop hardness is presented as a
function of the orientation of the long axis of the Knoop diamond indenter with
respect to the ⟨10$\bar{1}$0⟩ direction at 10° intervals. The hardness decreases smoothly
to a minimum value of about 2670 at a location near 30° from the ⟨10$\bar{1}$0⟩ direction.
Figure 2.8 indicates that the ⟨10$\bar{1}$0⟩ direction is the maximum hardness
direction and that the maximum hardness is about 2830. The hardness results are
consistent with those of other workers [2.21, 2.22].

Friction anisotropy in abrasion.—If a hard material is brought into contact with
a softer single crystal, the plastic deformation and accordingly the coefficient of
friction of the single crystal are anisotropic and relate to the crystal structure. In
crystals of comparable purity and crystallographic perfection, slip always begins
when the shear stress across the slip planes reaches a certain definite value known
as the critical resolved shear stress. The actual stress required to start
deformation depends on the orientation of the slip planes relative to the applied

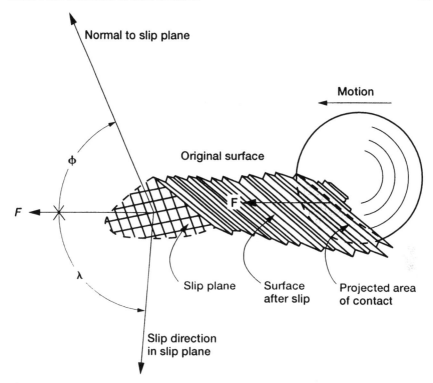

Figure 2.9.—Application of friction force to single crystal, causing slip to occur in definite direction. (The angle ϕ is measured between the normal to the slip plane and the axis of friction.)

stress. If the crystal is plowed or cut by a force F as in Fig. 2.9, the resolved stress on the slip plane in the slip direction is

$$\sigma = \frac{F}{A}\cos\phi\cos\lambda \tag{2.1}$$

where A is the projected area of contact and $\cos\phi\cos\lambda$ is the Schmid factor [2.23, 2.24]. The Schmid factor m is determined by the angles that the direction of the applied stress makes with the slip direction λ and with the normal to the slip plane ϕ. Therefore, the friction force F is

$$F = \frac{\sigma A}{m} \tag{2.2}$$

If the slip planes are nearly perpendicular to the sliding direction, ϕ approaches zero and a large force is required to initiate slip. This situation can arise in crystals of hexagonal-close-packed (hcp) metals and ceramics, such as zinc, magnesium, and Al_2O_3, where there is only one set of slip planes. In face-centered cubic (fcc) crystals there are four differently oriented sets of {111} slip planes so that, regardless of crystal axis orientation, there will be at least one set of slip planes more or less favorably oriented for slip.

When deformation occurs on a single set of parallel slip planes, it can continue until the crystal is literally sheared apart, with little stress increase above that needed to start deformation. If the crystal, as in the fcc case, has several sets of planes that can slip, then although deformation may begin on only one set, it will eventually start on others. The occurrence of slip on two or more sets of {111} planes means that active slip planes will interact with each other. This interaction hinders the glide on both sets, with the result that the stress must be continually increased to keep the deformation going. In single crystals of the hcp metals zinc and cadmium, the strain hardening—the rate of stress increase with strain—is small. In the fcc metals copper, aluminum, silver, and gold, it is much greater due to the intersecting slip systems.

As examples, anisotropic plastic deformation and fracture behavior of single-crystal ceramics in solid-state contact are discussed here as they relate to friction. Figures 2.10(a) and (b) show the coefficient of friction and the width of the permanent groove in plastic flow measured as functions of the crystallographic direction on the {0001}, $\left\{10\bar{1}0\right\}$, and $\left\{11\bar{2}0\right\}$ planes of single-crystal SiC in sliding contact with a spherical diamond pin in mineral oil. The results presented indicate that the coefficient of friction and the groove width are influenced by the crystallographic orientation. The $\left\langle11\bar{2}0\right\rangle$ direction on the basal {0001} plane has the larger groove, primarily as a result of plastic flow, and is the direction of high friction for this plane. The $\left\langle0001\right\rangle$ directions on the $\left\{10\bar{1}0\right\}$ and $\left\{11\bar{2}0\right\}$ planes have the greater groove widths and are the directions of high friction when compared with the $\left\langle11\bar{2}0\right\rangle$ directions on the $\left\{10\bar{1}0\right\}$ plane and $\left\langle10\bar{1}0\right\rangle$ directions on the $\left\{11\bar{2}0\right\}$ plane. The anisotropies of friction are $\mu\left\langle11\bar{2}0\right\rangle/\mu\left\langle10\bar{1}0\right\rangle = 1.2$ on {0001}, $\mu\langle0001\rangle/\mu\left\langle11\bar{2}0\right\rangle = 1.3$ on $\left\{10\bar{1}0\right\}$, and $\mu\langle0001\rangle/\mu\left\langle10\bar{1}0\right\rangle = 1.3$ on $\left\{11\bar{2}0\right\}$.

Figure 2.10(c) represents the contact pressure calculated from the groove width data in Fig. 2.10(b) and the Knoop hardness obtained by other workers [2.21]. The anisotropy of the contact pressure during sliding and the Knoop hardness clearly correlate with each other.

The explanation for SiC slip behavior is known [2.25–2.28]. Several slip systems have been observed in hexagonal, single-crystal SiC, such as {0001}$\left\langle11\bar{2}0\right\rangle$, $\left\{3\bar{3}01\right\}\left\langle11\bar{2}0\right\rangle$, and $\left\{10\bar{1}0\right\}\left\langle11\bar{2}0\right\rangle$ [2.25, 2.26]. The $\left\{10\bar{1}0\right\}\left\langle11\bar{2}0\right\rangle$ slip system observed on the sliding surface was responsible for the anisotropic friction and plastic deformation behavior of SiC during sliding in the $\left\langle10\bar{1}0\right\rangle$ and $\left\langle11\bar{2}0\right\rangle$ directions on the SiC {0001} surface. The experimental results on the {0001} plane generally agreed with Daniels and Dunn's [2.27] resolved shear stress analysis based on the slip system $\left\{10\bar{1}0\right\}\left\langle11\bar{2}0\right\rangle$. The minima and maxima of the resolved shear stress for the {0001} plane of a hexagonal crystal match the hard and soft directions on that plane (i.e., the $\left\langle10\bar{1}0\right\rangle$ and $\left\langle11\bar{2}0\right\rangle$ directions, respectively). The

35

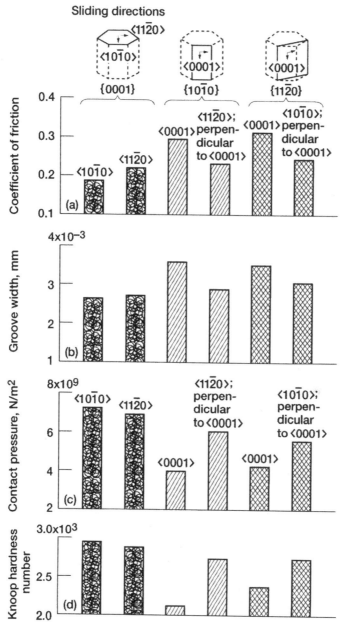

Figure 2.10.—Anisotropies on SiC {0001}, {10$\bar{1}$0}, and {11$\bar{2}$0} surfaces. (a) Coefficient of friction. (b) Groove width. (c) Contact pressure (load, 0.2 N). (d) Knoop hardness number (load, 1 N).

results shown in Fig. 2.10 cannot be explained by the resolved shear stress analysis based on the $\{0001\}\langle11\bar{2}0\rangle$ slip system [2.26]. Thus, the anisotropies of friction, contact pressure, and Knoop hardness on the $\{0001\}$ plane strongly correlate with the resolved shear stress based on the slip system $\{10\bar{1}0\}\langle11\bar{2}0\rangle$.

The anisotropies of friction, contact pressure, and Knoop hardness on the $\{10\bar{1}0\}$ and $\{11\bar{2}0\}$ planes correlate with the resolved shear stress [2.28] based on the $\{10\bar{1}0\}\langle0001\rangle$ and $\{10\bar{1}0\}\langle11\bar{2}0\rangle$ slip systems. They also correlate with the resolved shear stress [2.26] based on the $\{11\bar{2}0\}\langle10\bar{1}0\rangle$ slip system. However, the slip system actually observed in hexagonal SiC, the $\{10\bar{1}0\}\langle11\bar{2}0\rangle$ slip system, may be responsible for those anisotropies on the $\{10\bar{1}0\}$ and $\{11\bar{2}0\}$ planes.

Similar data for the Knoop hardness anisotropy of the tungsten carbide $\{0001\}$ and $\{10\bar{1}0\}$ surfaces have been reported by other workers [2.28], who used a resolved shear stress calculation involving the $\{10\bar{1}0\}\langle0001\rangle$ and $\{10\bar{1}0\}\langle11\bar{2}0\rangle$ slip systems to explain the data. Note that Fig. 2.10 suggests that the $\langle10\bar{1}0\rangle$ directions on the SiC basal plane would exhibit the lowest coefficient of friction and greatest resistance to abrasion resulting from plastic deformation.

Polycrystalline Materials

Both the weakest and strongest forms of metals such as iron are single crystals. In large single crystals the dislocations move easily; in microscopic whiskers the dislocations are immobile or perhaps absent. Extensive slip occurs more easily in single crystals than in polycrystalline materials with random grain orientations.

Important defects in polycrystalline materials are grain boundaries, or the interfaces between single-crystal grains. The boundaries are high-energy sites where they emerge on a solid surface, and they generally contain or act as an irregularity in the surface geometry because a cusp usually forms where the two grains meet. Grain boundaries are an atomic scale defect.

Slip will be blocked at every grain boundary in polycrystalline form; accordingly much greater stresses are required to deform a polycrystalline material, and much less deformation can be achieved before failure. A wide variety of materials, such as metals and ceramics, can be strengthened by decreasing the grain size (i.e., decreasing the mean free path for movement of dislocations). In a pure material the yield strength S is greatly increased by the presence of grain boundaries; this relationship can be derived from the Hall-Petch equation $S = S_i + k/d^{1/2}$, where d is the grain diameter and S_i and k are constants [2.23]. In solid solutions, solute segregation at the grain boundaries can further exaggerate the peculiarities of the regions. For example, a pronounced hardness peak occurs near grain boundaries in some solid solutions, such as a dilute solution of MgO in Al_2O_3. An important exception to the Hall-Petch relation occurs in certain high-strength materials at high temperatures (e.g., in MgO and in complex high-temperature alloys). In these materials deformation in the grain boundaries may become a major component of the deformation process; rupture strength then increases with increasing grain size and may be a maximum for a single crystal of the matrix phase.

2.2.2 Worked Layer

Mechanically abraded or polished metal, polymer, and ceramic surfaces are frequently used as lubricating coatings, self-lubricating substances, or wear-resistant substances. Such a surface is extremely rough on an atomic scale, and crystalline structure is distorted.

When a number of grits of hard abrasive material embedded in a resin matrix come into contact with a softer ceramic, the abrasive grits begin to cut or skive grooves in the ceramic surfaces (the so-called two-body condition). When abrasive-impregnated tapes rub on manganese-zinc (Mn-Zn) ferric oxide ceramic ($MnO–ZnO–Fe_2O_3$), for example, the abrasive grits (e.g., SiC and Al_2O_3) can cut into and remove material from the oxide ceramic. Grooves are formed plastically primarily by the plowing and microcutting of the abrasive grits. Figure 2.11(a) presents a surface abraded in this manner [2.29]. A large number of plastically deformed grooves formed on the ferrite surface in the sliding direction of the 14-μm Al_2O_3 abrasive-impregnated tape. The maximum height of irregularities on the wear surface as measured by surface profilometer was 0.2 mm.

Figure 2.11(b) presents a scanning electron micrograph of the surface of a tape impregnated with 14-μm Al_2O_3 abrasive after 10 passes in sliding contact with Mn-Zn ferrite. Particles of Al_2O_3 are held in the nonmetallic binder. Figure 2.11(b) clearly indicates that relatively few abrasive grits were in contact with the ferrite and that ferrite was removed by a microcutting and plowing action. Only the tips of such grits contained ferrite wear debris particles involved in the abrasive action. Most of the ferrite debris observed on the tape surface was powdery and irregular in shape, but some was curled, indicating a cutting action.

When a third particle harder than one or both of the surfaces in contact becomes trapped at the interface, abrasive wear, lapping, or polishing can also occur, removing material from one or both surfaces. This mode of wear is called three-body abrasion. For example, Fig. 2.12 presents replication electron micrographs and reflection electron diffraction patterns of single-crystal Mn-Zn ferrite wear surfaces abraded by 15- and 4-μm SiC grits in the three-body condition. The abrasion with 15-μm SiC grits resulted in brittle fractured facets on the Mn-Zn ferrite surface due to cleavage and quasi-cleavage. The 4-μm SiC grits mostly produced a large number of plastically deformed indentations and grooves, formed primarily by plowing and microcutting, as well as a few brittle fractured facets.

Reflection electron diffraction patterns taken on both abraded single-crystal Mn-Zn ferrite surfaces (Fig. 2.12) contain continuous arcs extending over nearly a semicircle that indicate a surface in a polycrystalline state. The rest were blocked out by the shadow of the solid specimen. The arcs for the abraded surface generated by the 4-μm SiC grits are much broader than those for the abraded surface

38

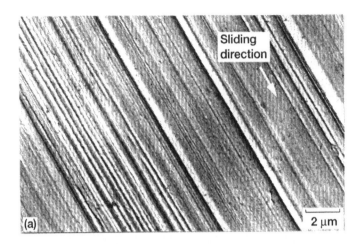

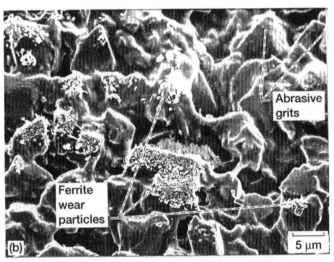

Figure 2.11.— Wear surfaces. (a) Replication electron micrograph of Mn-Zn ferrite surface after single-pass sliding of tape impregnated with 14-μm Al_2O_3 (sliding surface, {110}; sliding direction, ⟨110⟩; sliding velocity, 11.19 m/s; laboratory air; room temperature). (b) Scanning electron micrograph of tape impregnated with 14-μm Al_2O_3 after 10-pass sliding against Mn-Zn ferrite (sliding velocity, 0.52 m/s; laboratory air; room temperature).

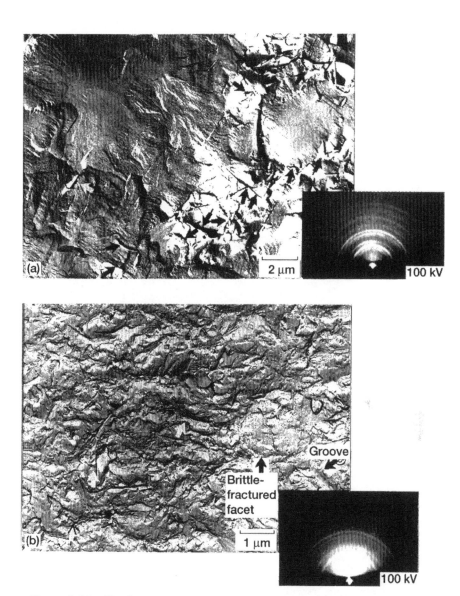

Figure 2.12.—Replication electron micrographs and reflection electron diffraction patterns of abraded single-crystal Mn-Zn ferrite {110} surface. Abrasion direction, ⟨011⟩; lapping disk, cast iron; lapping fluid, olive oil; sliding velocity, 0.5 m/s; abrasive-to-fluid ratio, 27 wt%. (Arrows denote cracks.) (a) After contact with 15-μm SiC grit at 3 N/cm². (b) After contact with 4-μm SiC grit at 8 N/cm².

generated by the 15-μm SiC grits. The broad arcs indicate the large extent of plastic deformation on the abraded surface generated by the 4-μm SiC grits. Thus, the abrasion mechanism for Mn-Zn ferrite oxide ceramics is drastically changed by the size of the abrasive grits. With 15-μm SiC grits ferrite abrasion is principally due to brittle fracture, whereas with 4-μm SiC grits it is due to plastic deformation.

The plastic deformation and microcracks subdivide a single-crystal Mn-Zn ferrite into polygonal subgrains, as described above. The nature of the strain or structural surface damage varies with depth from the surface. This variation can be revealed by combining chemical etching with reflection electron diffraction. For example, Fig. 2.13 shows a replication electron micrograph of an abraded single-crystal Mn-Zn ferrite {110} surface, similar to that in Fig. 2.11(a), and reflection electron diffraction patterns of the surface after chemical etching. The diffraction

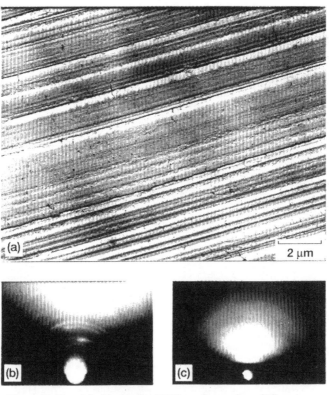

Figure 2.13.—Mn-Zn ferrite {110} surface after sliding on
Al$_2$O$_3$ lapping tape (1000 mesh). (a) Replication electron
micrograph. Electron diffraction patterns showing
(b) highly textured polycrystalline surface and (c) nearly
amorphous surface.

patterns indicate a surface severely distorted and in a polycrystalline state ranging from highly textured to nearly amorphous. The diffraction pattern from the ferrite surface etched to a depth of 0.2 μm (Fig. 2.14(a)) shows a large streak-spot pattern. The streaking indicates severe plastic deformation (i.e., a highly strained, mosaic, single-crystal structure). Large numbers of line defects can cause streaking in a diffraction pattern. Strain and structural damage decrease at greater depths below the abraded surface. At 0.4 μm below the surface (Fig. 2.14(b)) the diffraction pattern shows a relatively sharp spot pattern and no streaking. At depths of 0.8 and 1.0 μm (Figs. 2.14(c) and (d)) the patterns show Kikuchi lines (pairs of black and white lines), which indicate a bulk crystalline structure containing no mechanically stressed areas. Thus, reflection electron diffraction studies can reveal the type of surface damage as well as the total thickness of the deformed, worked layer.

Figure 2.15 schematically summarizes the crystal structure of the worked layer verified by reflection electron diffraction and chemical etching techniques. It also presents the etching rate for the abraded surface as a function of etching depth (distance from the abraded surface). This graph is called an etching rate–depth profile. The outermost layer, which is textured and polycrystalline, has the highest

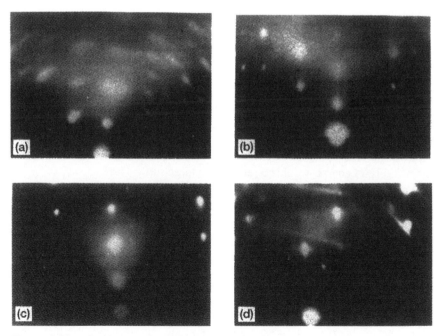

Figure 2.14.—Reflection electron diffraction patterns of surface in Fig. 2.13(a) etched at depths of (a) 0.2 μm, (b) 0.4 μm, (c) 0.8 μm, and (d) 1 μm.

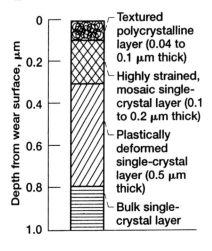 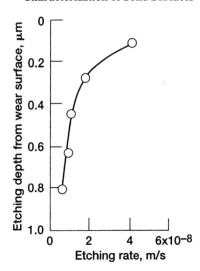

Figure 2.15.—Crystal structure of deformed layer and etching rate as function of etching depth. (Sliding materials same as in Fig. 2.13.)

etching rate, indicating that the layer contains severe defects. A decreasing etching rate is related to decreasing defects: the fewer the defects, the lower the etching rate.

The thickness of the worked layer and the degree of deformation in the layer are functions of (1) the amount of work, or energy, put into the deformation process and (2) the structure and composition of the solid. Figure 2.16(a) shows that the worked layer thickened as the normal load increased. Note that at all loads the greater part of the worked layer was composed of a plastically deformed single-crystal layer (Fig. 2.16(b)), which increased in extent as the load increased. The etching rate (Fig. 2.16(c)) also depended on the normal load applied during abrasion; that is, the number of defects in the worked layer increased with increasing normal load. Some solids are much more prone to deformation than are others. This, of course, would be reflected in these surface layers.

Worked layers are extremely important in tribology and in solid film lubricant deposition because their physical, chemical, and mechanical properties can be entirely different from those in the bulk solid. Further, these properties depend on the depth of the worked layers and on the concentration and nature of the defects, such as whether they are dislocations or microcracks. For example, extrinsic magnetic properties of Mn-Zn ferrites, such as initial permeability, coercive force, and magnetic loss, are influenced by both the surface and bulk crystalline states of the ferrite. On a videotape recorder the sliding of the magnetic tape abrades the ferrite head and scratches its surface. Considerable plastic flow occurs on the ferrite surface, and the large number of defects produced can drastically change the crystalline state of the ferrite head and produce a worked layer on its surface. The worked layer decreases readback signal amplitude and degrades the signal obtained

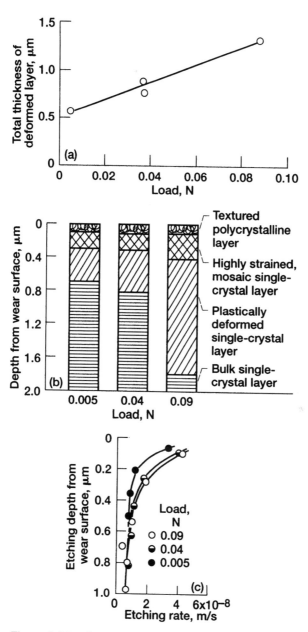

Figure 2.16.—Crystal structure of deformed layer as
function of load. (Sliding materials same as in Fig.
2.13.) (a) Layer thickness as function of load.
(b) Depth from wear surface as function of load.
(c) Etching rate as function of etching depth.

in short-wavelength recording. Figure 2.17(a) shows the readback signals from a new, chemically etched Mn-Zn ferrite magnetic head in sliding contact with a magnetic tape as a function of sliding distance. Figures 2.17(b) and (c) show the electron diffraction patterns taken from the magnetic head surface before and after the sliding. Clearly, the sliding action changed the crystalline state of the magnetic head surficial layer from a single-crystal structure to a nearly amorphous one. That crystallographic change in Mn-Zn ferrite is a critical factor in the readback signal losses shown in Fig. 2.17(a).

2.2.3 Surface Topography

All surfaces contain irregularities, or hills and valleys. Even in cleaved surfaces of single-crystal materials there are cleavage steps. Surface topography is a permanent record of the deformation and fracture process and provides valuable information on surface properties. The world of the engineer is made of solids whose surfaces acquire their texture from many processes, such as cutting, grinding, lapping, polishing, etching, peening, sawing, casting, molding, calendering, and coating [2.30]. Surface topography, or roughness, is an important parameter in characterizing engineering surfaces used in tribological applications.

The unaided eye, fingers, hand lens, optical microscope, optical (interferometry) profiler, stereo microscope, scanning electron microscope, transmission electron microscope with replicas (sometimes also used in optical microscopy), stylus profilometer, and scanning probe microscopes (e.g., scanning tunneling microscope and atomic force microscope) are typically used for studying surface topography [2.31]. Gross features, such as tool marks, isotropy of the surface texture, surface defects, and discoloration, are often best viewed with the unaided eye. Such viewing is rapid and versatile, allows large areas to be examined, and generally prevents details from obscuring the overall pattern. Also, surface texture and roughness are felt with the fingers as they touch a solid surface. The hand lens (e.g., 10X) extends the eye's capability without much loss of speed or surface inspection area. The optical microscope remains one of the most useful and cost-effective tools, in terms of initial cost as well as speed, effectiveness, and versatility of use.

The optical (interferometry) profiler can measure surface features without contact. Light reflected from the surface of interest interferes with light from an optically flat reference surface. Deviations in the fringe pattern produced by the interference are related to differences in surface height. If an imaging array is used, three-dimensional information can be provided. In general, optical profilers have some advantages—no specimen preparation and short analysis time—but also some disadvantages. If the surface is too rough (roughness greater than 1.5 µm), the interference fringes can be scattered to the extent that topography cannot be determined. If more than one matrix is involved (e.g., multiple thin films on a substrate, or if the specimen is partially or totally transparent to the wavelength of

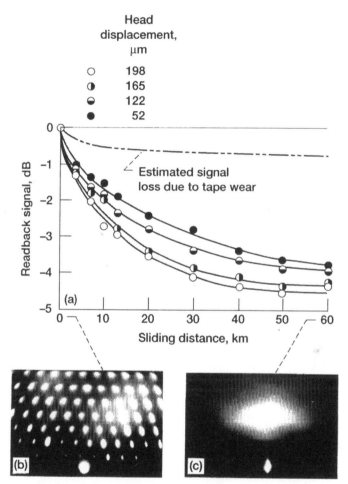

Figure 2.17.—Crystal structure and magnetic signal as function of sliding distance. Sliding materials, chemically etched Mn-Zn ferrite and magnetic tape. (a) Readback signal. Electron diffraction patterns of (b) highly strained single-crystal structure before sliding and (c) nearly amorphous structure after 60 km of sliding.

the measurement system), measurement errors can be introduced. Multiple-matrix specimens can be measured if coated with a layer that is not transparent to the wavelength of light used.

The stereo microscope allows for three-dimensional viewing of specimens themselves, of a stereogram, or of a pair of stereo pictures taken by an optical or scanning electron microscope. Stereo imaging consists of two images taken at different angles of incidence a few degrees apart. Stereo imaging, in conjunction with computerized frame storage and image processing, can provide three-dimensional images with the quality normally ascribed to optical microscopy.

The scanning electron microscope produces micrographs with sufficient resolution to reveal individual features (lateral resolution of 1 to 50 μm in the secondary electron mode at sampling depths from a few nanometers to a few micrometers, depending on the accelerating voltage and the analysis mode) and yet with a large enough field of view that the interrelation of many such features can be seen. In practice, however, the scanning electron microscope has three disadvantages: specimen size is limited to 10 cm or smaller, roughness cannot be quantified, and specimens must be vacuum compatible [2.30, 2.31]. Surface features can be measured best by cleaving the specimen and taking a cross-sectional view.

Transmission electron microscopy is always of replicas (sometimes also used in optical microscopy). It delineates surface features, such as fine pores, cracks, grooves, and especially cleavage steps, better than scanning electron microscopy (e.g., see Fig. 2.12). The transmission electron microscope has some disadvantages: the difficulty of locating specific areas limits it to general characterization and large features, such as large pores, and it fails to reveal cracks owing to replica breakage at these points.

The stylus profilometer (mechanical profilometer) is today most widely used for measuring surface roughness and analyzing surface topography. A diamond stylus with a tip radius of a few micrometers moves up and down as it is dragged across a specimen surface. This up-and-down motion effectively replicates the surface topography. Lateral resolution depends on the stylus radius. If the radius of curvature of the surface of interest is smaller than the radius of curvature of the stylus, the measurement will not satisfactorily reproduce the surface. The typical stylus radius is about 3 μm, but smaller radii down to submicrometer sizes are available. The load applied to the stylus is usually in the millinewton range, 1 to 40 mN. In spite of the light loads used, however, the contact pressure is on the order of gigapascals and is sufficiently large to damage surfaces. Sampling distance ranges from tens of micrometers to tens of millimeters. No specimen preparation is required, almost any specimen regardless of engineering material can be measured rapidly, and results can be obtained in seconds. The stylus profiler provides somewhat limited two-dimensional information. For three-dimensional topographical information, consecutive line scans are needed. This procedure can be quite time consuming.

Scanning probe microscopes [2.31, 2.32] can be considered as derivatives of the stylus profilometer. One popular variant is the scanning tunneling microscope (STM). In this technique a tip is brought to within one nanometer of the specimen surface, and a small bias voltage of typically 0.01 to 1 V is applied between them. Under these conditions electrons can penetrate the potential barrier between a specimen and a probe tip, producing an electron tunneling current I that varies exponentially with tip-to-surface spacing s as follows:

$$I \propto e^{-1.025\sqrt{\varphi}s} \tag{2.3}$$

where φ is the composite work function in electron volts (typically ~4 eV) and s is in angstroms (1 Å equals 0.1 nm). If the tip-to-surface spacing increases (or decreases) by 1 Å, the tunneling current decreases (or increases) by about a factor of 10. Therefore, the tunneling current is a sensitive function of tip-to-surface spacing s. The exponential dependence of the magnitude of I upon s means that, in most cases, a single atom on the tip will image the single nearest atom on the specimen surface. This tunneling current is the imaging mechanism for the scanning tunneling microscope. A piezoelectric scanner is usually used as an extremely fine positioning stage to move the probe over the specimen (or the specimen under the probe).

Clearly, this technique works only with conducting and semiconducting materials. With insulating materials one can put on conducting coatings. Alternatively, one can use a different form of scanning probe microscope, namely, the atomic force microscope (AFM), also called a scanning force microscope (SFM).

An atomic force microscope, instead of using the electron tunneling current to measure the tip-to-surface distance, can measure the force of interaction between a specimen surface and a sharp probe tip. The tip, a couple of micrometers long and often less than 10 nm in diameter, is located at the free end of a cantilever 100 to 200 μm long. When the tip comes within a few angstroms of the specimen's surface, repulsive van der Waals forces between the atoms on the tip and those on the specimen cause the cantilever to deflect, or bend. Figure 2.18 shows the dependence of the van der Waals force on the tip-to-surface spacing. A detector, such as the position-sensitive photodetector (PSPD), measures the cantilever deflection as the tip is scanned over the specimen or the specimen is scanned under the tip (Fig. 2.19). As a piezoelectric scanner gently traces the tip across the specimen (or the specimen under the tip), the contact force causes the cantilever to bend to accommodate changes in topography. The measured cantilever deflections allow a computer to generate a map of surface topography. Atomic force microscopes can be used to study insulating and semiconducting materials as well as electrical conducting materials.

Note that in Fig. 2.18 at the right side of the curve the atoms are separated by a large distance. As the atoms are gradually brought together, they will first weakly

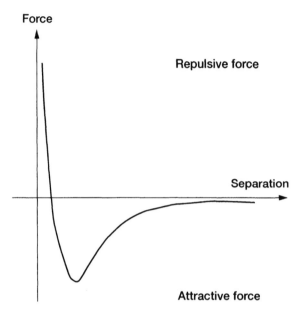

Figure 2.18.—Van der Waals force as function of tip-to-surface separation. (From [2.32].)

attract each other. This attraction increases until the atoms are so close together that their electron clouds begin to repel each other electrostatically. This electrostatic repulsion progressively weakens the attractive force as the interatomic separation continues to decrease. The force goes to zero when the distance between the atoms reaches a couple of angstroms, about the length of a chemical bond. When the total van der Waals force becomes positive (repulsive), the atoms are in contact. The slope of the van der Waals curve is steep in the repulsive or contact region. As a result, the repulsive van der Waals force will balance almost any force that attempts to push the atoms closer together. In the atomic force microscope this means that when the cantilever pushes the tip against the specimen, the cantilever will bend rather than force the tip atoms closer to the specimen atoms.

Most atomic force microscopes currently used detect the position of the cantilever with optical techniques. In the most common scheme, shown in Fig. 2.19, a laser beam bounces off the back of the cantilever onto the position-sensitive photodetector. As the cantilever bends, the position of the laser beam on the detector shifts. The position-sensitive photodetector itself can measure displacements of light as small as 1 nm. The ratio of the path length between cantilever and detector to the length of the cantilever itself produces a mechanical amplification. As a result, the system can detect even 0.1-nm vertical movement of the

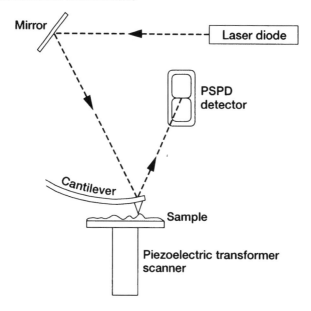

Figure 2.19.—Schematic diagram of optical
technique for detecting cantilever deflection
(also called bounce-beam detection).
(From [2.32].)

cantilever tip. Other methods of detecting cantilever deflection rely on optical
interference, a scanning tunneling microscope tip, or piezoresistive detection
(fabricating the cantilever from a piezoresistive material).

   The shape of a surface can be displayed by a computer-generated map developed
from digital data derived from many closely spaced parallel profiles taken by this
process. Such a map shows details of individual features and also the general
topography over an area and describes surfaces. Figures 2.20(a) and (b) are
examples of atomic force microscopy images of an ion-beam-deposited, diamondlike
carbon film and a chemical-vapor-deposited, fine-grain diamond film on mirror-
polished silicon substrates (DLC on silicon and CVD diamond on silicon). The
surface of the diamondlike carbon film has a smooth, flat morphology. The CVD
diamond surface has a granulated or spherulitic morphology: the surface contains
spherical asperities of different sizes. The surface roughness of the DLC on silicon
is 0.49 nm root-mean square (rms); the surface roughness of the CVD diamond on
silicon is 58.8 nm rms. Also, Fig. 2.20 shows actual height profiles, histograms, and
bearing ratios for the DLC on silicon and the CVD diamond on silicon. Height
profiles, histograms, and bearing ratios are explained in the following paragraphs.

   The fraction of a surface lying in each stratum can readily be obtained by
sampling its height at regular intervals or by tracing a profile. This procedure gives

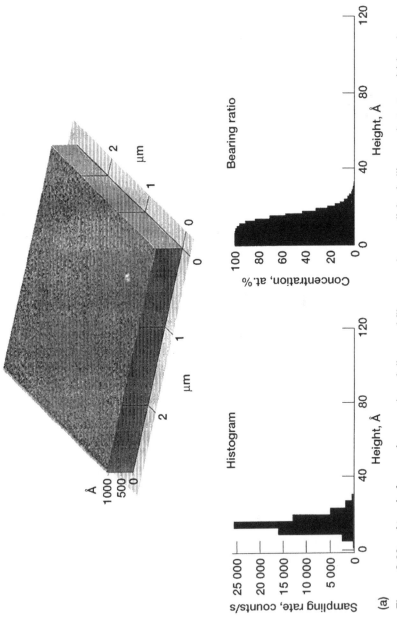

Figure 2.20.—Atomic force micrographs of diamond films on mirror-polished silicon substrates. (a) Ion-beam-deposited, diamondlike carbon film. (b) Chemical-vapor-deposited, fine-grain diamond film.

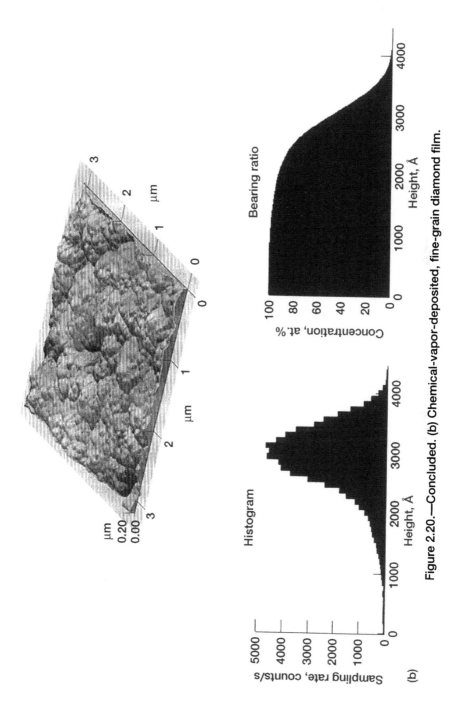

Figure 2.20.—Concluded. (b) Chemical-vapor-deposited, fine-grain diamond film.

the height profile (distribution curve of roughness), bearing ratio (Abbott's bearing curve) or bearing area curve, height histogram, and height distribution, as shown in Fig. 2.21. Many engineering surfaces have height distributions that are approximately Gaussian (i.e., they can be described by the normal probability function)

$$f(u) = \frac{1}{\sqrt{2\pi}\sigma} e^{-(1/2)(u/\sigma)^2} \tag{2.4}$$

where u is a deviation from the median line of the profile curve and σ is the standard deviation of the profile curve.

It is also useful to describe surfaces in terms of the integral of the distribution (bearing ratio in Fig. 2.21(b)), which gives the fraction of the surface at or below each height. The well-known Abbott's bearing curve, which gives the contact area that would exist if the hills were worn down to the given height by an ideally flat body, is the fraction of the surface at or above each height. Some modern surface analyzers provide chart or video displays of height histogram and bearing ratio as standard features. Examples are shown in Fig. 2.20.

Profiles of engineering surfaces usually contain three major components: roughness, waviness, and errors of form, as shown in Fig. 2.22 [2.30, 2.33]. Roughness shows closely spaced irregularities, the height, width, and direction of which create the predominant surface pattern. Roughness includes those surface features intrinsic to the production process. Waviness shows surface irregularities of greater spacing than roughness. It often results from heat treatment, machine or workpiece or specimen deflections, vibrations, or warping strains. Errors of form are gross deviations from the nominal or ideal shape. They are not normally considered part of the surface texture.

Surface roughness occurs at all length scales. Figure 2.23 presents length scales for various surface-related phenomena of interest in engineering. The size range of contaminant particles in the environment is enormous. Such particles include human particles (hair, skin flakes), combustion products (smoke, fly ash), and particles produced by abrasion (machining, car tires, sand). In general, any system involving fluid bearings is susceptible to contaminant particle damage. The dimensions of microelectromechanical systems (MEMS) and the flying heights of magnetic hard-disk drives range from tens of micrometers down to tens of nanometers. The scale of the world in tribology is essentially determined by the size of the contact areas between surfaces. The diameters of contact spots range from about 100 μm down to less than 1 μm, similar to the height range of surface features. The width of that range is large and so is the depth of the worked layers.

Let us now statistically consider the size of the contact areas between surfaces [2.34, 2.35]:

1. An ideal surface in contact with a rough surface
2. A rough surface in contact with a rough surface

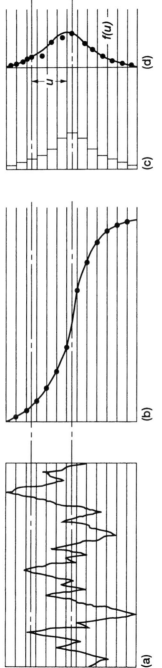

Figure 2.21.—Fraction of surface lying in each stratum. (a) Height profile (distribution curve of roughness). (b) Bearing ratio (Abbott's bearing curve) or bearing area curve. (c) Height histogram. (d) Height distribution.

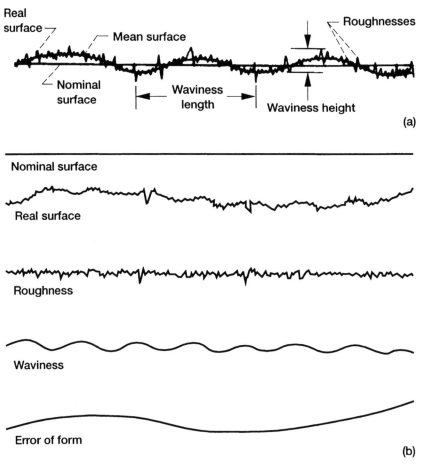

Figure 2.22.—Magnified surface profile (a) and its components (b). (From [2.30, 2.33].)

When a hard, ideal flat surface is brought into contact with a soft, rough surface having the Gaussian height distribution, the actual contact takes place over a small area, at the tips of the asperities or surface irregularities, as indicated in Fig. 2.24(a). The distance $m\sigma$ between the ideal surface and the median line of the profile curve for the rough surface is the distance between the tips of the highest asperities and the median line of the profile curve. The constant m can be determined from the peak-to-valley profile height. These asperity regions initially deform elastically, and then, if the load is sufficiently high, they deform plastically until the load can be supported. That is, the real contact area A continues to increase with deformation until it is sufficient to support the load W applied to the two surfaces in solid-state contact. In this case, A may be approximated by

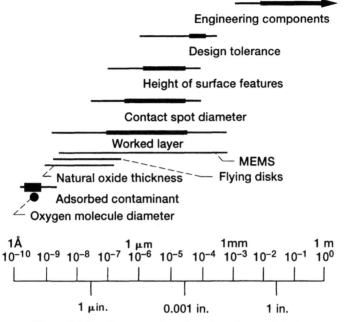

Figure 2.23.—Comparative size of surface-related
phenomena. (From [2.30].)

$$A = \frac{W}{p_m} \tag{2.5}$$

where

p_m flow pressure of soft surface
W applied load

For a normally distributed rough surface with the same profile curve in any section perpendicular to the y axis, as indicated in Fig. 2.24(b), A is given statistically by

$$A = \int_0^{L_y} L_x \frac{F(u)}{F(-m\sigma)} dy = L_x L_y \frac{F(u)}{F(-m\sigma)} = L_x L_y \frac{\Phi(t)}{\Phi(-m)} \tag{2.6}$$

where

$$F(u) = \int_u^{m\sigma} f(u) du$$

$$f(u) = \frac{1}{\sqrt{2\pi}\sigma} e^{-(1/2)(u/\sigma)^2}$$

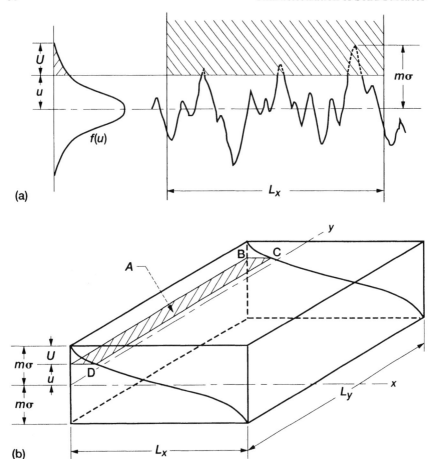

Figure 2.24.—Contact between ideal flat surface and rough surface (a), and Abbott's bearing curves on L_x and L_y and real area of contact A (b).

$$\Phi(t) = \int_t^m \varphi(t)dt$$

$$\varphi(t) = \frac{1}{\sqrt{2\pi}} e^{-t^2/2}$$

σ standard deviation

m constant depending on surface roughness and type of surface finish ($m\sigma$ is distance between tip of highest asperity and median line of profile curve on considerably wide surface; m is determined experimentally from peak-to-valley height of profile curve on considerably wide surface)

$L_x L_y$ apparent contact area between two surfaces

Combining Eqs. (2.5) and (2.6) gives

$$L_x L_y \frac{\Phi(t)}{\Phi(-m)} = \frac{W}{P_m} \qquad (2.7)$$

The distance (separation) between the ideal surface and the median line of the profile curve for the rough surface under applied load W can be given by

$$u = t\sigma \qquad (2.8)$$

and the penetration depth of the ideal flat surface U is given by

$$U = (m\sigma - u) = (m - t)\sigma \qquad (2.9)$$

When two rough surfaces having Gaussian height distribution are brought into contact, the contact also takes place over a small area, actually at the tips of the asperities or surface irregularities, as indicated in Fig. 2.25(a). In this case, the probability density $g(w)$ of the contact is given by

$$g(w) = \frac{1}{\sqrt{2\pi}\sqrt{\sigma_1^2 + \sigma_2^2}} e^{-(1/2)\left(w/\sqrt{\sigma_1^2 + \sigma_2^2}\right)^2} \qquad (2.10)$$

where σ_1 and σ_2 are standard deviations of the two profile curves. This relation coincides with the probability density of contact between an ideal flat surface and a rough surface having a standard deviation of

$$\sigma = \sqrt{\sigma_1^2 + \sigma_2^2} \qquad (2.11)$$

The separation w between two rough surfaces in contact is given by

$$w = t\sqrt{\sigma_1^2 + \sigma_2^2} \qquad (2.12)$$

The penetration depth U is given by

$$U = (m - t)\sqrt{\sigma_1^2 + \sigma_2^2} \qquad (2.13)$$

Thus, contact is limited to a relatively small area, and the rest of the surfaces are held apart. The real contact area and the interfacial gap can contribute significantly

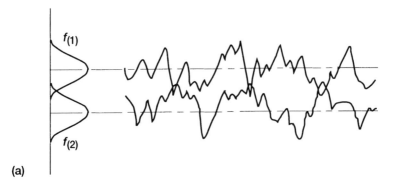

(a)

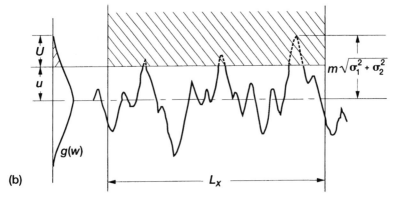

(b)

Figure 2.25.—Contact between two rough surfaces (a), and contact between ideal flat surface and rough surface having $\sqrt{\sigma_1^2 + \sigma_2^2}$ (b).

to all major aspects of tribology: adhesion, friction, lubrication, wear, corrosion, heat transfer and conductivity, electrical conductivity, seal or leakage, interference (shrinkage) fits, cylinder/cylinder liner tolerances, and more.

The interfacial gap formed is usually continuous, permitting gas and liquid access to the whole interface. Surface textures can be designed to provide a particular interfacial gap. The surfaces of mechanical seals, for example, can be smoothed by superfinishing, honing, or lapping to reduce gaseous or liquid leakage and further given noninterconnecting dimples to carry some lubricants to reduce wear. Conversely, the surfaces of heavy-duty sliding bearings can be made to have wide interfacial gaps to facilitate lubricant access and debris removal; with a second process they can also be given shallow-domed plateaus that can carry the load with little plastic deformation and hence little wear.

Another example, presented in Fig. 2.26, is a plain bearing having circumferential microgrooves that was developed for crankshaft and connecting-rod bearings

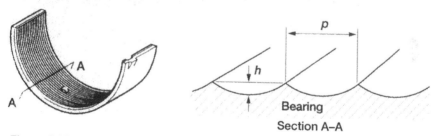

Figure 2.26.—Microgrooved bearing. Pitch, *p*, 0.20 to 0.25 mm; groove depth, *h*, 4.0 to 4.5 μm.

in gasoline engines [2.36]. The microgrooved bearing shows superior conformability relative to normal plane bearings, resulting in a shorter time to establish the hydrodynamic lubrication regime, increased oil flow resulting in lower bearing temperature, and superior oil-retaining properties that prevent seizure for a considerable time (more than 50 min) after cutting off the supply of lubricating oil.

The textures of engineering surfaces used in tribological applications vary widely. Even single-crystal surfaces have a variety of surface defects depending on bulk defects and preparation, such as ledges in crystal faces and steps where dislocations emerge as large-scale shape deviations [2.30]. Brittle, inorganic crystals that are cleaved have flat, atomically smooth surfaces between surface cleavage steps. The cleavage steps in these materials, such as SiC, Al_2O_3, MgO, and Mn-Zn ferrite, occur along cleavage planes. Some metals, such as zinc, undergo brittle cleavage at cryogenic temperatures [2.37]. Other alterations in surface geometry can occur from other processes, such as growth steps that develop during solidification from the liquid state and during crystal growth.

Mechanical preparation techniques produce, in addition to the microstructural defects discussed in Section 2.2.2, another level of surface topographic defects (e.g., grooves (scratches), smears, cavities, grain pullout, cracks, porosity, and contamination (embedded foreign elements)). Grooves can be generated when a hard-surface particle rubs against a softer surface (two-body abrasion) or when small, hard particles are entrapped between two solid surfaces (three-body abrasion). Figure 2.11(a) shows a single-crystal Mn-Zn ferrite surface after two-body abrasion, and Fig. 2.12 shows the same surface after three-body abrasion [2.29]. The surfaces are covered with hills and valleys. The replication electron micrographs show a large number of plastically deformed grooves (Fig. 2.11(a)), brittle-fractured facets (Fig. 2.12(a)), and plastically deformed indentations and grooves (Fig. 2.12(b)). When the material is plastically pushed (i.e., moved across the surface) instead of being cut, the resulting plastic deformation of a large area is called smearing. Cavities may be left after grains or particles are torn out of the solid surface. They are usually found in hard or brittle materials and in materials with inclusions. Hard or brittle materials can hardly be deformed plastically, so small parts of the surface material shatter and may fall out or be pulled out during machining or abrasion. Cracks occur in brittle materials and in materials with layers, such as coatings. Some materials have natural porosity (e.g., cast metals, spray coatings, and ceramics). Ceramics prepared by powder compaction and heat treatment are almost always porous. Deposits on a solid surface can serve as surface geometric irregularities. These deposits could include materials from a source other than the solid itself, such as wear and abrasive particles. Such contamination can occur on all types of material.

Another type of surface defect, the etch pit or line, can occur when the environment or a lubricant constituent reacts with the surfaces. Etch pits or lines can occur at high-energy sites (e.g., grain boundaries and subsurface defects, such as dislocation sites) after chemical, physical, or optical etching.

2.2.4 Contaminant and Oxide Layers

In addition to the presence of irregularities, commonly called asperities, as discussed in Section 2.2.3, the solid surface itself is covered with thin contaminant layers of atomic dimensions (~ 2 nm thick). These contaminant layers are unavoidably present on every surface of any solid matter that has been exposed to air. In other words, the simplest and most common occurrence recognized with real surfaces is that nearly all surfaces contain adsorbates, either physically

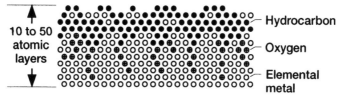

Figure 2.27.—Schematic model of metal exposed to air, showing contaminant layer.

adsorbed or chemically adsorbed material that has formed through interaction with the environment. Sometimes oxide layers of various depths are present beneath the surface contaminants. Knowledge of the contaminant and oxide layers is of great interest to materials researchers in tribology as modern technology tries to improve material properties, such as coefficient of friction and resistance against wear and corrosion in near-surface regions.

Figure 2.27 shows, for example, how the ambient atmosphere affects the dry surface of an elemental metal. A thin contaminant layer of adsorbates is present, as on the surface of any solid. It is made up of contaminants, such as adsorbed gases, water vapor, and hydrocarbons. Various hydrocarbons are detected if the component has been near operating machinery because lubricating or processing oils vaporize.

2.3 Surface Characterization Techniques

Although a wide range of physical surface analysis techniques is available, certain traits common to many of them can be classified from two viewpoints. Most techniques involve either electrons, photons (light), x rays, neutral species, or ions as a probe beam striking the material to be analyzed. The beam interacts with the material in some way. In some of these techniques the changes induced by the beam (energy, intensity, and angular distribution) are monitored after the interaction, and analytical information is derived from observing these changes. In other techniques the information used for analysis comes from electrons, photons, x rays, neutral species, or ions that are ejected from the specimen under the stimulation of the probe beam (Fig. 2.28). In many situations several connected processes may be going on more or less simultaneously, with a particular analytical technique picking out only one aspect (e.g., the extent of incident light absorption or the kinetic energy distribution of ejected electrons).

Table 2.1 briefly summarizes the analytical techniques available to the solid lubricants user and tribologist today in studying the properties and behavior of solid surfaces. The table allows quick access to the type of information provided by the techniques and their popularity. The reader will find the basic principles and instrumentation details for a wide range of analytical techniques in the literature

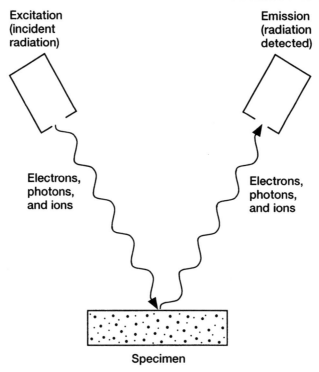

Figure 2.28.—Schematic diagram showing excitation
and emission processes on solid surface.

(e.g., [2.4–2.6, 2.9, 2.10]). However, the analytical instrumentation field is moving rapidly and within a year current spatial resolutions, sensitivities, imaging/mapping capabilities, accuracies, and instrument costs and size are likely to be out of date. Therefore, these references should be viewed with caution.

What types of information are provided by these analytical techniques? Elemental composition is perhaps the most basic information, followed by chemical state information, phase identification, the determination of structure (atomic sites, bond lengths, and angles), and defects.

The elemental and chemical state, phase, microstructure, and defects of a solid often vary as a function of depth into the material or spatially across the material, and many techniques specialize in addressing these variations down to extremely fine dimensions (on the order of angstroms in some cases). Requests are made for physical and chemical information as a function of depth to depths of 1 mm or so (materials have about 3 million atomic layers per millimeter of depth), as shown in Fig. 2.29. It is this region that affects a broad spectrum of properties: adhesion, bonding, friction, wear, lubrication, corrosion, contamination, chemical composition, and chemical activity. Knowledge of these variations is of great importance to the selection and use of solid lubricants and other tribological materials.

TABLE 2.1.—POPULAR ANALYTICAL TECHNIQUES FOR SURFACE, THIN FILM, INTERFACE, AND BULK ANALYSIS OF SOLID LUBRICANT ON SUBSTRATE

Technique	Main information	Use (popularity)
Light microscopy	Defects Image and morphology	Extensive
Scanning electron microscopy (SEM)	Defects Image and morphology	Extensive
Scanning tunneling microscopy (STM) and atomic force microscopy (AFM)	Structure Defects Image and morphology	Medium
Transmission electron microscopy (TEM)	Phase Structure Defects Image and morphology	Medium
Energy-dispersive x-ray spectroscopy (EDS)	Element composition Image	Medium
Electron energy-loss spectroscopy (EELS)	Element composition Chemical state Image	Not common
Cathodoluminescence (CL)	Element composition Defects	Not common
Electron probe x-ray microanalysis (EPMA)	Element composition Image	Medium
X-ray diffraction (XRD)	Phase Structure Defects	Extensive
Low-energy electron diffraction (LEED)	Structure Defects	Medium
Reflection high-energy electron diffraction (RHEED)	Structure Defects	Medium
X-ray photoelectron spectroscopy (XPS)	Element composition Chemical state Image	Extensive
Auger electron spectroscopy (AES)	Element composition Chemical state Image	Extensive
X-ray fluorescence (XRF)	Element composition	Extensive
Photoluminescence (PL), or fluorescence spectrometry	Chemical state Defects Image	Medium
Variable-angle spectroscopic ellipsometry (VASE)	Film thickness	Not common
Fourier transform infrared spectroscopy (FTIR)	Chemical state Defects	Extensive
Raman spectroscopy	Chemical state Defects	Medium
Solid-state nuclear magnetic resonance (NMR)	Chemical state Phase Structure	Not common
Rutherford backscattering spectrometry (RBS)	Element composition Structure Defects	Medium
Elastic recoil spectroscopy (ERS)	Hydrogen content	Not common

TABLE 2.1.—Concluded.

Technique	Main information	Use (popularity)
Ion-scattering spectroscopy (ISS)	Element composition	Not common
Secondary ion mass spectrometry (SIMS)	Element composition Chemical state Image	Extensive
Mass spectrometries: Sputtered neutral (SNMS) Laser ionization (LIMS) Spark source (SSMS) Glow discharge (GDMS) Inductively coupled plasma (ICPMS)	Element composition Chemical state	Extensive to medium
Profilometers: Stylus profiler Optical profiler	Surface roughness Image	Extensive
Adsorption	Surface areas Chemisorption	Medium

For surfaces, interfaces, and thin films there is often little material to analyze; hence, the presence of many microanalytical methods in Table 2.1. Within micro-analysis it is often necessary to identify trace components down to extremely low concentrations (parts per trillion in some cases), and a number of techniques specialize in this aspect. In other cases a high degree of accuracy in measuring the presence of major components might be the issue. Usually, the techniques that are good for trace identification do not accurately quantify major components. Most complete analyses require the use of multiple techniques, the selection of which depends on the nature of the specimen and the desired information.

Figure 2.30 demonstrates the importance of the sampling (information) depth by comparing the analysis of two different specimens. The first has an atomic layer of impurity atoms on the surface; the other has these atoms distributed homogeneously within the specimen. When a conventional x-ray microprobe with a sampling depth of about 1000 nm is used, the specimens produce signals of equal intensity and it is not possible to differentiate between them. One can sometimes get an indication of whether there is a bulk impurity or a surface segregation (or a thin film) by lowering the electron beam voltage or by measuring at grazing incidence. However, this is not possible with only a few atomic layers. On the other hand, because of their insufficient detection limits Auger electron spectroscopy (AES) or x-ray photoelectron spectroscopy (XPS) will produce two quite different spectra: in the first specimen, a strong signal from the impurity layer caused by the low sampling depth; in the second specimen, only the spectrum of the pure bulk material with no indication of the impurities.

Figure 2.31 illustrates the information depth capabilities of surface-sensitive characterization tools (ion scattering spectroscopy (ISS), secondary ion mass spectrometry (SIMS), AES, and XPS, which is also called electron spectroscopy for chemical analysis (ESCA)) and of bulk characterization tools (electron

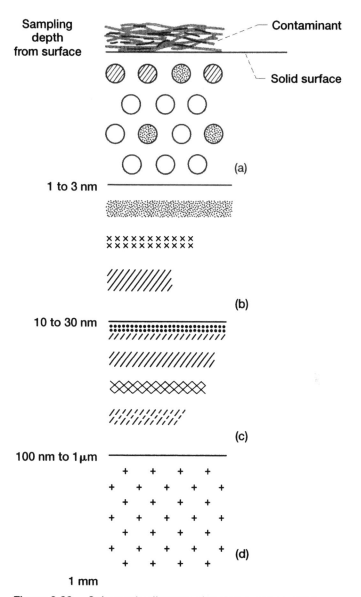

Figure 2.29.—Schematic diagram showing regimes of (a) surface analysis, (b) thin film analysis, (c) interface analysis, and (d) bulk substrate analysis.

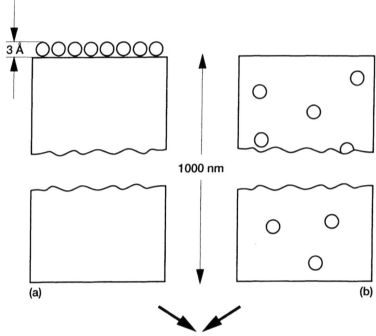

Electron microprobe gets <u>same</u> signal for both specimens

Figure 2.30.—Schematic diagram showing importance of information depth.
(a) Specimen 1: monolayer of foreign (impurity) atoms on surface.
(b) Specimen 2: same atoms with homogeneous distribution. (From [2.44].)

microprobe analysis (EMPA) and energy-dispersive x-ray analysis (EDX)). Thus, knowing both the structure of real surfaces and the capabilities of the various characterization techniques is of great importance.

It was during the 1960's that the amazing growth and diversification of surface analytical techniques began and evolved with the development of two types of ultra-high-vacuum electron spectroscopy—Auger electron spectroscopy (AES) closely followed by x-ray photoelectron spectroscopy (XPS). The combination of the all-encompassing definition of tribology with these surface analytical techniques, including a variety of electronic, photonic, and ionic spectroscopies and microscopies, reflects the trend of tribology and solid film lubricants today. In the following subsections characterization equipment commonly used in the fields of tribology and tribological coatings (including solid film lubricants) and examples of their use are described. Also, it is shown where surface science can play a role in advancing our knowledge of adhesion, friction, lubrication, and wear as well as the basics of tribological phenomena.

A number of techniques are now available for measuring the composition of any solid surface. The most widely used techniques for surface analysis are AES, XPS

67

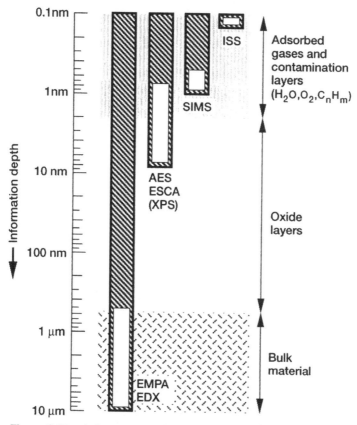

EMPA, electron microprobe analysis
EDX, energy-dispersive x-ray analysis
AES, Auger electron spectroscopy
ESCA, electron spectroscopy for chemical analysis
XPS, x-ray photoelectron spectroscopy
SIMS, secondary ion mass spectrometry
ISS, ion scattering spectroscopy

Figure 2.31.—Information depth capabilities of various surface
analysis techniques. (The unshaded areas within the bars
represent the variation of the information depths, which
depend on instrumental parameters and on the material
itself.) (From [2.44].)

(or ESCA), and SIMS. These techniques are well suited for examining extremely thin layers, including the contaminant layers and the oxide layers.

2.3.1 AES and XPS

The surface analytical techniques most commonly used in tribology are AES and XPS (or ESCA). Each can determine the composition of the outermost atomic layers of a clean surface or of surfaces covered with adsorbed gas films, oxide films, lubricants, reaction film products, and frictionally transferred films and can describe the surface [2.4–2.11].

AES and XPS are generally called "surface analysis" techniques, but this term can be misleading [2.5]. Although these techniques derive their usefulness from their intrinsic surface sensitivity, they can also be used to determine the composition of deeper layers. Such a determination is normally achieved through controlled surface erosion by ion bombardment. AES or XPS analyzes the residual surface left after a certain sputtering time with rare gas ions. In this way, composition depth profiles can be obtained that provide a powerful means for analyzing thin films, surface coatings, reaction film products, transferred films, and their interfaces. Clearly, this capability also makes AES and XPS ideal for studying wear-resistant coatings and solid film lubricants. There are, however, a number of practical differences between the two techniques (e.g., detection speed, background, and spatial resolution, as shown in Table 2.2) that are generally more advantageous in AES profiling. Table 2.2 lists the methods and capabilities of AES and XPS, the two principal solid-surface analysis techniques. Again, because the instrumentation field is moving rapidly, this table should be viewed only as a reference and with caution.

AES.—AES uses a focused electron beam to create secondary electrons near a solid surface, as schematically shown in Fig. 2.32. Some of these electrons (the Auger electrons) have energies characteristic of the elements. Figure 2.33, for example, presents an AES spectrum of the surface of a molybdenum disulfide (MoS_2) film. The MoS_2 films were deposited by magnetron radiofrequency sputtering to a nominal thickness of 110 nm on sputter-cleaned AISI 440C stainless steel disk substrates. The MoS_2 films were exposed to air for a short time (less than 15 min) prior to AES analysis. A carbon contamination peak is evident as well as an oxygen peak. The presence of these peaks indicates that the MoS_2 surface is covered with oxides as well as a simple adsorbed film of contaminants, including oxygen and carbon. Adsorption of common contaminants, such as hydrocarbon vapor and water vapor, and oxidation easily take place on MoS_2 surfaces exposed to air. The contaminants and oxide films exist in layers of atomic dimensions. Note that the surface of the MoS_2 film shows a highly dense, smooth, featureless appearance by scanning electron microscopy.

As stated above, AES can characterize the specimen in depth and provide elemental depth profiles when used in combination with sputtering (e.g., argon-ion

TABLE 2.2.—CHARACTERISTICS OF AES AND XPS (OR ESCA) FOR
PRACTICAL SURFACE ANALYSIS OF SOLID LUBRICANTS

Method and capabilities	Auger electron spectroscopy (AES)	X-ray photoelectron spectroscopy (XPS, or ESCA)
Excitation by–	Electrons	X rays
Detection of–	Electrons	Electrons
Destructive?	No, except to electron-beam-sensitive materials and during depth profiling	No, some beam damage to x-ray-sensitive materials
Element analysis?	Yes, semiquantitative without standards and quantitative with standards	Yes, semiquantitative without standards and quantitative with standards; not a trace element method
Range of elements	All except hydrogen and helium; no isotopes	All except hydrogen and helium; no isotopes
Sensitivity differences for range of elements	Factor of 10	Factor of 10
Detection limit	1000 ppm or 10^{-10} g/cm^3	1000 ppm or 10^{-10} g/cm^3
Chemical state information?	Yes, in many materials	Yes
Depth probed	0.5 to 10 nm	0.5 to 5 nm
Depth profiling?	Yes, in combination with ion beam sputtering	Yes, over the top 5 nm; greater depths require sputter profiling
Lateral resolution	30 nm for Auger analysis, even less for imaging	5 mm to 75 μm; down to 5 μm in special instruments
Imaging/mapping?	Yes, called scanning Auger microscopy (SAM)	Yes, called imaging XPS (or ESCA scope)
Sample requirements	Vacuum-compatible materials	Vacuum-compatible materials

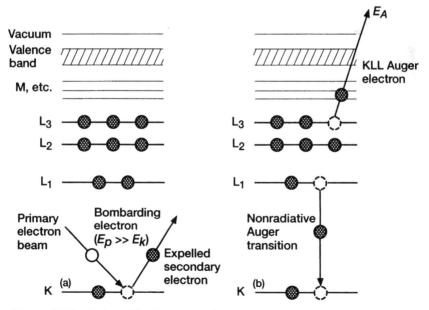

Figure 2.32.—Schematic diagram of Auger emission in a solid. (a) Excitation by electrons. (b) Emission of Auger electrons.

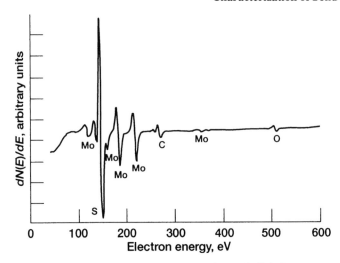

Figure 2.33.—AES spectrum from surface of slightly
contaminated MoS_2 film deposited on AISI 440C stainless
steel by magnetron radiofrequency sputtering.

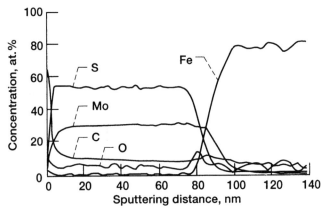

Figure 2.34.—AES depth profile of MoS_2 film on AISI 440C
stainless steel.

sputter etching) to gradually remove the surface. Figure 2.34 presents an
example of an AES depth profile, with contaminant concentration shown as a
function of sputtering distance from the MoS_2 film surface. The concentrations of
sulfur (S) and molybdenum (Mo) at first rapidly increase with an increase in
sputtering distance, whereas the concentrations of carbon (C) and oxygen (O)
initially decrease. They all remain constant thereafter. The MoS_2 film has a
sulfur-to-molybdenum ratio of approximately 1.7. The relative concentrations of
various constituents are determined by using manufacturers' sensitivity factors.

Although the contaminants are removed from the film surface by sputtering, the MoS_2 film contains small amounts of carbon and oxygen impurities in the bulk.

In addition to energies characteristic of the elements, as shown in Fig. 2.32, some of the Auger electrons detected have energies characteristic, in many cases, of the chemical bonding of the atoms from which they are released. Because of their characteristic energies and the shallow depths from which they escape without energy loss, Auger electrons can characterize the elemental composition and, at times, the chemistry of surfaces. The Auger peaks of many elements show significant changes in position or shape in different chemical conditions and environments. For example, AES can distinguish between amorphous carbon, carbide, and graphite. Figures 2.35 and 2.36 compare the survey AES spectra and high-resolution AES spectra for carbon in various forms present on single-crystal SiC exposed to various environments. These AES spectra were obtained from three different surfaces: (1) a surface polished at room temperature, (2) a surface heated to 800 °C in ultrahigh vacuum, and (3) a surface heated to 1500 °C in ultrahigh vacuum. AES peaks of amorphous carbon, carbide, and graphite were detected from single-crystal SiC {0001} surfaces. When the polished single-crystal SiC is placed in a vacuum, one of the principal contaminants on the surface is adsorbed carbon. The high-resolution AES spectrum for carbon on the as-received surface appears only as the single main carbon peak, labeled A_0 in Fig. 2.36(a), where A is used to denote an AES peak. The carbon peak is similar to that obtained for amorphous carbon but not to that obtained for carbide. The high-resolution AES peaks for carbon on the SiC surface heated to 800 °C indicate a carbide peak, similar to that obtained for an SiC surface cleaned by argon-ion sputtering at room temperature. The carbide AES peaks are characterized by the three peaks labeled A_0, A_1, and A_2 in Fig. 2.36(b). The high-resolution spectrum of the SiC surface heated above 1500 °C clearly reveals a graphite peak at 271 eV. The graphite form of carbon is characterized by a step labeled A in Fig. 2.36(c).

Furthermore, Fig. 2.37 presents a survey AES spectrum and a high-resolution AES spectrum for carbon obtained from a single-crystal {111} diamond surface after argon-ion sputter etching. The high-resolution spectrum has four peaks, as is characteristic of diamond. The peaks have been labeled A_0 to A_3. The energies of the peaks in this case are 267 to 269 eV for A_0, 252 to 254 eV for A_1, 240 eV for A_2, and 230 to 232 eV for A_3. The high-resolution AES spectrum is consistent with that obtained by other workers for a clean diamond surface [2.38].

Thus, AES has the attributes of high lateral resolution, relatively high sensitivity, and standardless semiquantitative elemental analysis. It also provides chemical bonding information in some cases. Further, the high spatial resolution of the electron beam and the sputter etching process allow microanalysis of three-dimensional regions of solid specimens. . .

XPS.—In XPS monoenergetic soft x rays bombard a specimen material, causing electrons to be ejected, as shown in Fig. 2.38. The elements present in the specimen can be identified directly from the kinetic energies of these ejected

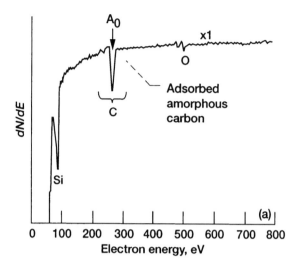

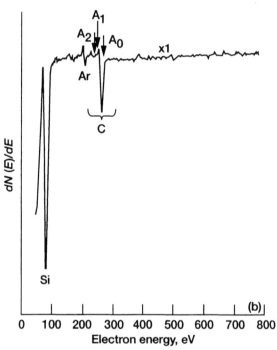

Figure 2.35.—Survey AES spectra of SiC {0001} surfaces treated in different conditions. (a) Polished surface at room temperature. (b) Surface heated to 800 °C in ultrahigh vacuum. (c) Surface heated to 1500 °C in ultrahigh vacuum.

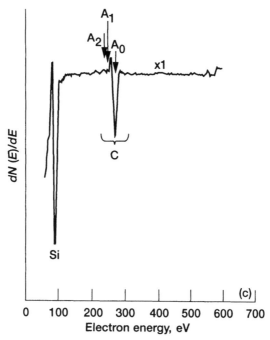

Figure 2.35.—Concluded. (c) Surface heated to
1500 °C in ultrahigh vacuum.

photoelectrons. Electron binding energies are sensitive to the chemical state of the atom. Although the XPS is designed to deal with solids, specimens can be gaseous, liquid, or solid. XPS is applicable to metals, ceramics, semiconductors, organic, biological, and polymeric materials. Although x-ray beam damage can sometimes be significant, especially in organic materials, XPS is the least destructive of all the electron or ion spectroscopy techniques. The depth of solid material sampled varies from the top 2 atomic layers to 15 to 20 layers. This surface sensitivity, combined with quantitative and chemical analysis capabilities, has made XPS the most broadly applicable general surface analysis technique used today, especially in the field of tribology. Like AES, XPS can also characterize the specimen in depth and provide elemental depth profiles when used in combination with sputtering (e.g., argon-ion sputter etching) to gradually remove the surface. Specific examples are given in the following paragraphs.

Figure 2.39 presents XPS survey spectra of the untreated and laser-annealed tungsten disulfide (WS_2) coatings taken with magnesium K_α radiation. The major elements present in both coatings were tungsten, sulfur, oxygen, and carbon. A small amount of nitrogen was present on the surface of the untreated specimen. XPS depth profiles showing the concentrations of the major elements as a function of depth from the surface were obtained from both coatings. The nitrogen on the untreated coating disappeared immediately upon sputtering and so is not

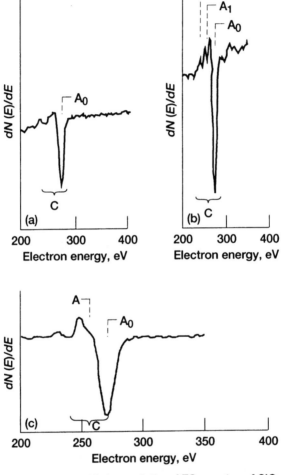

Figure 2.36.—High-resolution AES spectra of SiC
{0001} surfaces treated in different conditions.
(a) Amorphous carbon at room temperature.
(b) Carbide at 800 °C. (c) Graphitic carbon at
1500 °C.

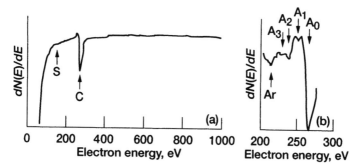

Figure 2.37.—AES spectrum for diamond. (a) Survey spectrum. (b) High-resolution spectrum.

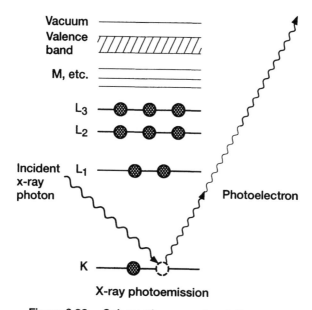

X-ray photoemission

Figure 2.38.—Schematic energy level diagram illustrating photoexcitation of K-shell electron by low-energy x-ray photon.

included in the profiles. The depth profiles were made by recording high-resolution spectra of the W_{4f}, S_{2p}, O_{1s}, and C_{1s} regions, sputtering for 50 s, and repeating the procedure until the total sputter time reached 300 s. Under the conditions used, the sputter etch rate was 0.021 nm/s as calibrated on a tantalum pentoxide (Ta_2O_5) standard. The depth profiles were quantified by using sensitivity factors supplied by the instrument manufacturer. No standards were run.

The major difference in the XPS analyses of the two coatings is evident from the depth profiles shown in Fig. 2.40. In the untreated coating there was almost

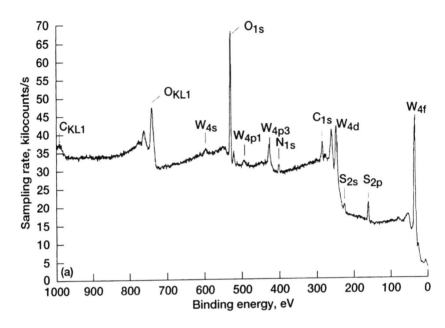

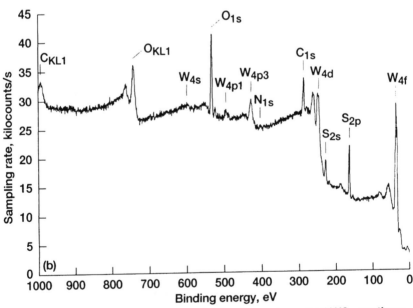

Figure 2.39.—Survey XPS spectra of pulse-laser-deposited WS$_2$ coatings. (a) Untreated. (b) Laser annealed.

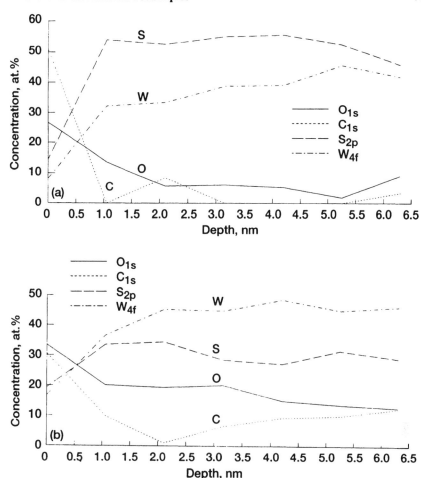

Figure 2.40.— XPS depth profiles of pulsed-laser-deposited WS_2 coatings.
(a) Untreated. (b) Laser annealed.

no carbon or oxygen below a depth of 3 nm. In the laser-annealed coating, on the other hand, carbon and especially oxygen persisted throughout the profile. The sulfur concentration in the laser-annealed coating was correspondingly less than that in the untreated coating, although the tungsten concentrations were about the same.

The high-resolution O_{1s} spectra taken after 300 s of sputtering confirmed the much greater oxygen concentration in the laser-annealed coating than in the untreated coating (Fig. 2.41). High-resolution S_{2p} spectra both from the surface (Fig. 2.42(a)) and from 6 nm deep (Fig. 2.42(b)) showed that sulfur was present as the sulfide at all levels in both coatings. In particular, any sulfate would give a peak at a much higher binding energy, as indicated on these spectra. The high-resolution

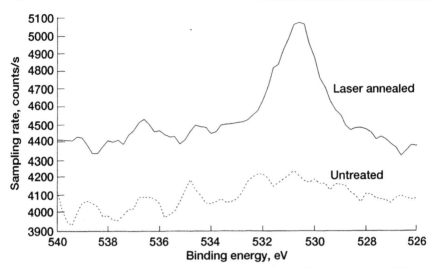

Figure 2.41.—O_{1s} spectra of pulsed-laser-deposited WS_2 coating after 300 s of sputtering.

W_{4f} spectra were composed of a poorly resolved 7/2 and 5/2 doublet in the 31-eV to 35-eV range and a $5p_{3/2}$ line around 37 eV (Fig. 2.43). The published positions of the 7/2 line for several possible tungsten compounds are indicated in Fig. 2.43(a). Each of these would have a corresponding 5/2 and $5p_{3/2}$ component.

In summary of XPS analysis the surface oxides, represented by the initial high oxygen concentrations in the depth profiles, were 1 to 2 nm deep. To a depth of 6 nm the untreated coating was relatively uncontaminated WS_2, but the laser-annealed coating contained high levels of oxygen and somewhat less carbon, which replaced sulfur in the coating. In all cases sulfur was present as the sulfide.

As presented above, on a finer scale the chemical state of the elements present can be identified from small variations in the kinetic energies. Another example is shown in Fig. 2.44. High-resolution photoelectron lines for C_{1s} of the SiC {0001} surface are split asymmetrically into doublet peaks. Three spectral features, which depend on the chemical nature of the specimen, can be observed: (1) two kinds of doublet peaks, (2) change of the peak vertical height, and (3) shift of the peaks. The XPS spectra of the as-received surface indicate distinguishable kinds of carbon (i.e., a carbon contamination peak and a carbide peak associated with the SiC). At room temperature the carbon contamination peak is higher than the carbide peak (Fig. 2.44(a)). At 800 °C the carbide peak is much higher than the graphite peak (Fig. 2.44(b)). At 1500 °C the spectra indicate a large graphite peak and a small carbide peak (Fig. 2.44(c)), as is typical of the surface heated above 800 °C.

The XPS carbon peaks (Fig. 2.44) are similar to the AES carbon peaks (Fig. 2.36). AES and XPS provide both elemental and chemical information for

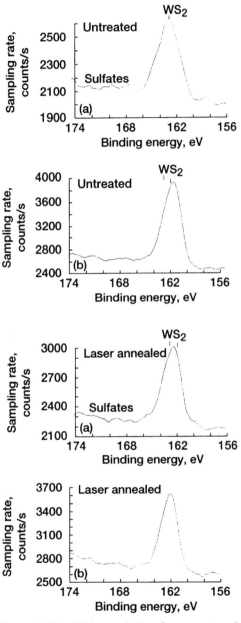

Figure 2.42.—High-resolution S_{2p} spectra of untreated and laser-annealed, pulsed-laser-deposited WS_2 coatings. (a) Surface. (b) After 300 s of sputtering from 6 nm deep.

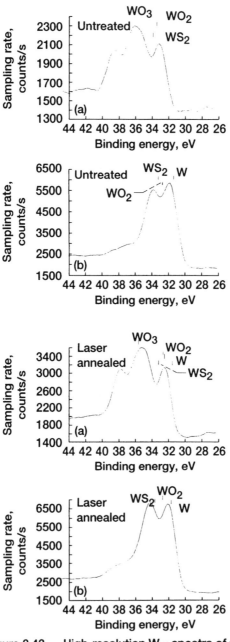

Figure 2.43.— High-resolution W_{4f} spectra of pulsed-laser-deposited WS_2 coatings. (a) Surface. (b) After 300 s of sputtering from 6 nm deep.

carbon in various forms present on single-crystal SiC. In general, AES provides elemental information only. The AES peaks of many elements, however, show significant changes in position or shape in different chemical environments. It is possible to distinguish among amorphous carbon, carbide, and graphite in AES, as demonstrated in Fig. 2.36. On the other hand, the main advantage of XPS is its ability to provide chemical information from the shifts in binding energy. The particular strengths of XPS are quantitative elemental analysis of surfaces (without standards) and chemical state analysis.

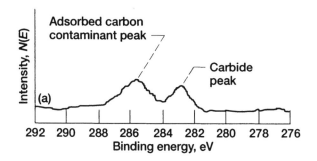

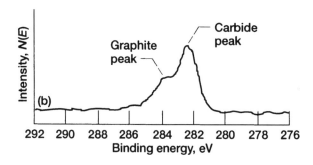

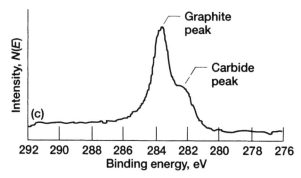

Figure 2.44.—Carbon XPS peaks on SiC.
(a) Room temperature. (b) 800 °C. (c) 1500 °C.

For a solid, AES and XPS probe 2 to 10 and 2 to 20 atomic layers deep, respectively, depending on the material, the energy of the photoelectron concerned, and the angle (with respect to the surface) of the measurement. Note that the AES analysis of SiC heated to 1500 °C (Fig. 2.36(c)) indicates that the carbon peak shown was only of graphite form. The XPS analysis (Fig. 2.44(c)), however, indicates that carbide as well as graphite was present on the SiC surface heated to 1500 °C. This difference can be accounted for by the fact that the analysis depth (sampling depth) with XPS is deeper.

The thickness of the outer graphite layer can be determined as follows: (1) by studying the attenuation by the graphite layer of photoelectrons originating in the bulk SiC material and (2) by studying the variation in intensity of photoelectrons emitted by the layer itself as a function of thickness. The photoelectron flux penetrating a layer thickness d is simply

$$I_d = I_0 \exp(-d/\lambda) \tag{2.14}$$

where I_d is the flux emerging at the surface (i.e., the XPS peak intensity from the layer of thickness d), I_0 is the flux emitted by the clean SiC, and λ is variously called the inelastic mean free path, the mean escape depth, or the attenuation length of an electron having an electron energy within the material.

On the other hand, the intensity of a photoelectron signal from the layer is

$$I_d = I_0\{1 - \exp(-d/\lambda)\} \tag{2.15}$$

In either case, the mean escape depth λ relates to the layer material and to electrons of a given kinetic energy. The potential value varies with the angle of electron emission, or takeoff angle θ, in the XPS analysis. The generalized theoretical prediction for intensity changes with θ for the bulk SiC plus the graphite overlayer system is given by

$$I_d = I_0\{1 - \exp[-d(\lambda \sin \theta)]\} \tag{2.16}$$

Table 2.3 shows inelastic mean free paths of various elements and the estimated thickness of the graphite layer formed on the SiC surface. These values have been estimated by using Eqs. (2.14) to (2.16) and the values of λ for silicon, carbon, and graphite [2.39, 2.40]. Table 2.3 suggests that carbon collapse in two or three successive SiC surficial layers after silicon evaporation is the most probable mechanism for SiC surface graphitization. The outermost surficial graphite layer on the SiC heated to 1500 °C was 1.5 to 2.4 nm thick. These thicknesses suggest that carbon collapse in three successive SiC layers was the most probable mechanism for the initial graphitization stages of the SiC basal planes. Similar results were also obtained with sintered polycrystalline SiC.

TABLE 2.3.—THICKNESS OF GRAPHITIC CARBON
LAYER ON SiC SURFACE PREHEATED TO 1500 °C

Element and photoelectron, Mg K_α	Electron inelastic mean free path, λ, nm	Layer thickness, d, nm
Si_{2p}	[a]4.7	2.0
	[a,b]3.9	1.7
C_{1s}	[b]4.4	1.8
Graphite C_{1s}	[a]2.1	1.5
	[a]3.1	2.3
	[a]3.4	2.4

[a]From [2.39]
[b]From [2.40]

2.3.2 SEM and EDS

The single most useful tool available today to tribologists and lubrication engineers interested in studying the morphology, defects, and wear behavior of tribological surfaces is undoubtedly the scanning electron microscope (SEM). Especially, the combination of the SEM and x-ray analysis utilizing either energy-dispersive x-ray spectroscopy (EDS) or wavelength-dispersive x-ray spectroscopy (WDS) provides a powerful tool for local microchemical analysis [2.10, 2.41]. The use of electron microprobe techniques in the SEM is now a well-established procedure. The two techniques, EDS and WDS, differ only in the use of an energy-dispersive, solid-state detector versus a wavelength-dispersive crystal spectrometer. Successful studies have been carried out to characterize the surfaces of bearings, seals, gears, magnetic recording components, and other manufactured sliding or rolling surfaces and to select tribological materials and solid lubricants for mechanisms with improved reliability and life.

The SEM is often the first analytical instrument used when a quick look at a material is required and the light microscope no longer provides adequate spatial resolution or depth of focus. The SEM provides the investigator with a highly magnified image of a material surface. Its resolution can approach a few nanometers and it can be operated at magnifications from about 10X to 300 000X. The SEM produces not only morphological and topographical information but information concerning the elemental composition near surface regions.

In the SEM an electron beam is focused into a fine probe and subsequently raster scanned over a small rectangular area. As the electron beam interacts with the specimen, it creates various signals, such as secondary electrons, internal currents, and photon emissions, all of which can be collected by appropriate detectors. The SEM–EDS produces three principal images: secondary electron images, backscattered electron images, and elemental x-ray maps (Fig. 2.45). Secondary and backscattered electrons are conventionally separated according to their energies. They are produced by different mechanisms. When a high-energy primary electron interacts with an atom, it undergoes either inelastic scattering

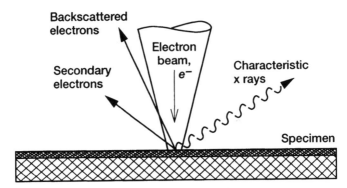

Figure 2.45.—Signals generated when focused electron
beam interacts with specimen in scanning electron
microscope.

with atomic electrons or elastic scattering with the atomic nucleus. In an inelastic collision with an electron some amount of energy is transferred to the other electron. If the energy transferred is extremely small, the emitted electron will probably not have enough energy to exit the surface. If the energy transferred exceeds the work function of the material, the emitted electron will exit the solid. When the energy of the emitted electron is less than about 50 eV, by convention it is referred to as a secondary electron (SE), or simply a secondary. Most of the emitted secondaries are produced within the first few nanometers of the surface. Secondaries produced much deeper in the material suffer additional inelastic collisions, which lower their energy and trap them in the interior of the solid.

Higher energy electrons are primary electrons that have been scattered without loss of kinetic energy (i.e., elastically) by the nucleus of an atom, although these collisions may occur after the primary electron has already lost some energy to inelastic scattering. Although backscattered electrons (BSE's) are, by definition, electrons that leave the specimen with only a small loss of energy relative to the primary electron beam energy, BSE's are generally considered to be the electrons that exit the specimen with an energy greater than 50 eV, including Auger electrons. The BSE imaging mode can be extremely useful for tribological applications, since the energy, spatial distribution, and number of BSE's depend on the effective atomic number of the specimen, its orientation with respect to the primary beam, and the surface condition. The backscatter coefficient, or relative number of electrons leaving the specimen, increases with increasing atomic number, as shown in Fig. 2.46 [2.41]. The higher the atomic number of a material, the more likely it is that backscattering will occur. Thus, as a beam passes from a low-Z (atomic number) to high-Z area, the signal due to backscattering and consequently the image brightness will increase. There is a built-in contrast caused by elemental differences. BSE images can therefore be used to distinguish different phases, transferred films, coatings, and foreign species of the specimen

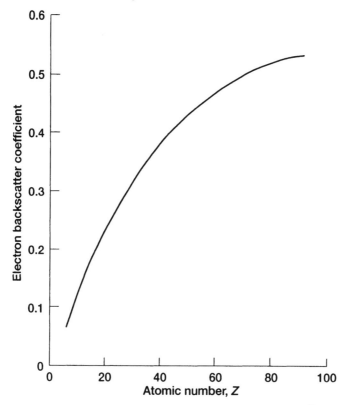

Figure 2.46.—Electron backscatter coefficient as function of atomic number.

having different mean atomic numbers (atomic number contrast). For most specimens examined in SEM, except for those that are flat or polished, the specimen both varies in chemistry from area to area and exhibits a varying rough surface. As a result, both atomic number and topographic contrast are present in the BSE signal (as well as in the SE signal). In general, if the high-energy BSE's are collected from the specimen at a relatively high takeoff angle, atomic number information is emphasized. Conversely, if the high-energy electrons leaving the specimen are collected at a relatively low takeoff angle, topographic information is emphasized. For nearly all BSE applications the investigator is interested in the atomic number contrast and not in the topographic contrast. Note that the backscatter coefficient is defined as the number of BSE's emitted by the specimen for each electron incident on the specimen. Because of the relatively deep penetration of the incident electron beam combined with the extensive range of the BSE's produced, spatial resolution in the BSE mode is generally limited to about 100 nm in bulk specimens under the usual specimen/detector configurations.

Both energy-dispersive and wavelength-dispersive x-ray detectors can be used for element detection in the SEM. When the atoms in a material are ionized by high-energy radiation, usually electrons, they emit characteristic x rays (Fig. 2.45). The detectors produce an output signal that is proportional to the number of x-ray photons in the area under electron bombardment. EDS is a technique of x-ray spectroscopy that is based on the collection and energy dispersion of characteristic x rays. Most EDS applications are in electron column instruments like the SEM, the electron probe microanalyzer (EPMA), and the transmission electron microscope (TEM). X rays entering a solid-state detector, usually made from lithium-drifted silicon, in an EDS spectrometer are converted into signals that can be processed by the electronics into an x-ray energy map or an x-ray energy histogram. A common application of the x-ray systems, such as EDS and WDS, involves x-ray mapping, in which the concentration distribution of an element of interest is displayed on a micrograph. The detectors can be adjusted to pass only the pulse range corresponding to a particular element. This output can then be used to produce an x-ray map or an elemental image. Higher concentrations of a particular element yield higher x-ray photon pulse rates, and the agglomeration of these pulses, which appear as dots in the image, generate light and dark areas relating to the element concentration distribution. In x-ray spectroscopy the x-ray spectrum consists of a series of peaks that represent the type and relative amount of each element in the specimen. The number of counts in each peak can be further converted into elemental weight concentration either by comparison with standards or by standardless calculations. Three modes of analysis are commonly used: spectrum acquisition; spatial distribution or dot mapping of the elements; and element line scans.

Figures 2.47 and 2.48 present examples of secondary electron images, backscattered electron images with both atomic number contrast and topographic contrast, and elemental x-ray maps. These examples were taken from the wear surfaces of a WS_2-film lubricant deposited on an AISI 440C stainless steel disk in sliding contact with a bare AISI 440C stainless steel ball in ultrahigh vacuum for approximately 500 000 disk revolutions. Figures 2.47(a) and 2.48(a) show secondary electron images taken with a primary energy of 20 keV on the disk wear track and the ball wear scar, respectively. The surfaces of the wear track and the wear scar are generally smooth. Likewise, most of the loose wear debris has accumulated on the outside of both the wear track and the wear scar. The darkest areas are essentially AISI 440C stainless steel. The light gray areas are essentially WS_2 wear debris. Darker gray areas, later identified as WS_2-film lubricant, are interveined. Secondary electron imaging generally yields better resolution than backscatter imaging (Figs. 2.47(b) and (c) and 2.48(b) and (c)) and x-ray maps (Figs. 2.47(d) and 2.48(d)). Clearly, chemistry plays an important role in developing image contrast or varying signal when imaging secondary electrons. In other words, a major contribution to the secondary electron image is generated by BSE's.

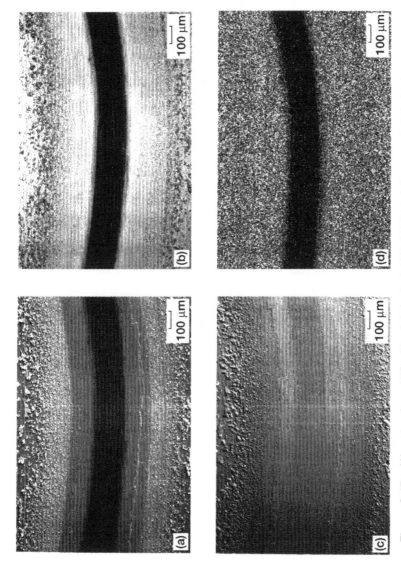

Figure 2.47.—Wear track on WS$_2$-film-lubricated AISI 440C stainless steel disk. (a) Secondary electron images. (b) Backscattered electron images with atomic number contrast. (c) Backscattered electron images with topographic contrast. (d) X-ray maps.

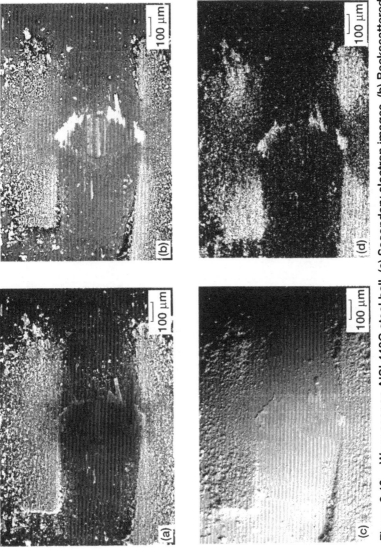

Figure 2.48.—Wear scar on AISI 440C steel ball. (a) Secondary electron images. (b) Backscattered electron images with atomic number contrast. (c) Backscattered electron images with topographic contrast. (d) X-ray maps.

Figures 2.47(b) and (c) and 2.48(b) and (c) show atomic-number-contrast and topographic-contrast BSE images of the ball and disk, respectively. Atomic number contrast can be used to distinguish two materials, AISI 440C stainless steel and WS_2, because the actual BSE signal increases somewhat predictably with the concentration of the heavier element of the pair, as described above. In other words, higher Z (atomic number) areas should always be bright owing to the higher backscatter coefficient. In the present example (Figs. 2.47(b) and 2.48(b)) the light areas in the backscatter photomicrographs show where the most WS_2-film lubricant is present, and the dark areas show the AISI 440C stainless steel where no or less WS_2-film lubricant is present.

The topographic contrast information in Figs. 2.47(c) and 2.48(c) clearly indicates WS_2 wear debris on both wear track and wear scar, smeared WS_2 film on both sides of the wear track, and transferred WS_2 film on the wear scar. Even though for nearly all BSE applications the investigator is interested in the atomic number contrast and not in the topographic contrast, the topographic contrast of the BSE imaging mode can be extremely useful in solid lubricant investigations, especially wear analysis of solid lubricants. The tungsten (M_α, L_α, and L_β) and sulfur (K_α and K_β) x-ray maps in Figs. 2.47(d) and 2.48(d), respectively, are quite revealing. Areas of the WS_2 wear debris, WS_2 transferred films, and WS_2-film lubricant yield a high x-ray count. Major chemistry differences are quite obvious.

With EDS the output signal is displayed as a histogram of counts versus x-ray energy displays. The x-ray energy spectrum shown in Fig. 2.49(a) was produced by allowing the electron beam to dwell on one of the WS_2-wear-debris-rich areas deposited on the WS_2 film coating the disk specimen. The spectrum shows the presence of peaks corresponding to tungsten and sulfur and a small amount of iron (Fe) and chromium (Cr) from the AISI 440C stainless steel substrate material. Because this specimen was slightly contaminated with carbon and oxygen during deposition of the WS_2 film onto the substrate, the small carbon and oxygen peaks were expected. The spectrum shown in Fig. 2.49(b) was produced by allowing the electron beam to dwell on one of the WS_2-wear-debris-rich areas deposited on the counter material (the AISI 440C stainless steel ball specimen). The spectrum shows the presence of peaks corresponding to tungsten and sulfur and a large amount of iron and chromium from the AISI 440C stainless steel ball material. Small amounts of carbon and oxygen are also present in the spectrum.

Note that when examining a micrograph it is important to remember that the average human eye may perceive a spatial resolution on the order of 0.2 mm. Figure 2.50 presents the size of a 1-mm feature in a micrograph as a function of magnification. The size of the smallest feature resolvable by the human eye (0.2 mm) is also given [2.41]. The magnification relates to the final magnification (including enlargement if done) of the micrograph. Figure 2.50 indicates that for a photomicrograph to appear perfectly sharp to the eye at 20 000X magnification, the instrument or electron optical resolution must be 10 nm or better.

90

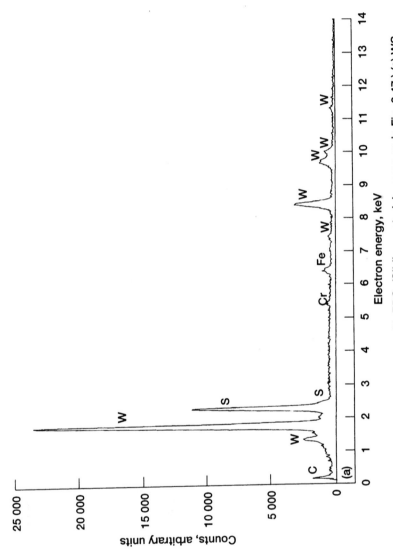

Figure 2.49.—X-ray energy spectra with EDS. (Sliding materials same as in Fig. 2.47.) (a) WS$_2$-film-coated AISI 440C stainless steel disk. (b) AISI 440C stainless steel ball.

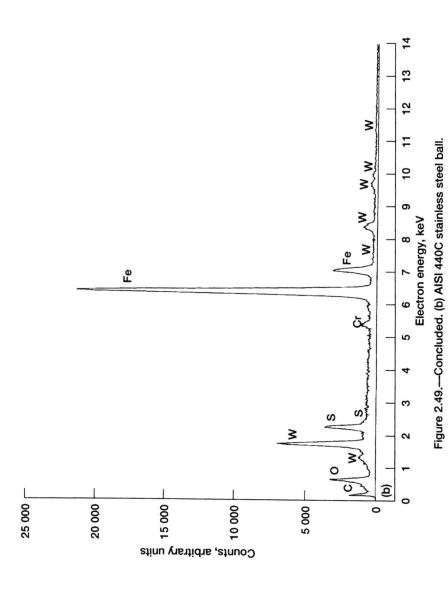

Figure 2.49.—Concluded. (b) AISI 440C stainless steel ball.

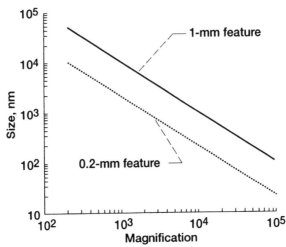

Figure 2.50.—True sizes of 1-mm image feature and 0.2-mm image feature (size of smallest feature resolvable by human eye) in photomicrograph as function of scanning electron microscope's magnification.

2.4 Surrounding Environments

The environment may be a conventional, normal atmospheric condition, such as air in the terrestrial environment and vacuum in space, or some artificially imposed environment, such as inert gas, reactive gas, or liquid lubricant.

2.4.1 Substances in Environments

Substances in the surrounding environment have physical, chemical, and mechanical effects on the composition, structure, chemical state, and electronic state of the surfaces of tribological components and solid lubricating films in their formations, reactions, and combinations. Most substances in the surrounding environment fall under the categories of organic materials, inorganic materials, radiation, and/or vacuum ([2.42] and Table 2.4).

Organic materials contain the elements carbon and (usually) hydrogen as a key part of their structures, and they are usually derived from living things. Polymers, natural resins, soils, some foods, and living organisms are organic solids. Liquid lubricants, fuels, chemicals, oils, paints, and some foods are organic liquids. Hydrocarbons and carbon oxides are organic gases.

Inorganic materials are those substances not derived from living things. Solids such as metals, ceramics, composites, glasses, clays, cements, sand, and rock; liquids such as water, acids, bases, and drugs; and gases such as chlorine, argon, and helium are all inorganic.

TABLE 2.4.—SUBSTANCES IN SURROUNDING ENVIRONMENT

Organic substances			Inorganic substances		
Solids	Liquids	Gases	Solids	Liquids	Gases
Polymers	Lubricants	Hydrocarbons	Metals	Water	Oxygen
Natural resins	Chemicals	Carbon oxides	Ceramics	Acids	Nitrogen
Living	Fuels		Composites	Bases	Argon
organisms	Oils		Glasses	Drugs	Helium
Soils	Paints		Stone		Sulfur
Foods	Foods		Lubricants[a]		Chlorine
					Atomic
					oxygen
					Ozone

[a]For example, MoS_2 and graphite.

Radiation	Vacuum
Gamma rays	Industrial vacuum processes
X rays	Low vacuum ($>10^2$ to 10^5 Pa)
Ultraviolet	Medium vacuum (10^{-1} to 10^2 Pa)
Visible spectra	High vacuum (10^{-5} to 10^{-1} Pa)
Infrared	Ultrahigh vacuum ($<10^{-5}$ Pa)
Hertzian waves	Outer space (10^{-4} to 10^{-9} Pa)

TABLE 2.5.—WAVELENGTHS OF VARIOUS TYPES
OF RADIATION

Type of radiation	Wavelength, μm
Gamma rays	0.005 to 0.140
X rays	0.01 to 10
Ultraviolet[a]	<400
Visible spectrum	400 to 700
Violet (representative, 410 μm)	400 to 424
Blue (representative, 470 μm)	424.0 to 491.2
Green (representative, 520 μm)	491.2 to 575.0
Maximum visibility	556
Yellow (representative, 580 μm)	575.0 to 585.0
Orange (representative, 600 μm)	585.0 to 647.0
Red (representative, 650 μm)	647.0 to 700.0
Infrared	>700
Hertzian waves	$>0.22 \times 10^6$

[a]Limit of Sun's ultraviolet at Earth's surface, 292.

Wavelengths of various radiations vary from less than 1 nm (high-energy radiation such as gamma rays and x rays) to a few hundred nanometers or more (e.g, infrared), as shown in Table 2.5. In nuclear plants, space systems such as communication satellites, and bridges solid lubricants are exposed to radiation over long times. The resistance of various solid lubricants to extreme temperatures has significantly expedited their development. Various types of solid lubricant perform well against thermal radiation at temperatures in the cryogenic region and up to 1000 °C.

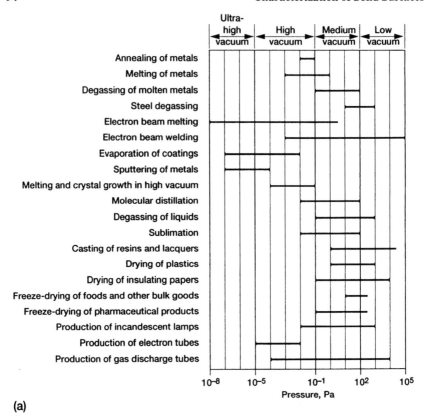

(a)

Figure 2.51.—Fields of application of high vacuum. (a) Industrial applications. (b) Research and development applications.

Vacuum ranges involved in industrial vacuum processes vary from low to ultrahigh vacuum (Fig. 2.51). For example, a steel degassing process requires a vacuum range of 10 to 10^3 Pa, and a metal sputtering process requires a vacuum range of 10^{-7} to 10^{-4} Pa. Solid lubricants offer one of the most promising methods of space system lubrication in the vacuum range 10^{-4} to 10^{-9} Pa.

2.4.2 Particulates and Fine Dusts

The environment can be polluted by particulates and fine dusts created by people's activities and by nature [2.43]. In 1974 the U.S. Environmental Protection Agency set a total suspended particulates standard of 75 mg/m^3 as a primary standard for the protection of public health. A secondary standard of 60 mg/m^3 was established as necessary to protect the public welfare. Actual measurements at background sites have consistently shown counts of about 35 mg/m^3. Measurements at national monitoring sites run close to the tolerance level of

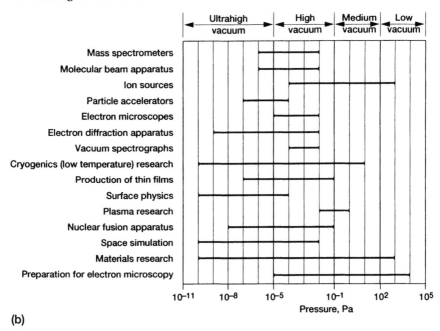

(b)

Figure 2.51.—Concluded. (b) Research and development applications.

75 mg/m^3. Small particles in polluted air are controlled hardly at all by gravitational forces. Once they enter the atmosphere, their residence time is likely to range from a month to several years. The suspended atmospheric particles are of four general types. The first is the light ions produced in the air by cosmic rays and radioactivity. They consist of small aggregates of molecules having dimensions up to a few molecular diameters. The second important type of particle consists of the so-called Aitken nuclei. These particles range in radius from 2 nm up to 2 mm. They are prevalent near the Earth's surface, particularly near cities. As a rule of thumb one anticipates finding 100 000 of these particles per cubic centimeter in a large city, 10 000/cm^3 in the country, and 1000/cm^3 at sea. The numbers decrease with increasing altitude and only 10% of the surface populations are found at an altitude of 7000 m. The fine-particle pollution of the air is largely composed of these nuclei. The third type of atmospheric particle is the cloud droplet having a radius from 1 to 50 mm. Finally, cloud droplets associate to form raindrops or snowflakes that fall at velocities dependent on their size.

Particles in the environment are involved in virtually all tribological situations that produce contamination, abrasive wear, corrosion, and erosion. They adsorb on surfaces and change surface properties and tribological behavior. Fine particles in the atmosphere may also influence lifetime and reliability in all mechanical systems with moving parts. Despite their importance they have not received the same attention as other areas of tribology.

References

2.1 P.F. Kane and G.B. Larrabee, *Characterization of Solid Surfaces*, Plenum Press, New York, 1974.

2.2 F.P. Bowden and D. Tabor, *The Friction and Lubrication of Solids—Part 1*, Clarendon Press, Oxford, UK, 1954.

2.3 F.P. Bowden and D. Tabor, *The Friction and Lubrication of Solids—Part 2*, Clarendon Press, Oxford, UK, 1964.

2.4 *Surface Analysis and Pretreatment of Plastics and Metals* (D.M. Brewis, ed.), Macmillan Publishing Co., New York, 1982.

2.5 D. Briggs and M.P. Seah, *Practical Surface Analysis: By Auger and X-Ray Photo-Electron Spectroscopy*, Vol. 1, John Wiley & Sons, New York, 1983.

2.6 D.H. Buckley, *Surface Effects in Adhesion, Friction, Wear, and Lubrication*, Elsevier Scientific Publishing Co., New York, 1981.

2.7 Y.W. Chung and H.S. Cheng, *Advances in Engineering Tribology*, STLE SP–31, Society of Tribologists and Lubrication Engineers, Park Ridge, IL, 1991.

2.8 Y.W. Chung, A.M. Homola, and G.B. Street, *Surface Science Investigations in Tribology: Experimental Approaches*, Developed From the 201st National Meeting of the American Chemical Society Symposium, American Chemical Society, Washington, DC, 1992.

2.9 W.A. Glaeser, *Characterization of Tribological Materials*, Butterworth-Heinemann, Boston, MA, 1993.

2.10 C.R. Brundle, C.A. Evans, Jr., and S. Wilson, Jr., *Encyclopedia of Materials Characterization: Surfaces, Interfaces, Thin Films*, Butterworth-Heinemann, Boston, MA, 1992.

2.11 K. Miyoshi and Y.W. Chung, *Surface Diagnostics in Tribology: Fundamental Principles and Applications*, World Scientific Publishing Co., River Edge, NJ, 1993.

2.12 K.F. Dufrane and W.A. Glaeser, Rolling-contact deformation of MgO single crystals, *Wear 37*: 21–32 (1976).

2.13 R.P. Steijn, On the wear of sapphire, *J. Appl. Phys. 32, 10*: 1951–1958 (1961).

2.14 K. Tanaka, K. Miyoshi, Y. Miyao, and T. Murayama, Friction and deformation of Mn-Zn ferrite single crystals, *Proceedings of the JSLE–ASLE International Lubrication Conference* (T. Sakurai, ed.), Elsevier Scientific Publishing Co., Amsterdam, 1976.

2.15 K. Miyoshi and D.H. Buckley, Friction, deformation, and fracture of single-crystal silicon carbide, *ASLE Trans. 22*: 79–90 (1979).

2.16 H. Ishigaki, K. Miyoshi, and D.H. Buckley, Influence of corrosive solutions on microhardness and chemistry of magnesium oxide {001} surface, NASA TP–2040 (1982).

2.17 K. Miyoshi, D.H. Buckley, G.W.P. Rengstorff, and H. Ishigaki, Surface effects of corrosive media on hardness, friction, and wear of materials, *Ind. Eng. Chem. Prod. Res. Dev. 24, 3*: 425–431 (1985).

2.18 S. Hattori, K. Miyoshi, D.H. Buckley, and T. Okada, Plastic deformation of a magnesium oxide {001} surface produced by cavitation, *Lubric. Eng. 44, 1*: 53–58 (1988).

2.19 J. Narayan, The characterization of the damage introduced during micro-erosion of MgO single crystals, *Wear 25*: 99–109 (1973).

2.20 J. Narayan, Physical properties of a ⟨100⟩ dislocation in magnesium oxide, *J. Appl. Phys. 57, 8*: 2703–2716 (1985).

2.21 P.T.B. Shaffer, Effect of crystal orientation on hardness of silicon carbide, *J. Am. Ceram. Soc. 47, 9*: 466 (1964).

2.22 O.O. Adewoye, G.R. Sawyer, J.W. Edington, and T.F. Page, Structural studies of surface deformation in MgO, SiC, and Si_3N_4, Annual Technical Report AD–A008993, Cambridge University Press, Cambridge, England, 1974.

2.23 A.G. Guy, Deformation of materials, *Introduction to Materials Science*, McGraw-Hill, New York, 1972, pp. 399–451.

2.24 R.M. Brick, A.W. Pense, and R.B. Gordon, Strengthening mechanisms; deformation hardening and annealing, *Structure and Properties of Engineering Materials*, McGraw-Hill, New York, 1977, pp. 69–92.

2.25 S. Amelinckx, G. Strumane, and W.W. Webb, Dislocations in silicon carbide, *J. Appl. Phys. 31, 8:* 1359–1370 (1960).

2.26 C.A. Brookes, J.B. O'Neill, and B.A.W. Redfern, Anisotropy in the hardness of single crystals, *Proc. R. Soc. London, A 322:* 73–88 (1971).

2.27 F.W. Daniels and C.G. Dunn, The effect of orientation on Knoop hardness of single crystals of zinc and silicon ferrite, *Trans. Am. Soc. Met. 41:* 419–442 (1949).

2.28 D.N. French and D.A. Thomas, Hardness anisotropy and slip in WC crystals, *Trans. Metall. Soc. AIME 233:* 950–952 (May 1965).

2.29 D.H. Buckley and K. Miyoshi, Friction and wear of ceramics, *Wear 100:* 333–353 (1984).

2.30 J.B.P. Williamson, The shape of surfaces, *CRC Handbook of Lubrication* (E.R. Booser, ed.), CRC Press Inc., Boca Raton, FL, Vol. II, 1984, pp. 3–16.

2.31 Y.W. Chung, Characterization of topography of engineering surfaces, *Surface Diagnostics in Tribology: Fundamental Principles and Applications* (K. Miyoshi and Y.W. Chung, eds.), World Scientific Publishing, River Edge, NJ, 1993, pp. 33–46.

2.32 R.S. Howland, J. Okagaki, and L. Mitobe, *How To Buy a Scanning Probe Microscope*, Park Scientific Instruments, 1993.

2.33 T. Tsukizoe, *Precision Metrology*, Yokkendo Publishing, Tokyo, Japan, 1970, pp. 180–199.

2.34 T. Hisakado, On the mechanism of contact between solid surfaces, *Bull. JSME 12, 54:* 1519–1527, 1528–1536 (1969).

2.35 T. Tsukizoe, T. Hisakado, and K. Miyoshi, Effects of surface roughness on shrinkage fits, *Bull. JSME 17, 105:* 359–366 (1974).

2.36 Y. Kumada, K. Hashizume, and Y. Kimura, Performance of plain bearings with circumferential microgrooves, *Tribol. Trans 39:* 81–86 (1996).

2.37 D.H. Buckley and K. Miyoshi, Tribological properties of structural ceramics, *Structural Ceramics* (J.B. Wachtman, Jr., ed.), Academic Press, Boston, MA, 1989, pp. 203–365.

2.38 P.G. Lurie and J.M. Wilson, The diamond surface, *Surf. Sci. 65:* 476–498, North-Holland Publishing Co. (1976).

2.39 S. Evans, R.G. Pritchard, and J.M. Thomas, Escape depths of x-ray (Mg K_α)-induced photoelectrons and relative photoionization cross sections for the 3p subshell of the elements of the first long period, *J. Phys. C. 10, 13:* 2483–2498 (July 1977).

2.40 P. Cadman, S. Evans, J.D. Scott, and J.M. Thomas, Determination of relative electron inelastic mean free paths (escape depths) and photoionization cross sections by x-ray photoelectron spectroscopy, *J. Chem. Soc., Faraday Trans. II 71, 10:*1777–1784 (1975).

2.41 J.T. Norton and G.T. Cameron, Sr., eds., *Electron optical and x-ray instrumentation for research, product assurance and quality control*, Amray Technical Bulletins, Vol. 2, No. 1, Amray, Inc., Bedford, MA, Jan. 1986.

2.42 K.G. Budinski, *Engineering Materials Properties and Selection*, Prentice Hall, Englewood Cliffs, NJ, 1992.

2.43 M. Sittig, *Particulates and Fine Dust Removal Processes and Equipment*, Noyes Data Corp., Park Ridge, NJ, 1977.

2.44 H. Hantsche, Comparison of basic principles of the surface-specific analytical methods AES/SAM, ESCA(XP), SIMS and ISS with x-ray microanalysis, *Scanning 11, 6:* 257–280, FACM, Inc., Mahwah, NJ (1989).

Chapter 3
Properties of Clean Surfaces: Adhesion, Friction, and Wear

3.1 Introduction and Approach

As described in Chapter 2, a contaminant layer may form on a solid surface either by the surface interacting with the environment or by the bulk contaminant diffusing through the solid itself. Thin contaminant layers, such as adsorbed gases, water vapor, and hydrocarbons of atomic dimensions (approximately 2 nm thick), are unavoidably present on every surface of any solid that has been exposed to air. Surface analysis techniques, particularly x-ray photoelectron spectroscopy (XPS) and Auger electron spectroscopy (AES), are well suited for examining these thin contaminant layers. However, contaminant surface layers can affect the spectrum by attenuating the electron signal from the underlying surface, thereby masking spectral features related to the bulk material [3.1–3.3].

Contamination is an important factor in determining such solid surface properties as adhesion and friction. Contaminant layers can greatly reduce adhesion and friction and, accordingly, provide lubrication. The adhesion, friction, and wear behaviors of contaminated surfaces will be discussed in Chapter 4.

Because contaminants are weakly bound to the surface, physically rather than chemically, they can be removed by bombarding them with rare gas ions (e.g., argon ions) or by heating to say 700 °C [3.4, 3.5]. Contaminant surface layers can also be removed by repeated sliding, making direct contact of the fresh, clean surfaces unavoidable [3.3, 3.6]. This situation applies in some degree to contacts sliding in air, where fresh surfaces are continuously produced by a counterfacing material. It also applies in vacuum tribology to wear-resistant components used in aerospace mechanisms, semiconductor-processing equipment, machine tool spindles, etc. Obviously, understanding the behavior of clean surfaces in metal-ceramic couples is of paramount practical importance.

This chapter presents the fundamental tribology of clean surfaces (i.e., the adhesion, friction, and wear behaviors of smooth, atomically clean surfaces of solid-solid couples, such as metal-ceramic couples, in a clean environment). Surface and bulk properties, which determine these behaviors, are described. The primary emphasis is on the nature and character of the metal, especially its surface energy and ductility. Also, the friction and wear mechanisms of clean, smooth surfaces are stated.

To understand the adhesion and friction behaviors of atomically clean solid surfaces, a simple experimental approach has been taken to control and characterize as carefully as possible the materials and environment in tribological studies [3.1, 3.4, 3.7–3.11]. High-purity metals are used, as much as possible, in an ultrahigh vacuum (Fig. 3.1) that contains an XPS or AES spectrometer [3.7–3.11]. Adsorbed contaminant layers (water vapor, carbon monoxide and dioxide, hydrocarbons, and oxide layers) are removed by argon sputtering. Surface cleanliness is verified by AES or XPS (see Chapter 2). Adhesion and friction are measured by a pin-on-flat configuration, as shown in Fig. 3.1. Removing contaminant films from

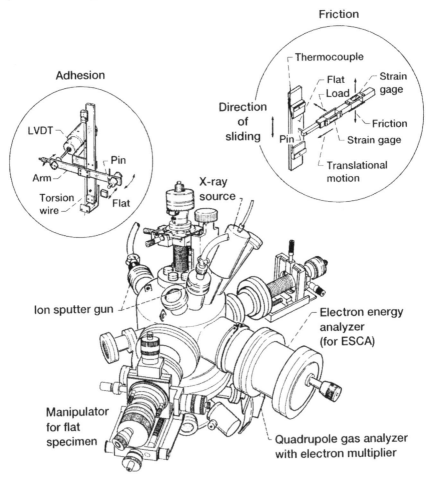

Figure 3.1.—Apparatus for measuring adhesion and friction in ultrahigh vacuum.

the surfaces of solids has enabled us to better understand the surface and bulk properties that influence adhesion and friction when two such solids are brought into contact in an ultrahigh vacuum.

3.2 Adhesion Behavior

When smooth, atomically clean solid surfaces are brought into contact under a normal load, the atoms must be in contact at some points. Thus, interatomic forces will come into play [3.1, 3.2, 3.12] and cause some adhesion at these points.

Adhesion, a manifestation of mechanical strength over an appreciable area, has many causes, including chemical bonding, deformation, and the fracture processes involved in interface failure [3.1, 3.2, 3.13–3.21]. Adhesion undoubtedly depends on the area of real contact, the micromechanical properties of the interface, and the modes of junction rupture. However, there is no satisfactory theory or experimental method for determining the area of real contact. Vibration, which may cause junction (contact area) growth in the contact zone, and the environment also influence the adhesion and deformation behaviors of solids. There are many unknown and unresolved problems. Therefore, adhesion studies of solids are best performed only through refined experiments under carefully controlled laboratory conditions, such as in an ultrahigh vacuum or an inert gas, to reduce secondary effects.

In practical cases, adhesion develops in the film formation processes of joining, bonding, and coating. Beneficially, it is a crucial factor in the structural performance of engineering materials, including monolithics, composites, and coatings, used in engines, power trains, gearboxes, and bearings [3.22–3.27]. The joining of solid to solid, fiber to matrix, and coating to substrate is determined by adhesion. Destructively, adhesion occurs during friction and wear in solid-state contacts, causing high friction and heavy surface damage.

We can use a variety of methods to quantify bonding forces. Some, such as pull-off force measurements, involve tensile pulling on the interface. Others, such as friction force measurements, are based on tangential shearing of the junction [3.3]. The stronger the interfacial bond strength, the greater the resistance to separate (pull off) or to move one surface relative to the other normally or tangentially. Such measurements are sufficiently sensitive that the adhesive bond forces for different-material couples, when two atomically clean material surfaces are brought into solid-state contact, can be readily quantified.

A torsion balance was used to measure adhesion in this study. The balance was adapted from the principle of the Cavendish balance used to measure gravitational forces in 1798–99 and also from a similar balance invented by Coulomb in 1784–85 for studying electrical attraction and repulsion [3.28, 3.29].

The adapted torsion balance [3.10, 3.11] consists of a solid (A) and a displacement sensor, such as an electromechanical transducer, mounted at opposite ends of a horizontal arm (Fig. 3.2). The arm is supported at its center by a vertical wire,

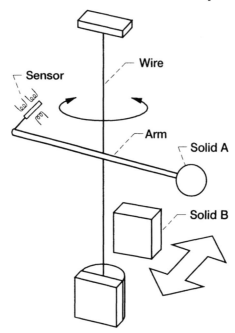

Figure 3.2.—Schematic diagram of torsion
balance adapted from Cavendish balance.

perhaps a single strand of music wire. Another solid (B) is moved horizontally toward solid A, presses against it, and twists the wire through a small angle with normal force—the normal loading process—thereby moving the sensor. Solid B is then gradually moved horizontally backward until the two solids are pulled apart in a normal direction—the unloading process. If the adhesive force between the two solids is zero, solid A separates from solid B at its original position and untwists the wire, thereby moving the sensor back to its original position. If an adhesive force is present between the two solids, the force twists the wire as solid B moves backward until the wire develops sufficient force to separate the surfaces of solids A and B in the normal direction.

In this system the attractive force of adhesion and the force required to pull the surfaces of two solids apart (the pull-off force) act along a horizontal direction and are not affected by gravity and buoyancy. The axis of weight and buoyancy for all the components (the arm, sensor, and wire) is different from that of the pull-off microforce to be measured and is in the vertical direction because of gravity.

Because the pull-off force is measured by the torsional moment acting on the torsion wire, the force can be calibrated in three ways:

1. By calculation from the geometrical shape of the torsion wire, such as its length and area of section

2. By calculation from measured natural periods of the arm's harmonic motion when it is freely oscillating

3. By direct comparison of microforce with standard weight when the arm and torsion wire are held horizontally

The pull-off forces determined by all three methods of calibration were nearly the same [3.10].

For the actual balance shown in Fig. 3.1 the pin specimen (corresponding to solid A in Fig. 3.2) was mounted at one end of a movable arm. A free-moving, rod-shaped magnetic core was mounted on the other end of the arm. The coils of a linear variable differential transformer (LVDT) were mounted on a stationary arm. There was no physical contact between the movable magnetic core and the coil structure. The movable arm was supported by a single strand of music wire acting as a torsion spring. The flat (corresponding to solid B in Fig. 3.2) was mounted on a specimen attached to a manipulator, allowing electron beam specimen heating in a vacuum. Therefore, measurements could be made in ultrahigh vacuum, even to temperatures as high as 1200 °C.

For in situ pull-off force (adhesion) measurements in vacuum the flat specimen was brought into contact with the pin specimen by moving the micrometer headscrew forward manually. Contact was maintained for 30 s; then the pin and flat specimen surfaces were pulled apart by moving the micrometer headscrew backward. An LVDT monitored the displacement of the pin specimens. Figure 3.3 shows a typical force-time trace resulting from such adhesion experiments. Contact occurred at point A. The line A–B represents the region where load was being applied. The displacement B–X corresponds to the normal load. The line B–C represents the region where the contact was maintained at the given load and the specimen surfaces were stationary. The line C–D represents the region where both

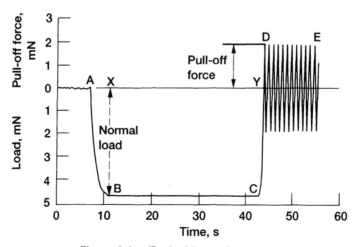

Figure 3.3.—Typical force-time trace.

the unloading point and the separation forces were being applied on the adhesion joint. The onset of separation occurred at point D. The displacement D–Y corresponds to the pull-off force. After the pin specimen separated from the flat, the pin fluctuated back and forth, as represented by the D–E region.

Atomically clean solids will exhibit strong adhesive bonds when brought into solid-state contact. A number of bulk and surface properties of solids have been shown to affect the nature and magnitude of the adhesive bond forces that develop for solids. Surface properties include electronic surface states, ionic species present

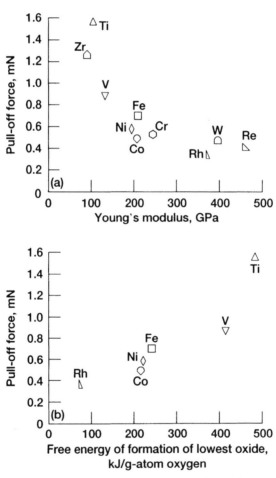

Figure 3.4.—Pull-off force (adhesion) for various metals in contact with ferrites (MnO–ZnO–Fe$_2$O$_3$) in ultrahigh vacuum. (a) As function of Young's modulus of metal. (b) As function of free energy of formation of lowest metal oxide.

TABLE 3.1.—CONDITIONS OF EXPERIMENTS IN ULTRA-HIGH-
VACUUM ENVIRONMENT

[Hemispherical pin (0.79-mm radius) and flat specimens were polished
with 3-μm diamond powder and 1-μm sapphire powder, respectively.
Both specimens were argon sputter cleaned.]

Condition	Adhesion (pull-off force) measurements	Friction measurements
Load, N	0.0002 to 0.002	0.05 to 0.5
Vacuum, Pa	10^{-8}	10^{-8}
Temperature, °C	23	23
Motion	Axial	Unidirectional sliding
Sliding velocity, mm/min	---------------	1
Total sliding distance, mm	---------------	2.5 to 3

at the surface, chemistry, and the surface energy of the contacting materials. Bulk
properties include elasticity, plasticity, fracture toughness, cohesive bonding energy,
defects, and the crystallography of the materials.

Figure 3.4 presents the pull-off force, which reflects interfacial adhesion, measured for various argon-ion-sputter-cleaned metals in contact with argon-ion-
sputter-cleaned ferrites in ultrahigh vacuum. Table 3.1 (from [3.10]) gives the
conditions of the adhesion experiments. As Fig. 3.4(a) shows, pull-off force
decreased as the Young's modulus E (also known as the elastic modulus) of the
metal increased. Thus, the bulk properties of the metal, such as Young's modulus,
affect the magnitude of the adhesive bond forces that develop at the metal-ceramic
interface. Similar pull-off force (adhesion) results were obtained for clean metal-
silicon nitride (Si_3N_4) couples [3.10, 3.11].

Figure 3.4(b) shows that the pull-off forces for clean metal-ferrite couples
increased as the free energy of formation of the lowest metal oxides increased. This
correlation suggests that the adhesive bond at the metal-ceramic interface is a
chemical bond between the metal atoms on the metal surface and the large oxygen
anions on the ferrite ($MnO-ZnO-Fe_2O_3$) surface. Further, Fig. 3.4(b) indicates that
the strength of this chemical bond is related to the oxygen-to-metal bond strength
in the metal oxide. Similar adhesion behavior has been noted with other oxide
ceramics, such as nickel-zinc ferrite ($NiO-ZnO-Fe_2O_3$) and sapphire (Al_2O_3)
[3.10, 3.30].

3.3 Friction Behavior

In situ friction experiments were conducted with the friction device shown in
Fig. 3.1. Table 3.1 gives the conditions of the friction experiments.

Figure 3.5 presents the coefficient of friction, which reflects interfacial adhesion,
measured for various argon-ion-sputter-cleaned metals in contact with
argon-ion-sputter-cleaned ferrites in ultrahigh vacuum. As Fig. 3.5(a) shows, the
coefficient of friction decreased as the shear modulus of the metal G increased.

Thus, the bulk properties of the metal, such as shear modulus (also known as the torsion modulus or the modulus of rigidity), play an important role in the friction behavior of clean metal-ferrite couples. The similar shapes of Figs. 3.4(a) and 3.5(a) are not surprising because $E \approx 2.6G$ (as discussed in [3.31] and briefly later).

Figure 3.5(b) shows that the coefficient of friction increased as the free energy of formation of the lowest metal oxides increased. This correlation suggests that the adhesive bond at the metal-ceramic interface is a chemical bond between the metal atoms on the metal surface and the large oxygen anions on the ferrite (MnO–ZnO–Fe_2O_3) surface (as shown in Fig. 3.6). It also suggests that the strength of this chemical bond is related to the oxygen-to-metal bond strength in the metal oxide.

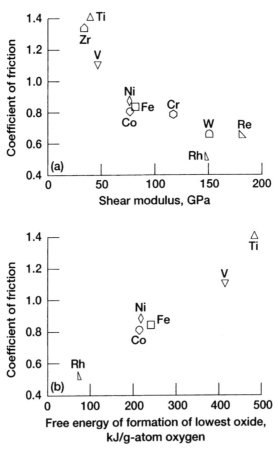

Figure 3.5.—Coefficient of friction for various metals in contact with ferrites (MnO–ZnO–Fe$_2$O$_3$) in ultrahigh vacuum. (a) As function of shear modulus of metal. (b) As function of free energy of formation of lowest metal oxide.

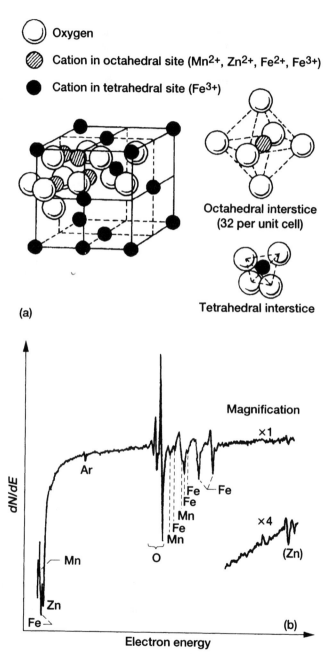

Figure 3.6.—Structure and surface of Mn-Zn ferrite.
(a) Spinel structure. (b) AES spectrum for {110}
surface after sputter cleaning.

Similar relationships have been observed with $NiO–ZnO–Fe_2O_3$ [3.9]. This dependence of friction on the shear modulus and chemical activity of the metal is analogous to the adhesion behavior described in the previous section.

Figure 3.6(a) illustrates the spinel crystal structure of manganese-zinc (Mn-Zn) ferrite. In the unit cell, which contains 32 oxygen ions, there are 32 octahedral sites and 64 tetrahedral sites. Sixteen of the octahedral sites are filled with equal amounts of divalent (Mn^{2+}, Zn^{2+}, and Fe^{2+}) and trivalent (Fe^{3+}) ions, and eight of the tetrahedral sites are filled with trivalent (Fe^{3+}) ions [3.32, 3.33]. The Auger peaks in Fig. 3.6(b) indicate that, in addition to oxygen and iron, small amounts of manganese and zinc occur on a clean Mn-Zn ferrite surface. The surface accommodated slightly more oxygen with the {110} plane than with the {211}, {111}, and {100} planes, in that order.

The values of the Young's and shear moduli used in this investigation of bulk polycrystalline metal were those reported by Gschneidner [3.31]. Young's modulus varies from 3.538 GPa (0.0361×10^6 kg/cm^2) for potassium to 1127 GPa (11.5×10^6 kg/cm^2) for diamond. Estimated values, however, would indicate that the lower limit is probably 1.6 GPa (0.017×10^6 kg/cm^2) for francium. A recent calculation for a hypothetical material, carbon nitride in β-C_3N_4 structure, predicted a bulk modulus comparable to that for diamond ($\beta = 410$ to 440 GPa) [3.34, 3.35]. Gschneidner reported that the ratio of Young's modulus to shear modulus is essentially constant (at nearly 2.6) and that the shear modulus, like Young's modulus, markedly depends on the metal's electron configuration (i.e., the group in which it lies). The maximum value encountered in a given period of the periodic table is associated with the elements having the most unpaired d electrons. The minimum near the end of each period occurs for the elements having an s^2p^1 configuration.

The adhesion and friction behaviors described here for oxide ceramics, such as ferrites, in contact with metals are not unique to oxide ceramics. Analogous behaviors occur for metals in contact with other nonmetallic materials.

3.4 Wear Behavior

Inspection of all the metal and ceramic surfaces after sliding contact revealed that the metal deformation was principally plastic and that the cohesive bonds in the metal fractured [3.36–3.38]. All the metals that were examined failed by shearing or tearing and were transferred to the ceramic during sliding. Because the interfacial bond between the metal and the ceramic is generally stronger than the cohesive bond within the metal, separation generally took place in the metal when the junction was sheared. Pieces of the metal were torn out and transferred to the ceramic surface. For example, when an atomically clean silicon carbide (SiC) surface was brought into contact with a clean aluminum surface, the interfacial adhesive bonds that formed in the area of real contact were so strong that shearing or tearing occurred locally in the aluminum. Consequently, aluminum wear debris particles were transferred to the SiC surface during sliding, as verified by a scanning electron micrograph and an aluminum K_α x-ray map (Fig. 3.7).

The morphology of metal transfer to ceramic revealed that metals with a low shear modulus exhibited much more wear and transfer than those with a higher shear modulus. Further, the more chemically active the metal, the greater was the metal wear and transfer to the ceramic.

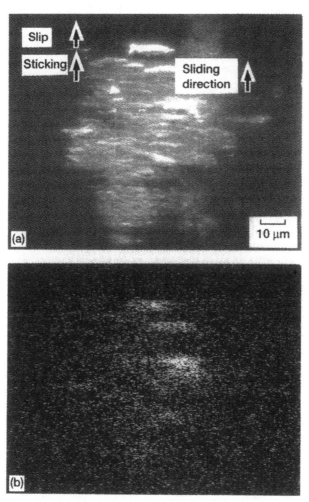

Figure 3.7.—Aluminum transferred to SiC {0001} surface before and after single-pass sliding in ultrahigh vacuum. (a) Initial contact area. (b) Aluminum K_α x-ray map (1.5×10^4 counts).

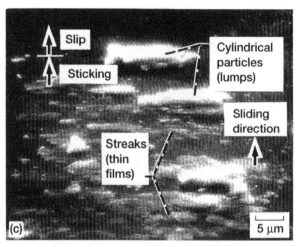

Figure 3.7.—Concluded. (c) Aluminum wear debris.

TABLE 3.2.—METALS TRANSFERRED TO SiC {0001} SURFACES AFTER
10 SLIDING PASSES IN ULTRAHIGH VACUUM

Metal	Form (size) of metal transferred				Extent of metal transfer	Shear modulus , GPa
	Small particle[a]	Piled-up particles[b]	Multilayer agglomeration	Large lump particle[b]		
Al	Yes	Yes	Yes	No	Most	27
Zr	Yes	Yes	Yes	No		34
Ti	Yes	Yes	Yes	No		39
Ni	Yes	Yes	No	No		75
Co	Yes	Yes	No	No		76
Fe	Yes	Yes	No	No		81
Cr	Yes	Yes	No	No		117
Rh	Yes	No	No	Yes		147
W	Yes	No	No	Yes		150
Re	Yes	No	No	Yes	Least	180

[a]Submicrometer.
[b]Several micrometers.

Table 3.2 summarizes the type of metal transfer to single-crystal SiC that was observed after multipass sliding. Generally, the metals at the bottom of the table had a higher shear modulus and less chemical affinity for silicon and carbon. Therefore, those metals exhibited less wear and transferred less metal to the SiC.

Note that sometimes the strong adhesion and high friction between a metal and a ceramic can locally damage the ceramic surface if that surface contains imperfections, such as microcracks or voids [3.36–3.38].

3.5 Relationship of Material Properties to Adhesion, Friction, and Wear

The tribological properties of clean, smooth, solid surfaces depend on the physical, mechanical, and metallurgical properties of the surface. As discussed in Sections 3.2 to 3.4, the physical properties, such as the Young's and shear moduli, influence observed adhesion, friction, and wear behaviors.

3.5.1 Mechanical Properties

Theoretical tensile strength.—A clean metal in sliding contact with a clean nonmetal or with itself will fail either in tension or in shear because some of the interfacial bonds are generally stronger than the cohesive bonds in the cohesively weaker metal. The failed metal subsequently transfers to nonmetallic material or to the other contacting metal (Fig. 3.7, [3.36–3.40]). Therefore, friction, metal wear, and metal transfer should be related to the metal's chemical, physical, and metallurgical properties and strength.

As greater and greater mechanical strengths are obtained from engineering materials, it is only logical to ask what the upper limit may be to the strength of a solid. This upper limit, or maximum strength, has come to be referred to as "the theoretical tensile strength." Therefore, let us consider first the relationship between theoretical tensile strength and tribological properties [3.41].

The generally accepted thinking on solid fracture is that the ideal elastic solid exhibits elastic response to a load until the interatomic forces are overcome and atomic separation takes place on a plane. At the atomistic level, fracture occurs when the bonds between atoms are broken across a fracture plane, creating a new surface. Bonds can be broken perpendicular to the fracture plane (Fig. 3.8(a)) or sheared across the fracture plane (Fig. 3.8(b)). Such behavior is expected for an ideal crystalline solid with no defects. Under such conditions the criteria for fracture are simple: fracture occurs when the local stress builds up either to the theoretical cohesive strength or to the theoretical shear strength.

The theoretical cohesive strength of an ideal elastic solid is calculated on the basis of all the energy used in separation being available for creating the two new surfaces. The surface energy is assumed to be the only energy expended in creating these surfaces. If the atoms A and A' in Fig. 3.8(a) are pulled apart, the stress required to separate the plane is the theoretical uniaxial tensile strength σ_{max}. When that strength is reached, the bonds are broken. That strength is given by the well-known equation

$$\sigma_{max} = \sqrt{\frac{E\gamma}{d}} \qquad (3.1)$$

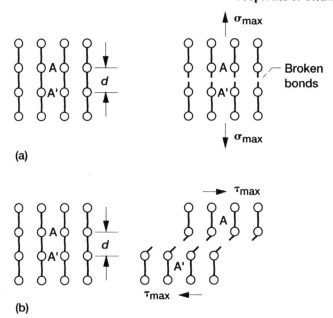

Figure 3.8.—Fracture viewed at atomistic level in terms
of breaking of atomic bonds. (a) Tensile fracture.
(b) Shear fracture.

where E is the appropriate Young's modulus, γ is the surface energy per unit area,
and d is the interplanar spacing of the planes perpendicular to the tensile axis [3.42–
3.46]. In this equation the theoretical tensile strength of the solid is directly related
to other macroscopic physical properties.

The foregoing approach is equally applicable to any solid. Frenkel used a similar
method to estimate the theoretical shear strength τ_{max} of a solid subjected to a simple
shear mode of deformation [3.42, 3.47]. He assumed that, for any solid, the stress
required to shear any plane a distance x over its neighbor is given by

$$\tau = \kappa \sin \frac{2\pi x}{b} \qquad (3.2)$$

where b is the appropriate repeat distance in the shear direction (the planes are
assumed to be undistorted by the shear) and κ is chosen to give the correct shear
modulus G. It is then easily shown that

$$\tau_{max} = \frac{Gb}{2\pi d} \qquad (3.3)$$

where d is the interplanar spacing of the shearing planes.

Figure 3.9 presents the coefficient of friction as a function of the theoretical tensile strength σ_{max}. (The values of σ_{max} can be obtained from Eq. (3.1).) There generally appears to be a strong correlation between friction and σ_{max}, with the friction decreasing as the theoretical tensile strength of the metals increased. The higher the tensile strength, the lower the friction.

When metallic and nonmetallic materials in sliding contact separate, fracture occurs in the metal as well as shear at the adhesive bonds in the interface. The morphology of metal transfer to nonmetal revealed that metals with low tensile strength exhibited much more transfer than those with higher tensile strength. For example, examination of wear tracks on SiC after single-pass sliding with titanium revealed evidence that both thin films and lump particles of titanium had transferred to the SiC. On the other hand, examination of the SiC surface after multipass sliding with titanium indicated the presence of thin transfer films, multilayer transfer films, small particles, and pileup of particles. Table 3.2 summarizes the metal transfer to single-crystal SiC observed after multipass sliding. Generally, metals closer to the bottom of Table 3.2 have less chemical affinity for silicon and carbon and greater resistance to tensile and shear fracture and, accordingly, lower coefficients of friction. Therefore, less transfer to SiC was observed with these metals.

Such dependency of metal transfer on the theoretical tensile strength arises from the adhesion and fracture properties of the metal. Thus, theoretical tensile strength, which is a function of surface energy, Young's modulus, and interplanar spacing in the crystal, plays a role in the adhesion, friction, and transfer of metals contacting metals or nonmetals. Surface energies of solid metals have been reported in the literature [3.48–3.56]. Investigators have sought correlations between surface energy and other physical properties [3.48, 3.49, 3.56]. The most successful and widely accepted of these correlations for elemental solids occurs where the heat of sublimation has been considered. A good correlation between surface and cohesive energy was also, however, found by Tyson and Jones [3.48, 3.49]. The correlation between surface energy and tribological properties will be sought later, in Section 3.6.

Theoretical shear strength.—Theoretical shear strength values were obtained from Eq. (3.3) and are presented in Table 3.3. It was assumed that the slip plane is in the slip direction, as indicated in Table 3.3.

Figure 3.10 presents the coefficient of friction as a function of the theoretical shear strength τ_{max}. These data show that the friction decreased as the theoretical shear strength of the metal bond increased. The theoretical shear strength generally correlated with the coefficient of friction for metals in contact with such nonmetals as diamond, pyrolytic boron nitride (BN), SiC, and Mn-Zn ferrite, as shown in Figs. 3.10(a) to (d). The coefficients of friction for metals in contact with themselves correlated with the metal's shear strength, except for platinum and palladium, as indicated in Fig. 3.10(e). In creating these figures the shear strength values for face-centered cubic metals from Table 3.3 were used. The shear strength values for the body-centered cubic metals were average values calculated from the shear strengths for three dominant slip systems. Those for the hexagonal metals

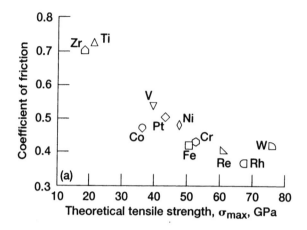

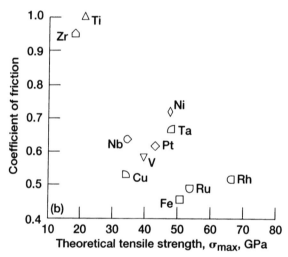

Figure 3.9.—Coefficient of friction as function of theoretical tensile
strength for metals in sliding contact with nonmetals and themselves. Tensile direction: ⟨111⟩ for face-centered cubic, ⟨110⟩ for
body-centered cubic, and ⟨0001⟩ for hexagonal metals; room
temperature; vacuum pressure, 10^{-8} Pa. (a) Sliding material,
single-crystal diamond {111} surface; sliding direction, ⟨1̄10⟩;
sliding velocity, 3 mm/min; load, 0.05 to 0.3 N. (b) Sliding material,
pyrolitic BN surface; sliding velocity, 0.77 mm/min; load, 0.3 N.
(c) Sliding material, single-crystal SiC {0001} surface; sliding
direction, ⟨101̄0⟩; sliding velocity, 3 mm/min; load, 0.05 to 0.5 N.
(d) Sliding material, single-crystal Mn-Zn ferrite {110} surface;
sliding direction, ⟨1̄10⟩; sliding velocity, 3 mm/min; load, 0.3 N.
(e) Sliding materials, metals against themselves; sliding velocity,
0.7 mm/min; load, 0.01 N.

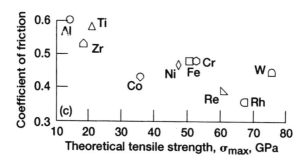

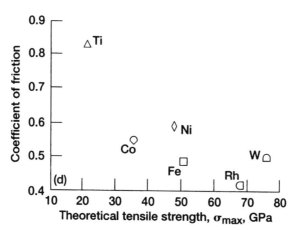

Figure 3.9.—Continued. (c) Sliding material, single-crystal SiC {0001} surface; sliding direction, ⟨10$\bar{1}$0⟩; sliding velocity, 3 mm/min; load, 0.05 to 0.5 N. (d) Sliding material, single-crystal Mn-Zn ferrite {110} surface; sliding direction, ⟨1$\bar{1}$0⟩; sliding velocity, 3 mm/min; load, 0.3 N.

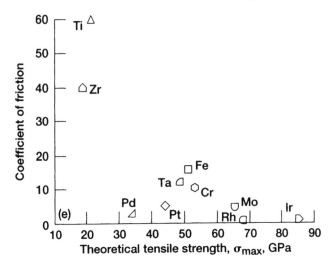

Figure 3.9.—Concluded. (e) Sliding materials, metals against themselves; sliding velocity, 0.7 mm/min; load, 0.01 N.

were average values calculated from the shear strengths for two dominant slip systems (i.e., the $\{10\bar{1}0\}\langle11\bar{2}0\rangle$ and $\{0001\}\langle11\bar{2}0\rangle$).

Thus, tensile and shear properties were shown to play important roles in the adhesion and friction of metals contacting nonmetals or metals contacting themselves. However, these simple calculations of the theoretical strength and the correlation between friction and strength can be criticized on several grounds. The extent of slip in a crystal varies with the magnitude of the shear stresses produced by the applied forces and the crystal's orientation with respect to these applied forces. This variation can be rationalized by the concept of the crystal's resolved shear stress for slip. Despite the foregoing, the relationship between the coefficient of friction and the theoretical shear strength may lead to an appreciation of how the physical properties of materials determine their tribological properties and mechanical behavior.

A good correlation between the coefficient of friction and the shear modulus (e.g., Fig. 3.5(a)) was also found for metals contacting nonmetals. The correlation is similar to that between the coefficient of friction and the shear strength (refer to Fig. 3.10). This similarity is to be expected because, as shown in Table 3.3, the ratio τ_{max}/G is essentially constant.

Actual shear strength.—The theoretical shear and tensile strengths are much greater than commonly found experimentally. In the previous sections the relationships between these theoretical strengths and the friction properties of metals in contact with nonmetals and with themselves were discussed. There is, in addition, an obvious need to compare the actual observed strengths of metals with their friction properties.

TABLE 3.3.—SIMPLE CALCULATIONS OF THEORETICAL
SHEAR STRENGTH

(a) Face-centered cubic structure;
shear plane and direction, $\{111\}, \langle 110 \rangle$

Metal	Shear strength, τ_{max}, GPa	Strength to modulus ratio, τ_{max}/G
Al	2.6	0.096–0.098
Ni	7.3	
Cu	4.4	
Rh	15	
Pd	5.0	
Ir	21	
Pt	5.9	↓

(b) Body-centered cubic structure

Metal	Shear plane and direction					
	$\{110\} \langle 111 \rangle$		$\{112\} \langle 111 \rangle$		$\{123\} \langle 111 \rangle$	
	τ_{max}, GPa	τ_{max}/G	τ_{max}, GPa	τ_{max}/G	τ_{max}, GPa	τ_{max}/G
V	3.1	0.65–0.66	5.3	0.11	8.1	0.17
Cr	7.6		13		20	
Fe	5.3		9.2		14	
Nb	2.4		4.2		6.4	
Mo	7.5		13		20	
Ta	4.5		7.8		12	
W	9.8	↓	17	↓	26	↓

(c) Close-packed hexagonal structure

Metal	Shear plane and direction					
	$\{0001\} \langle 11\bar{2}0 \rangle$		$\{10\bar{1}0\} \langle 11\bar{2}0 \rangle$		$\{10\bar{1}1\} \langle 11\bar{2}0 \rangle$	
	τ_{max}, GPa	τ_{max}/G	τ_{max}, GPa	τ_{max}/G	τ_{max}, GPa	τ_{max}/G
Ti	3.9	0.098–0.10	7.2	0.18–0.19	8.2	0.21
Co	7.5		14		16	
Y	2.6		4.8		5.5	
Zr	3.4		6.3		7.1	
Ru	16		29		34	
Re	18	↓	33	↓	38	↓

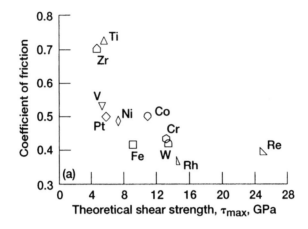

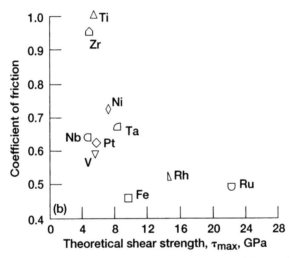

Figure 3.10.—Coefficient of friction as function of theoretical shear strength for metals in sliding contact with nonmetals and themselves. Room temperature; vacuum pressure, 10^{-8} Pa. (a) Sliding material, single-crystal diamond {111} surface; sliding direction, $\langle 1\bar{1}0 \rangle$; sliding velocity, 3 mm/min; load, 0.05 to 0.3 N. (b) Sliding material, pyrolytic BN surface; sliding velocity, 0.77 mm/min; load, 0.3 N. (c) Sliding material, single-crystal SiC {0001} surface; sliding direction, $\langle 10\bar{1}0 \rangle$; sliding velocity, 3 mm/min; load, 0.05 to 0.5 N. (d) Sliding material, single-crystal Mn-Zn ferrite {110} surface; sliding direction, $\langle 1\bar{1}0 \rangle$; sliding velocity, 3 mm/min; load, 0.3 N. (e) Sliding materials, metals against themselves; sliding velocity, 0.7 mm/min; load, 0.01 N.

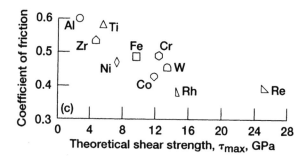

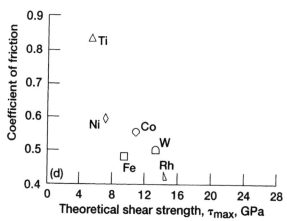

Figure 3.10.—Continued. (c) Sliding material, single-crystal SiC {0001} surface; sliding direction, $\langle 10\bar{1}0 \rangle$; sliding velocity, 3 mm/min; load, 0.05 to 0.5 N. (d) Sliding material, single-crystal Mn-Zn ferrite {110} surface; sliding direction, $\langle 1\bar{1}0 \rangle$; sliding velocity, 3 mm/min; load, 0.3 N.

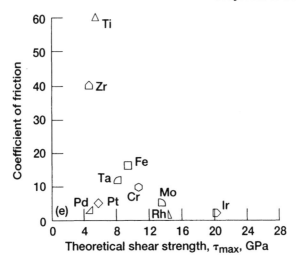

Figure 3.10.—Concluded. (e) Sliding materials,
metals against themselves; sliding velocity,
0.7 mm/min; load, 0.01 N.

The actual shear strengths of metals were estimated from Bridgman's experimental data [3.57]. The shear phenomena and strengths were studied at high hydrostatic pressures (to 4.9 GPa). The shear strength of a metal strongly depended on the hydrostatic pressure acting on it during shear, increasing as applied hydrostatic pressure increased. The actual shear strengths were estimated by extrapolating from the contact pressure during sliding experiments by using Bridgman's relationships between hydrostatic pressure and shear strength. The contact pressures for various metals in contact with nonmetals were calculated by using Hertz's classical equations [3.58].

Figure 3.11 presents the coefficient of friction for metals in contact with clean diamond, SiC, and Mn-Zn ferrite as a function of the actual shear strength. Generally, friction decreased as the actual shear strength increased. This correlation seems to indicate that the ratio of actual to theoretical shear strength does not vary greatly from one elemental metal to another.

Thus, the coefficients of friction for clean metals in contact with clean diamond, BN, SiC, Mn-Zn ferrite, and metals in ultrahigh vacuum can be generally related to the theoretical tensile, theoretical shear, and actual shear strengths of the metals. The stronger the metal, the lower the coefficient of friction.

Hardness.—In general, hardness implies resistance to deformation [3.59, 3.60]. With organic materials, such as rubber, the elastic properties play an important role in assessing hardness. With inorganic materials, such as metals and ceramics, however, the position is different, for although their elastic moduli are generally large, metals and ceramics deform elastically over a relatively small range, predominantly outside the elastic range. Consequently, considerable plastic or

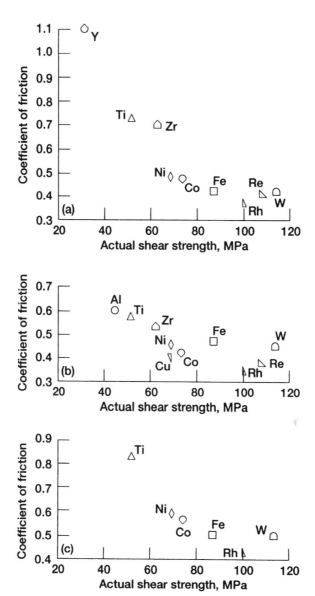

Figure 3.11.—Coefficient of friction as function of actual
metal shear strength for single-crystal metals in
sliding contact with various polycrystalline materials.
Sliding direction, $\langle 1\bar{1}0 \rangle$; sliding velocity, 3 mm/min;
room temperature; vacuum pressure, 10^{-8} Pa.
(a) Diamond {111} surface. (b) SiC {0001} surface.
(c) Mn-Zn ferrite {110} surface.

permanent deformation often occurs. For this reason the hardness of metals and ceramics is bound up primarily with their plastic strength properties and only to a secondary extent with their elastic properties. In ceramics the fracture properties may be as important as the plastic properties, particularly at high loads. Thus, hardness is another way of determining the plastic yield strength of a material, namely the amount of plastic deformation, produced mainly in compression, by a known force.

Hardness, like other mechanical properties, such as tensile and shear strength, is closely related to the Young's and shear moduli, as shown in Fig. 3.12. All the plotted Vickers hardness data were measured on polished metal surfaces by using a diamond pyramid indenter at a load of 0.25 N. Because the Vickers hardness of the metals increased as the shear modulus increased, their adhesion and friction are expected to be related to their hardness. Figure 3.13 shows the relationship between coefficient of friction and hardness for several metal-SiC couples. Friction decreased as hardness increased.

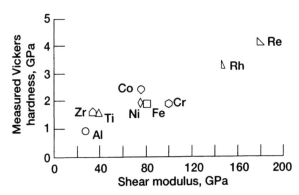

Figure 3.12.—Vickers hardness as function of shear modulus for various metals. Load, 0.25 N.

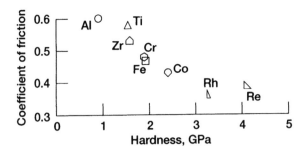

Figure 3.13.—Coefficient of friction as function of hardness for various metal-SiC couples.

3.5.2 Chemical Properties

Today, almost all the known elements are used to make ceramic materials and products. Probably the most widely used class of ceramic materials, however, is the oxides, such as sapphire (Al_2O_3) and zirconia (ZrO_2).

All but a handful of metals, alloys, and nonoxide ceramics (e.g., SiC, Si_3N_4, and molybdenum disulfide (MoS_2)) will form surface oxide films in air by their surface chemical reactivities. The thickness of the reaction oxide products varies depending on the material's reactivity to the environment, crystallographic orientation, grain boundary, impurities, dislocations, defects, surface topography, and mechanical stresses. However, 100 nm might be considered a typical thickness for such oxide layers [3.61, 3.62].

The surface reactivity required to form oxides is related to the mechanical properties of the parent material (Fig. 3.14). The free energy of formation of the lowest metal oxide correlates with the shear modulus of the metal. The higher the shear modulus, the lower the free energy of formation.

A close relationship exists between the coefficient of friction of a clean metal-ferrite contact and the free energy of formation of the lowest metal oxide (the strength of the chemical bond). The higher the free energy of formation, the greater the adhesion and friction (Figs. 3.4 and 3.5).

In 1948, Linus Pauling formulated a resonating valence bond theory of metals and intermetallic compounds in which numerical values could be placed on the bonding character of the various transition elements [3.63]. Because the d valence bonds are not completely filled in transition metals, they are responsible for such physical and chemical properties as cohesive energy, shear modulus, chemical stability, and magnetic properties. The greater the amount or percentage of d bond character that a metal possesses, the less active is its surface. Although there have been critics of this theory, it appears to be the most plausible explanation for the interfacial interactions of transition metals in contact with ceramics as well as with themselves [3.1].

When a transition metal is placed in contact with a ceramic material in an atomically clean state, the interfacial bonds formed between the metal and the ceramic depend heavily on the character of the bonding in the metal. Figure 3.15 shows, for example, the coefficients of friction for some transition metals in contact with a single-crystal diamond {111} surface as a function of both the shear modulus and d valence bond character of the metals [3.37]. The data for these sputter-cleaned surfaces indicate that adhesion and friction decreased as d valence bond character increased, as Pauling's theory predicts. Titanium and zirconium, which are chemically very active, exhibited strong adhesive bonding to the ceramic. In contrast, rhodium and rhenium, which have a high percentage of d bond character, had relatively low adhesion and friction. Thus, the more chemically active the metal, the higher the coefficient of friction.

124

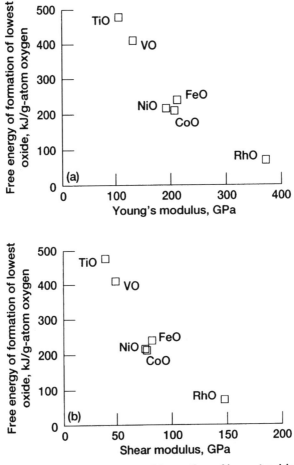

Figure 3.14.—Free energy of formation of lowest oxide. (a) As function of Young's modulus of metals. (b) As function of shear modulus of metals.

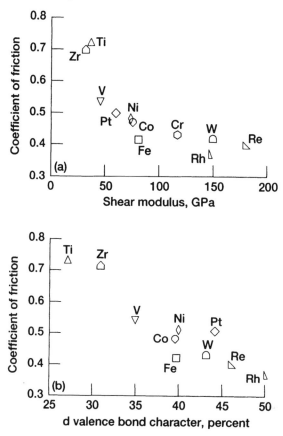

Figure 3.15.—Coefficient of friction for various metals in sliding contact with diamond {111} surface in ultrahigh vacuum. (a) As function of metal's shear modulus. (b) As function of percent of metal's d valence bond character.

3.5.3 Metallurgical Properties

There is little doubt that a solid's structure plays an important role in its mechanical behavior [3.64], particularly tribological behavior. Structure depends first on chemical composition and then on mechanical and thermal processing (sintering, casting, hot working, machining, and heat treatments of all kinds). For example, solid-solution alloying is a major mode of metal strengthening. Such chemical composition and processing steps influence tribological properties by their effect on phase, concentration of ingredients and their gradients, inclusions, voids, metastable phases, dispersed phases, and lattice imperfections of different kinds [3.64].

Alloying element effects.—Figure 3.16 (from [3.65]) shows the coefficients of friction for six iron-base binary alloys in contact with single-crystal SiC as a function of solute concentration (given in Table 3.4). The coefficient of friction initially increased markedly with the presence of any alloying element and then continued to increase more gradually as the concentration of alloying element increased. The rate of increase in the coefficient of friction strongly depended on the alloying element.

The average coefficient of friction for pure iron in sliding contact with single-crystal SiC is approximately 0.5 [3.38]. This value was obtained under identical experimental conditions to those of this investigation. The coefficients of friction were about 0.6 for pure titanium, 0.5 for pure nickel and tungsten, and 0.4 for pure rhodium. The coefficients of friction for the alloys were generally much higher, as much as twice those for pure metals.

Figure 3.17 presents the average coefficients of static friction for the various alloys of Fig. 3.16 as a function of solute-to-iron atomic radius ratio. The maximum solute concentration extended to approximately 16 at.%. The good agreement between the coefficient of static friction and the solute-to-iron atomic radius ratio differed for two cases: first, alloying with manganese and nickel, which have smaller atomic radii than iron; and second, alloying with chromium, rhodium, tungsten, and titanium, which have larger atomic radii than iron. The coefficients of static friction increased generally as the solute-to-iron atomic radius ratio increased or decreased from unity. The rate of increase was much greater for the first case than for the second case. Atomic size ratios reported herein are from [3.66] and [3.67]. The correlations indicate that the atomic size of the solute is an important factor in controlling the friction in iron-base binary alloys as well as the abrasive wear and friction reported in [3.66] and the alloy hardening reported in [3.67]. The mechanism controlling alloy friction may be raising the Peierles stress and/or increasing the lattice friction stress, by solute atoms, thus resisting the shear fracture of cohesive bonds in the alloy.

More detailed examination of Fig. 3.17 indicates that the correlation for manganese, nickel, and chromium was better than that for titanium, tungsten, and rhodium. The coefficient of friction for rhodium was relatively low, and that for titanium was relatively high. The relative chemical activity of the transition metals (metals with partially filled d shell) as a group can be ascertained from their percentage of d valence bond character after Pauling [3.63]. It has already been determined [3.38] that the coefficient of friction for SiC in contact with various transition metals is related to the d valence bond character (i.e., the chemical activity) of the metal. The more active the metal, the higher the coefficient of friction. Table 3.5 shows the reciprocal d valence bond character of metals calculated from the data of [3.62]. The greater the reciprocal d valence bond character, the more active the metal and the higher the coefficient of friction [3.38].

Rhodium-iron alloys in contact with SiC showed relatively low friction, but titanium-iron alloys showed relatively high friction. The results seem to be related

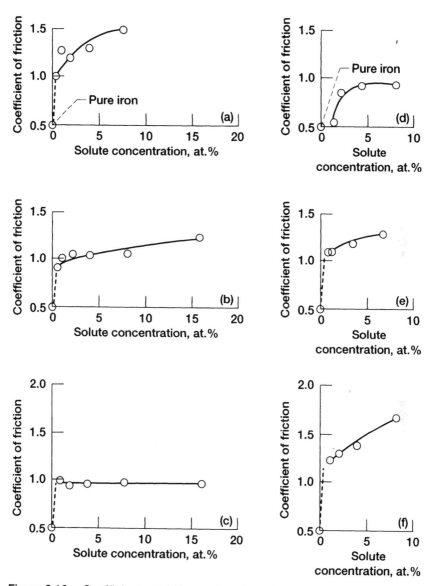

Figure 3.16.—Coefficient of friction as function of solute concentration for various iron-base binary alloys after single-pass sliding on single-crystal SiC {0001} surface. Sliding direction, $\langle 10\bar{1}0 \rangle$; sliding velocity, 3 mm/min; load, 0.2 N; room temperature; vacuum pressure, 10^{-8} Pa.

128

TABLE 3.4.—CHEMICAL ANALYSIS AND SOLUTE-TO-IRON
ATOMIC RADIUS RATIOS FOR IRON-BASE BINARY ALLOYS

Solute element	Analyzed solute concentration, at.%	Analyzed interstitial content, ppm by weight			Solute-to-iron atomic radius ratio
		C	O	P	
Ti	1.02	56	92	7	1.1476
	2.08	—	—	—	↓
	3.86	87	94	9	
	8.12	—	—	—	
Cr	0.99	—	—	—	1.0063
	1.98	50	30	12	↓
	3.92	—	—	—	
	7.77	40	85	10	
	16.2	—	—	—	
Mn	0.49	—	—	—	0.9434
	.96	39	65	6	↓
	1.96	—	—	—	
	3.93	32	134	8	
	7.59	—	—	—	
Ni	0.51	—	—	—	0.9780
	1.03	28	90	6	↓
	2.10	—	—	—	
	4.02	48	24	5	
	8.02	—	—	—	
	15.7	38	49	7	
Rh	1.31	—	—	—	1.0557
	2.01	20	175	22	↓
	4.18	—	—	—	
	8.06	12	133	19	
W	0.83	30	140	12	1.1052
	1.32	—	—	—	↓
	3.46	23	61	21	
	6.66	—	—	—	

TABLE 3.5.—AMOUNT AND
RECIPROCAL OF d VALENCE
BOND CHARACTER FOR
TRANSITION ELEMENTS

Metal	Amount of d character, percent	Reciprocal of d character
Fe	39.7	0.68
Mn	40.1	0.67
Ni	40.0	0.68
Cr	39	0.69
Rh	50	0.54
W	43	0.63
Ti	27	1

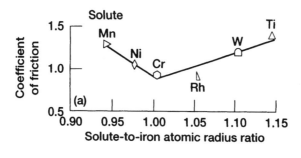

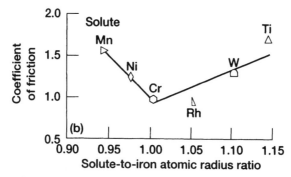

Figure 3.17.—Coefficient of friction as function of solute-to-iron atomic radius ratio for various iron-base binary alloys after single-pass sliding on single-crystal SiC {0001} surface. Sliding direction, $\langle 10\bar{1}0 \rangle$; sliding velocity, 3 mm/min; load, 0.2 N; room temperature; vacuum pressure, 10^{-8} Pa.

to the chemical activity of the alloying elements (i.e., rhodium is less active than iron, and titanium is more active), as indicated in Table 3.5. The good correlation for manganese, nickel, and chromium in Fig. 3.17 is due to their reciprocal d valence bond characters being almost the same as that for iron.

Figure 3.18 presents a scanning electron micrograph and an x-ray energy dispersive map of a wear track on SiC generated by an 8.12-at.%-titanium–iron alloy pin. In the x-ray map (Fig. 3.18(b)) the concentration of white spots corresponds to those locations in the micrograph (Fig. 3.18(a)) where copious amounts of alloy have transferred. Obviously, a large amount of alloy transferred to the SiC surface. The light area in Fig. 3.18(a), where alloy transfer is evident, was the contact area before sliding of the pin. In this area the surfaces of the titanium alloy and the SiC stuck together and strong interfacial adhesion occurred. Here, both the loading and tangential (shear) forces were applied to the specimen. All single-crystal SiC surfaces after sliding contact with the alloys whose analysis is shown in Table 3.4 contained metallic elements, indicating alloy transfer to the SiC. Alloys

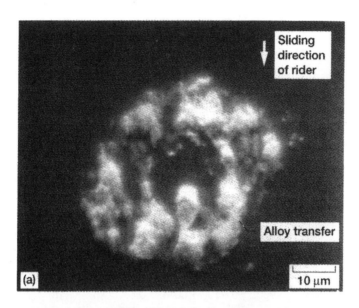

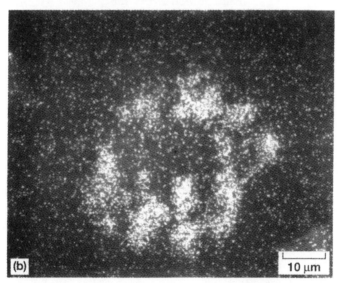

Figure 3.18.—Transfer of titanium-iron binary alloy (8.12-at.% Ti) to single-crystal SiC {0001} surface at start of sliding. Sliding direction, $\langle 10\bar{1}0 \rangle$; sliding velocity, 3 mm/min; load, 0.2 N; room temperature; vacuum pressure, 10^{-8} Pa.

having high solute concentrations produced more transfer than did alloys having low solute concentrations.

Figure 3.19 shows a typical pin wear scar on an iron-base binary alloy (in this case, 8.12-at.%-Ti–Fe alloy). The size of the wear scar (Fig. 3.19(a)) is comparable to the alloy transfer area shown in Fig. 3.18(a). The wear scar reveals a large number of small grooves and microcracks formed primarily by interface shearing and shearing in the alloy bulk. Close examination of Fig. 3.19(b) indicates that the cracks were small, were in the wear scar, and propagated nearly perpendicular to the sliding direction.

In summary, the atomic size misfit and the concentration of the alloying element are important factors in controlling the adhesion and friction of iron-base binary alloys in contact with SiC. The mechanism controlling alloy adhesion and friction may be raising the Peierles stress and/or increasing the lattice friction stress by solute atoms, thus resisting the shear fracture of cohesive bonds in the alloy. The coefficient of friction generally increased markedly with the presence of any concentration of alloying element in the pure metal and then increased more gradually as the concentration of alloying element increased. The coefficient of friction generally increased as the solute-to-iron atomic radius ratio increased or decreased from unity. The atomic size misfit and the concentration of alloying element were factors in controlling both friction and alloy transfer to SiC during multipass slidings.

Crystallographic orientation (anisotropy) effects.—Metals and ceramics exhibit anisotropic behavior in many of their mechanical properties. The friction and wear behaviors of ceramics are also anisotropic under adhesive conditions.

Anisotropy results can be of two kinds:

1. The observed variation in friction and wear when the sliding surface is changed from one crystal plane to another for a given material
2. The variation in friction and wear observed when the orientation of the sliding surface is changed with respect to a specific crystallographic direction on a given crystal plane

For example, the differences in the coefficients of friction with respect to the mating crystallographic planes and directions are significant under adhesive conditions, as indicated in Table 3.6 (from [3.68]). The data of Table 3.6 were obtained in vacuum with clean ferrite–ferrite oxide ceramics. The mating of preferred slip plane with highest atomic density plane and direction, such as {110}⟨110⟩ and {111}⟨110⟩ for Mn-Zn ferrite, gave the lowest coefficients of friction. In other words, the lowest coefficients of friction were obtained on the preferred slip plane when sliding in the preferred slip direction. Similar results have been obtained with SiC and Al_2O_3 (sapphire). Table 3.7 shows their anisotropic friction. Again, the coefficients of friction were lowest on the preferred basal slip plane when sliding in the preferred ⟨11$\bar{2}$0⟩ slip direction [3.69, 3.70]. The coefficient of friction reflects the force

132

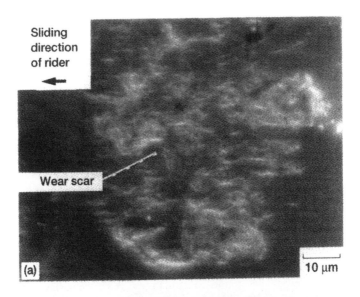

Sliding direction of rider

Wear scar

(a)

10 μm

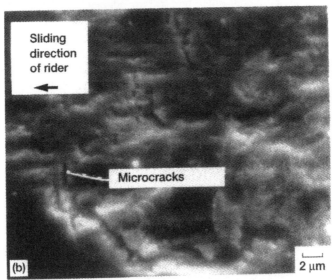

Sliding direction of rider

Microcracks

(b)

2 μm

Figure 3.19.—Wear scar on titanium-iron binary alloy (8.12-at.% Ti) showing grooves and cracks after single-pass sliding on single-crystal SiC {0001} surface. Sliding direction, $\langle 10\bar{1}0 \rangle$; sliding velocity, 3 mm/min; load, 0.2 N; room temperature; vacuum pressure, 10^{-8} Pa.

TABLE 3.6.—ANISOTROPIC FRICTION FOR Mn-Zn FERRITE
CONTACTING SiC UNDER ADHESIVE CONDITIONS
[Load, 0.05 to 0.5 N; sliding velocity, 3 mm/min; vacuum
pressure, 10^{-8} Pa; room temperature.]

Mated plane	Mated direction	Sliding direction	Coefficient of friction
Effect of crystallographic plane			
{110} on {110}	⟨110⟩ on ⟨110⟩	⟨110⟩	0.21
{111} on {111}	⟨110⟩ on ⟨110⟩	⟨110⟩	.21
{100} on {100}	⟨110⟩ on ⟨110⟩	⟨110⟩	.24
{110} on {100}	⟨110⟩ on ⟨110⟩	⟨110⟩	.27
{110} on {111}	⟨110⟩ on ⟨110⟩	⟨110⟩	.29
{110} on {211}	⟨110⟩ on ⟨110⟩	⟨110⟩	.29
Effect of crystallographic direction			
{110} on {110}	⟨110⟩ on ⟨110⟩	⟨110⟩	0.21
	⟨110⟩ on ⟨100⟩	⟨110⟩	.43

TABLE 3.7.—ANISOTROPIC FRICTION FOR
SAPPHIRE AND SiC CONTACTING
THEMSELVES UNDER
ADHESIVE CONDITIONS

Plane	Direction	Coefficient of friction
Sapphire sliding on sapphire[a]		
Prismatic $\{10\bar{1}0\}$	$\langle 11\bar{2}0\rangle$	0.93
	⟨0001⟩	1.00
Basal {0001}	$\langle 11\bar{2}0\rangle$	0.50
	$\langle 10\bar{1}0\rangle$	.96
SiC sliding on SiC[b]		
Basel {0001}	$\langle 11\bar{2}0\rangle$	0.54
	$\langle 10\bar{1}0\rangle$	.68

[a]Load, 10 N; sliding velocity, 7.8 mm/min;
vacuum pressure, 10^{-8}; room temperature.
[b]Load, 0.3 N; sliding velocity, 3 mm/min;
vacuum pressure, 10^{-8} Pa; room temperature.

required to shear at the interface when the SiC or Al_2O_3 basal planes are parallel to the interface. The results presented in Table 3.7 indicate that lower force is required to resist shear fracture of the adhesive bond at the interface in the preferred crystallographic $\langle 11\bar{2}0\rangle$ direction than in the $\langle 10\bar{1}0\rangle$ direction.

When the SiC {0001} surface was in contact with iron, as shown in Fig. 3.20, from room temperature to 800 °C in vacuum, the coefficient of friction was lower in the $\langle 11\bar{2}0\rangle$ direction than in the $\langle 10\bar{1}0\rangle$ direction over the entire temperature range [3.69]. The coefficient of friction generally increased with increasing temperature from about 0.5 in the $\langle 10\bar{1}0\rangle$ sliding direction and 0.4 in the $\langle 11\bar{2}0\rangle$ sliding direction at room temperature to 0.75 and 0.63, respectively, at about 800 °C.

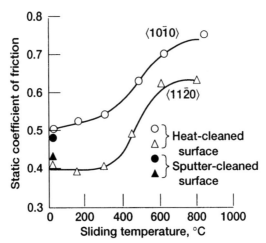

Figure 3.20.—Static coefficient of friction as function of sliding temperature and crystallographic orientation for SiC surface sliding against iron. Normal load, 0.2 N; vacuum pressure, 10^{-8} Pa. (The surfaces were heat cleaned at 800 °C before the friction experiments. The coefficient of friction was obtained by averaging three to five measurements.)

Although the coefficient of friction remained low below 300 °C, it increased rapidly with increasing temperature from 300 to 600 °C. There was, however, little further increase in friction above 600 °C.

The data of Fig. 3.20 indicate that the friction behavior of SiC in contact with iron is highly anisotropic over the entire range from room temperature to 800 °C. Several slip systems have been observed in α-SiC, including the $\{0001\}\langle11\bar{2}0\rangle$, $\{3\bar{3}01\}\langle11\bar{2}0\rangle$, and $\{10\bar{1}0\}\langle11\bar{2}0\rangle$ systems [3.69, 3.71]. The preferred crystallographic slip direction, or the shear direction for the basal $\{0001\}$ plane, is the $\langle11\bar{2}0\rangle$ direction. The coefficient of friction on the basal plane was lower in the $\langle11\bar{2}0\rangle$ direction than in the $\langle10\bar{1}0\rangle$ direction.

The coefficient of friction reflects the force required to shear at the interface when the SiC basal planes are parallel to the interface. The results presented in Fig. 3.20 indicate that less force is required to resist shear fracture of the adhesive bond at the interface in the $\langle11\bar{2}0\rangle$ direction than in the $\langle10\bar{1}0\rangle$ direction.

SiC $\{0001\}$ surfaces that were argon sputter cleaned or heat cleaned in situ revealed no significant differences in coefficient of friction. The frictional anisotropy was also similar (i.e., the coefficient of friction was lower in the $\langle11\bar{2}0\rangle$ direction than in the $\langle10\bar{1}0\rangle$ direction).

Sliding a metal or SiC pin on an SiC flat {0001} surface resulted in cracks along cleavage planes of $\langle 10\bar{1}0\rangle$ orientation. Figure 3.21 shows scanning electron micrographs of the wear tracks generated by 10 passes of rhodium and titanium pins on the SiC {0001} surface along the $\langle 10\bar{1}0\rangle$ direction. The cracks observed in the wear tracks propagated primarily along cleavage planes of the $\langle 10\bar{1}0\rangle$ orientation. Figure 3.21(a) reveals a hexagonal light area, which is the beginning of a wear track, and a large crack. Cracks were generated primarily along the $\{10\bar{1}0\}$ planes, propagated, and then intersected during loading and sliding of the rhodium pin over the SiC surface. It is anticipated from Fig. 3.21(a) that substrate cleavage cracking of the {0001} planes, which are parallel to the sliding surface, also occurs. Figure 3.21(b) reveals a hexagonal pit surrounded by a copious amount of thin titanium film. The hexagonal fracturing is caused primarily by cleavage cracking along the $\{10\bar{1}0\}$ planes and subsurface cleavage cracking along the {0001} planes. The smooth surface at the bottom of the hexagonal pit is due to cleavage of the {0001} planes.

Figure 3.22 illustrates the SiC wear debris produced by 10-pass sliding of aluminum pins on an SiC surface. The scanning electron micrographs reveal evidence of multiangular SiC wear debris particles with transferred aluminum wear debris on the SiC wear track. These multiangular wear debris particles had crystallographically oriented sharp edges and were nearly hexagonal, rhombic, parallelogramic, or square [3.71]. These shapes may be related to surface and subsurface cleavage of $\{10\bar{1}0\}$, $\{11\bar{2}0\}$, and {0001} planes.

Similar hexagonal pits and multiangular wear debris with crystallographically oriented sharp edges were also observed with single-crystal SiC in contact with itself. Figure 3.23 clearly reveals the gross hexagonal pits on the wear scar of the SiC pin and a nearly fully hexagonal and flat wear particle. The wear debris had transferred to the flat SiC specimen. Thus, crystallographically oriented cracking and fracturing of SiC resulted from both sliding of the metal pin and sliding of the SiC pin.

In summary, it has been shown that the friction and wear characteristics of single crystals are anisotropic. In general, the lowest coefficient of friction was observed when sliding was in the preferred slip direction on the preferred slip plane. Wear and fracture due to adhesion of clean surfaces behave with respect to crystallographic orientation in the same way as does friction.

3.6 Friction Mechanism of Clean, Smooth Surfaces

All the clean metal-ceramic couples, including the metal-diamond couples, exhibited a correlation between the surface and bulk properties of the metal (e.g., its elastic (Young's) and shear moduli, its bond strength, and the chemistry of the contacting materials) and the adhesion, friction, and wear behaviors of the metal. All the following properties decreased as the metal's Young's and shear moduli

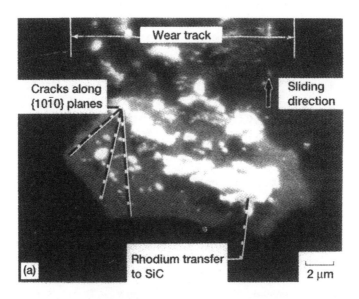

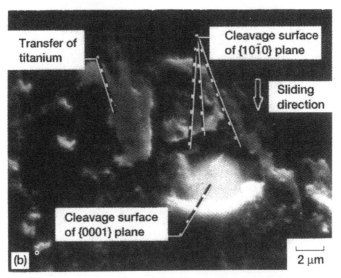

Figure 3.21.—Scanning electron micrographs of wear tracks
   on single-crystal SiC {0001} surface after 10 passes of
   rhodium and titanium pins in vacuum. Sliding direction,
   ⟨10$\bar{1}$0⟩; sliding velocity, 3 mm/min; load, 0.3 N; room
   temperature; vacuum pressure, 10^{-8} Pa. (a) Rhodium pin;
   hexagonal cracking. (b) Titanium pin; hexagonal pit.

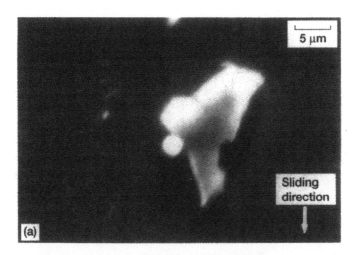

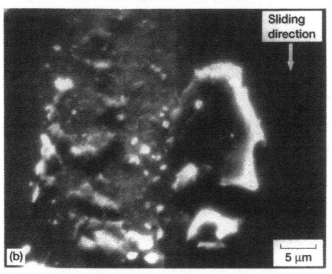

Figure 3.22.—Scanning electron micrographs of wear
tracks on and multiangular wear debris of flat single-
crystal SiC {0001} surface after 10 passes of aluminum
pin in vacuum. Sliding direction, $\langle 10\bar{1}0 \rangle$; sliding velocity,
3 mm/min; load, 0.2 N; room temperature; vacuum
pressure, 10^{-8} Pa.

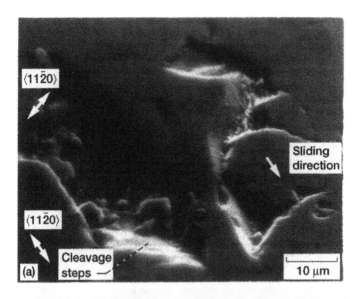

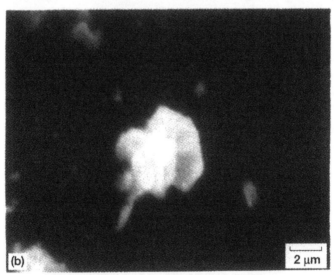

Figure 3.23.—Scanning electron micrographs of wear debris on single-crystal SiC {0001} surface after 10 passes of SiC pin in vacuum. Sliding direction, $\langle 10\bar{1}0\rangle$; sliding velocity, 3 mm/min; load, 0.5 N; room temperature; vacuum pressure, 10^{-8} Pa.

increased or its chemical activity decreased: adhesion, coefficient of friction, metal wear, and metal transfer to the ceramic. Perhaps the metal's bulk properties depend on the magnitude of its surface properties. It is interesting to consider then the role that the metal's basic surface and bulk properties, as found in the literature (such as its surface energy per unit area γ and its ductility) play in the adhesion, friction, wear, and transfer of metal-ceramic couples.

The surface energy per unit area γ of a metal is directly related to the interfacial bond strength per unit area at the metal-ceramic interface [3.61]. Figure 3.24 presents the γ values suggested in [3.48] and [3.56] for various metals at room temperature as a function of the shear modulus of the metal. As γ increased, so did the shear modulus. A comparison with Figs. 3.5(a) and 3.15(a) shows that γ (the surface or bond energy) behaved in the opposite manner from the coefficient or friction, which decreased with an increase in γ. Obviously, γ alone does not explain the friction trend shown in Figs. 3.5(a) and 3.15(a). Certainly, if γ is low, the interfacial bond strength per unit area is weak, but that does not mean that a low interfacial bond strength per unit area gives a mechanically weak interface in the real area of contact between the metal and ceramic surfaces.

A metal's ductility influences the real area of contact and accordingly the adhesion and friction at the metal-ceramic interface. Ceramics such as Si_3N_4 and SiC, unlike metals, are not considered to be ductile; these materials behave in a ductile manner only when subjected to high compressive stresses. Because of the marked difference in the ductilities of ceramics and metals, solid-state contact between the two materials can result in considerable plastic deformation of the softer metal. The real area of contact then for such a couple must be calculated from

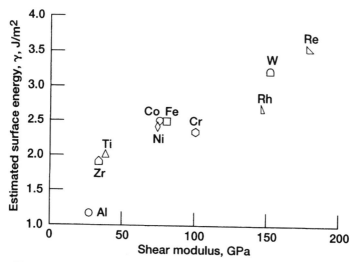

Figure 3.24.—Estimated surface energy as function of shear modulus for various metals.

the experimentally measured Vickers hardness of the metal. In this calculation the yield pressure of the surface asperities on the metal is assumed to be approximately the same as that of the bulk metal. Furthermore, no consideration is given to the growth of the real area of contact, known as junction growth, under both normal and shear (tangential) stresses acting at the interface. The real area of contact A is simply determined from the ratio of normal load to hardness. The hardness data were obtained from micro-Vickers indentation measurements of wear scars on metal pin specimens at a load of 0.25 N in an air environment. The calculated value of A depended strongly on the shear modulus of the metal (Fig. 3.25), decreasing as the shear modulus of the metal increased. The real area of contact obviously behaves in the same way as the coefficient of friction (see Figs. 3.5(a), 3.15(a), and 3.25).

A metal's total surface energy in the real area of contact is the product of the surface energy per unit area γ and the real area of contact A. It too decreased as the shear modulus of the metal increased. This relationship is brought out clearly in Fig. 3.26.

Comparing Fig. 3.26 with Figs. 3.5(a) and 3.15(a) shows that γA is associated with tribological behavior; the higher the value of γA, the greater the adhesion and friction. In addition, Fig. 3.27 clearly shows that the coefficient of friction for metal-SiC {0001} couples increased as γA increased. Comparing Table 3.2 with Fig. 3.26 indicates that γA is also related to metal wear and transfer to the ceramic (i.e., SiC); the higher the value of γA, the greater the metal wear and transfer.

The evidence from the adhesion and friction experiments reported herein points to the establishment of strong interfacial bonds in the real area of contact when clean metal-ceramic surfaces are brought into contact.

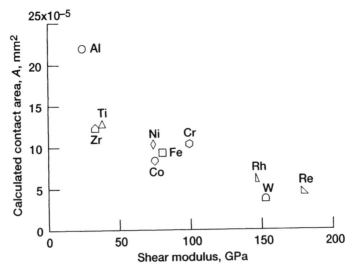

Figure 3.25.—Calculated contact area as function of shear modulus for various metals.

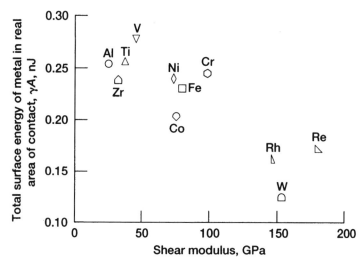

Figure 3.26.—Total surface energy in real area of contact as function of shear modulus for various metals.

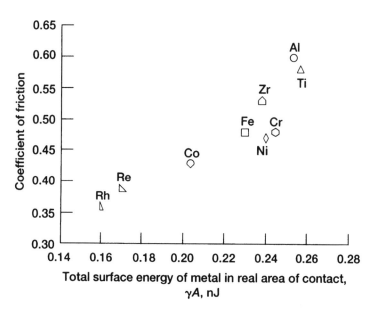

Figure 3.27.—Coefficient of friction for various metals in sliding contact with SiC {0001} surface in ultrahigh vacuum as function of total surface energy of metal in real area of contact.

References

3.1 D.H. Buckley, *Surface Effects in Adhesion, Friction, Wear and Lubrication*, Elsevier Book Series, Elsevier Scientific Publishing Co., Vol. 5, 1981.

3.2 D.H. Buckley, The use of analytical surface tools in the fundamental study of wear, *Wear 46,1*: 19–53 (1978).

3.3 D. Tabor, Status and direction of tribology as a science in the 80's—understanding and prediction, *New Directions in Lubrication, Materials, Wear, and Surface Interactions* (W.R. Loomis, ed.), Noyes Publications, Park Ridge, NJ, Vol. 1, 1985, pp. 1–17. (Also NASA CP–2300.)

3.4 K. Miyoshi and D.H. Buckley, Considerations in friction and wear, *New Directions in Lubrication, Materials, Wear, and Surface Interactions* (W.R. Loomis, ed.), Noyes Publications, Park Ridge, NJ, Vol. 1, 1985, pp. 291–320. (Also NASA CP–2300.)

3.5 K. Miyoshi and D.H. Buckley, Surface chemistry and wear behavior of single-crystal silicon carbide sliding against iron at temperatures to 1500 °C in vacuum, NASA TP–1947 (1982).

3.6 D. Tabor, Adhesion and friction, *The Properties of Diamond* (J.E. Field, ed.), Academic Press, New York, 1979, pp. 325–350.

3.7 K. Miyoshi, Uses of Auger and x-ray photoelectron spectroscopy in the study of adhesion and friction, *Advances in Engineering Tribology, Proceedings of the First International Symposium on Industrial Tribology* (Y.-W. Chung and H.S. Cheng, eds.), STLE SP–31, Society of Tribologists and Lubrication Engineers, 1991, pp. 3–12.

3.8 K. Miyoshi, Uses of AES and XPS in adhesion, friction, and wear studies, *Surface Diagnostics in Tribology* (K. Miyoshi and Y.-W. Chung, eds.), World Scientific Publishing Co., Inc., Vol. 1, 1993, pp. 93–134.

3.9 K. Miyoshi and D.H. Buckley, Properties of ferrites important to their friction and wear behavior, *Tribology and Mechanics of Magnetic Storage Systems* (Including the Symposium Proceedings of the Annual ASME–ASLE Lubrication Conference) (B. Bhushan, ed.), ASLE SP–16, American Society of Lubrication Engineers, 1984, pp. 13–20.

3.10 K. Miyoshi, C. Maeda, and R. Masuo, Development of a torsion balance for adhesion measurements, *Instrumentation for the 21st Century* (Including the Proceedings of the 11th Triennial World Congress of the International Measurement Confederation (IMEKO XI)), Instrument Society of America, 1988, pp. 233–248.

3.11 K. Miyoshi, Design, development, and applications of novel techniques for studying surface mechanical properties, *Interfaces Between Polymers, Metals, and Ceramics* (Including the Proceedings of the Materials Research Society Symposium), Materials Research Society, Vol. 153, 1989, pp. 321–330.

3.12 J.M. Georges, ed., *Microscopic Aspects of Adhesion and Lubrication: Proceedings of the 34th International Meeting of the Societe de Chimie Physique*, Elsevier Scientific Publishing Co., 1982.

3.13 H. Czichos, *Tribology—A Systems Approach to the Science and Technology of Friction, Lubrication and Wear*, Elsevier Scientific Publishing Co., New York, 1978.

3.14 J. Ferrante and J.R. Smith, A theory of adhesion at a bimetallic interface: overlap effects, *Surf. Sci. 38*: 77–92 (1973).

3.15 D.V. Keller, Jr., and R.G. Aldrich, Adhesion of metallic bodies initiated by physical contact, *J. Adhesion 1, 2*: 142–156 (1969).

3.16 J.B. Pethica and D. Tabor, Contact of characterized metal surfaces at very low loads: deformation and adhesion, *Surf. Sci. 89*: 182–190 (1979).

3.17 M.D. Pashley and D. Tabor, Adhesion and deformation properties of clean and characterized metal micro-contacts, *Vacua 31, 10–12*: 619–623 (1981).

3.18 J.E. Ingelsfield, Adhesion between Al slabs and mechanical properties, *J. Phys. F. Met. Phys. 6, 5*: 687–701 (1976).

3.19 D.D. Eley, *Adhesion*, Oxford University Press, London, 1961.

3.20 G.A.D. Briggs and B.J. Briscoe, Surface roughness and the friction and adhesion of elastomers, *Wear 57, 2*: 269–280 (1979).

3.21 N. Gane, P.F. Pfaelzer, and D. Tabor, Adhesion between clean surfaces at light loads, *Proc. R. Soc. London A340, 1623*: 495–517 (1974).

3.22 N.T. Saunders, Impact and promise of NASA aeropropulsion technology, *Aeropropulsion 1987*, NASA CP–10003, Vol. 1, 1988, pp. 1–30.

3.23 S.J. Grisaffe, Lewis materials research and technology: an overview, *Aeropropulsion 1987*, NASA CP–10003, Vol. 1, 1988, pp. 31–38.

3.24 M.A. Meador, High temperature polymer matrix composites, *Aeropropulsion 1987*, NASA CP–10003, Vol. 1, 1988, pp. 39–53.

3.25 J. Gayda, Creep and fatigue research efforts on advanced materials, *Aeropropulsion 1987*, NASA CP–10003, Vol. 1, 1988, pp. 55–72.

3.26 P.K. Brindley, Development of a new generation of high-temperature composite materials, *Aeropropulsion 1987*, NASA CP–10003, Vol. 1, 1988, pp. 73–87.

3.27 J.D. Kiser, S.R. Levine, and J.A. DiCarlo, Ceramics for engines, *Aeropropulsion 1987*, NASA CP–10003, Vol. 1, 1988, pp. 103–120.

3.28 W.E. Hazen and R.W. Pidd, *Physics*, Addison-Wesley Publishing Co., Reading, MA, 1965.

3.29 F.W. Sears and M.W. Zemansky, *College Physics—Mechanics, Heat, and Sound*, Second ed., Addison-Wesley Press Inc., Cambridge, MA, 1952.

3.30 S.V. Pepper, Shear strength of metal-sapphire contacts, *J. Appl. Phys. 47*: 801–808 (1976).

3.31 K.A. Gschneidner, Jr., Physical properties and interrelationships of metallic and semimetallic elements, *Solid State Physics, Advances in Research Applications* (F. Seitz and D. Turnbull, eds.), Academic Press, Vol. 16, 1964, pp. 275–426.

3.32 A.R. Von Hippel, *Dielectrics and Waves*, John Wiley & Sons, New York, 1954, pp. 219–228.

3.33 W.D. Kingery, H.K. Bowen, and D.R. Uhlmann, *Introduction to Ceramics*, Second ed., John Wiley & Sons, Inc., 1976, pp. 25–88, 975–1015.

3.34 A.Y. Liu and M.L. Cohen, Prediction of new low compressibility solids, *Science 245*: 841–842 (1989).

3.35 A.Y. Liu and M.L. Cohen, Structural properties and electronic structure of low-compressibility materials: β-Si_3N_4 and hypothetical β-C_3N_4, *Phys. Rev. B 41, 15*: 10727–10734 (1990).

3.36 K. Miyoshi and D.H. Buckley, Friction and wear of single-crystal manganese-zinc ferrite, *Wear 66, 2*: 157–173 (1981).

3.37 K. Miyoshi and D.H. Buckley, Adhesion and friction of single-crystal diamond in contact with transition metals, *Appl. Surf. Sci. 6, 2*: 161–172 (1980).

3.38 K. Miyoshi and D.H. Buckley, Friction and wear behavior of single-crystal silicon carbide in sliding contact with various metals, *ASLE Trans. 22, 3*: 245–256 (1979).

3.39 D.H. Buckley, The metal-to-metal interface and its effect on adhesion and friction, *J. Colloid Interface Sci. 58, 1*: 36–53 (1977).

3.40 D.H. Buckley, Friction and transfer behavior of pyrolytic boron nitride in contact with various metals, *ASLE Trans. 21, 2*: 118–124 (1978).

3.41 K. Miyoshi and D.H. Buckley, Correlation of tensile and shear strengths of metals with their friction properties, *ASLE Trans. 27, 1*: 15–23 (1984).

3.42 N.H. Macmillan, Review: the theoretical strength of solids, *J. Mater. Sci. 7, 2*: 239–254 (1972).

3.43 M. Polanyi, Uber die Natur des zerrei Bvorganges, *Z. Physik 7*: 323–327 (1921).

3.44 E. Orowan, Mechanical cohesion properties and the "real" structure of crystals, *Z. Kristallog, Kristallgeom, Kristallphys, Kristallchem 89*:327–343 (Oct. 1934).

3.45 E. Orowan, Fracture and strength of solids, *Rep. Phys. Soc. Prog. Phys. 12*: 185–232 (1948–49).

3.46 E. Orowan, Energy criteria of fracture, *Weld J. 34, 3*: 157S–160S (1955).

3.47 J. Frenkel, Zur Theorie der Elastizitatsgrenze and der festigkeit kristallinischer Korper, *Z. Phys. 37*: 572–609 (1926).

3.48 W.R. Tyson, Surface energies of solid metals, *Can. Metall. Q. 14, 4*: 307–314 (1975).

3.49 H. Jones, The surface energy of solid metals, *Met. Sci. J. 5*: 15–18 (1971).

3.50 L.E. Murr, *Interfacial Phenomena in Metals and Alloys*, Addison-Wesley, Reading, MA, 1975.

3.51 S.H. Overbury, P.A. Bertrand, and G.A. Somorjai, Surface composition of binary systems: prediction of surface phase diagrams of solid solutions, *Chem. Rev. 75, 5*: 547–560 (1975).

3.52 N. Eustathopoulos, J.C. Joud, and P. Desre, Interfacial tension of pure metals, Part I—Estimation of the liquid-vapor and solid-vapor surface tensions from the cohesion energy, *J. Chim. Phys.— Phys. Chim. Biol. 70, 1*: 42–48 (1973).

3.53 N. Eustathopoulos, J.C. Joud, and P. Desre, Interfacial tension of pure metals, Part II—Estimation of the solid-vapor and solid-liquid interfacial tensions from the surface tension of liquid metals, *J. Chim. Phys.—Phys. Chim. Biol. 70, 1*: 49–53 (1973).

3.54 R.G. Linford, Surface thermodynamics of solids, *Solid State Surface Science II* (M. Green, ed.), Marcel Dekker, 1973, pp. 1–152.

3.55 T.A. Roth, Surface and grain-boundary energies of iron, cobalt, and nickel, *Mater. Sci. Eng. 18, 2*: 183–192 (1975).

3.56 A.R. Miedema, Surface energies of solid metals, *Z. Metallk. 69, 5*: 287–292 (1978).

3.57 P.W. Bridgman, Shearing phenomena at high pressures, particularly in inorganic compounds, *Proc. Am. Acad. Arts Sci. 71*: 387–459 (1937).

3.58 S. Timoshenko and J.N. Goodier, *Theory of Elasticity*, Second ed., McGraw-Hill, New York, 1951.

3.59 D. Landau, Hardness: A Critical Examination of Hardness, Dynamic Hardness, and an Attempt To Reduce Hardness to Dimensional Analysis, The Nitralloy Corporation, New York, 1943.

3.60 D. Tabor, *The Hardness of Metals*, Clarendon Press, Oxford, 1951.

3.61 E. Rabinowicz, *Friction and Wear of Materials*, John Wiley & Sons, New York, 1965.

3.62 D.H. Buckley, *Friction, Wear, and Lubrication in Vacuum*, NASA SP-277, 1971.

3.63 L. Pauling, A resonating-valence-bond theory of metals and intermetallic compounds, *Proc. R. Soc. London A 196*: 343–362 (1949).

3.64 W. Hayden, W.G. Moffatt, and J. Wulff, Mechanical Behavior, *The Structure and Properties of Materials*, John Wiley & Sons, New York, Vol. 3, 1965.

3.65 K. Miyoshi and D.H. Buckley, The adhesion, friction, and wear of binary alloys in contact with single-crystal silicon carbide, *J. Lubr. Technol. 103*: 180–187 (1981).

3.66 K. Miyoshi and D.H. Buckley, The friction and wear of metals and binary alloys in contact with an abrasive grit of single-crystal silicon carbide, *ASLE Trans. 23, 4*: 460–477 (1980).

3.67 J.R. Stephens and W.R. Witzke, Alloy softening in binary iron solid solutions, *J. Less Common Met. 48*: 285–308 (Aug. 1976).

3.68 D.H. Buckley and K. Miyoshi, Tribological properties of structural ceramics, *Treatise on Materials Science and Technology* (John B. Wachtman, Jr., ed.), Academic Press, Boston, 1989, pp. 293–365.

3.69 K. Miyoshi and D.H. Buckley, Anisotropic tribological properties of SiC, *Wear 75, 2*: 253–268 (1982).

3.70 D.H. Buckley, Friction and wear behavior of glasses and ceramics, *Mater. Sci. Res. 7*: 101–126 (1974).

3.71 K. Miyoshi and D.H. Buckley, Wear particles of single-crystal silicon carbide in vacuum, NASA TP-1624 (1980).

Chapter 4
Properties of Contaminated Surfaces: Adhesion, Friction, and Wear

4.1 Introduction

When atomically clean, unlubricated surfaces are brought together under a normal load, the atoms at the surfaces must, at some points, be in contact. Then, the basic material properties of the solids themselves become extremely important in the adhesion, friction, and wear behavior of the materials, as described in Chapter 3. In most practical situations, however, contaminant layers, such as carbon compounds and water, are ubiquitous and are present on any solid surface that has been exposed to air. Even a supposedly "clean" material surface will show a significant carbon, oxygen, and water contribution to the Auger electron spectroscopy (AES) and x-ray photoelectron spectroscopy (XPS) spectrum because one or more layers of adsorbed hydrocarbons and oxides of carbon are present [4.1–4.4]. These contaminant layers mask the surface features of the solids in tribological contact. For example, a major characteristic of material wear is that for unlubricated surfaces the wear rate covers an enormous range (say 10^{-2} to 10^{-10} mm^3/N·m) while the coefficient of friction varies relatively little (0.01 to 1.5 in air). The small coefficient-of-friction range occurs because the solid surfaces in dry contact are masked by the contaminant layers. The friction between unlubricated surfaces is due to shearing in the adsorbed contaminant films, although these films may be partially destroyed by the sliding process.

This chapter presents the fundamental tribology of contaminated surfaces (i.e., the adhesion, friction, and wear behavior of smooth but contaminated surfaces of solid-solid couples, such as metal-ceramic couples). The subjects to be addressed include the effects of (1) surface contamination by the interaction of a surface with the environment, (2) surface contamination by diffusion of bulk elements or compounds, and (3) surface chemical changes with selective thermal evaporation [4.3, 4.4].

4.2 Effects of Surface Contamination by Environment

4.2.1 Ubiquitous Contamination From Atmosphere

Nonoxide ceramics.—Pull-off forces (adhesion) and coefficients of friction for hot-pressed polycrystalline silicon nitride (Si_3N_4) in contact with metals were examined in an ultra-high-vacuum environment [4.3, 4.4] by using the method described in Chapter 3.

Figure 4.1 shows the marked difference in adhesion for two surface conditions, contaminated (as received) surfaces and sputter-cleaned surfaces, in ultrahigh vacuum. Chemical interactions normally play an important role in the adhesion of silicon nitride–metal couples. With contaminated surfaces, however, the chemical activity or inactivity of the metal did not appear to play a role in adhesion (Fig. 4.1). Adhesion for the various as-received metals in contact with Si_3N_4 generally remained constant. In contrast, the adhesion properties for the sputter-cleaned surfaces were related to the relative chemical activity (percentage of d valence bond character) of the transition metals as a group, and adhesion was higher than for the as-received surfaces. According to Pauling's theory [4.5] the greater the percentage of d valence bond character, the less active the metal and the lower the pull-off force required to break the bonds. Thus, the adhesion results (Fig. 4.1) show that the more active the metal, the higher the adhesion. This conclusion is consistent with the

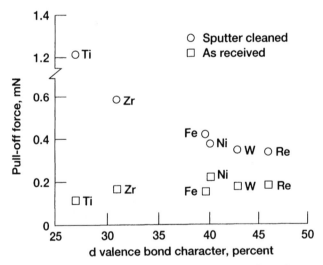

Figure 4.1.—Pull-off force (adhesion) as function of percentage of d valence bond character for transition metals in contact with monolithic Si_3N_4 in vacuum. From [4.5].

friction studies conducted on the surfaces of other nonmetal-metal couples (see Chapter 3). But why does this appreciable difference in adhesion occur for the two surface conditions, contaminated and sputter cleaned? Analyzing the surface chemistry of the foregoing Si_3N_4 specimens will assist us in understanding adhesion behavior.

XPS survey spectra of an as-received Si_3N_4 surface (Fig. 4.2(a)) reveal an adsorbed carbon peak as well as an oxygen peak. An adsorbate layer on the surface consisted of hydrocarbons and water vapor that may have condensed from the environment and become physically adsorbed to the Si_3N_4 surface. The contamination layer could be removed by ion etching just before analysis. But, since the ion beam itself can induce compositional changes in the specimen surface, it had to be used with care. After the Si_3N_4 surface was cleaned by argon ion sputtering, the carbon and oxygen contamination peaks became very small (Fig. 4.2(b)), and the peak intensity of both the silicon and nitrogen associated with Si_3N_4 increased markedly. The small amount of carbon and oxygen was associated with bulk contaminants of the Si_3N_4.

Comparing Fig. 4.1 with Fig. 4.2 shows that a prerequisite for the sameness in adhesion of as-received surfaces of ceramic-metal couples is that the surfaces be covered with a stable layer of contaminants. Thus, contaminant films on the surfaces of ceramics and metals can greatly reduce adhesion.

Friction results for boron nitride (BN) coatings in contact with metals in ultrahigh vacuum (Fig. 4.3) are analogous to the adhesion results (Fig. 4.1). The similar shapes of Figs. 4.1 and 4.3 are not surprising because XPS survey spectra of the as-received BN film surface reveal a carbon contaminant peak as well as an adsorbed

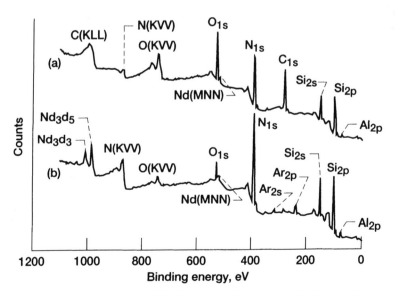

Figure 4.2.—XPS spectra of Si_3N_4 before (a) and after (b) ion sputtering.

Properties of Contaminated Surfaces

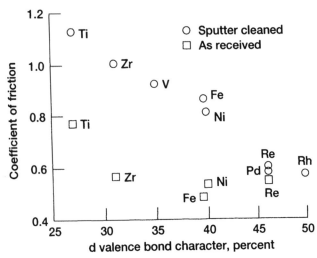

Figure 4.3.—Coefficient of friction as function of d valence bond character of metals in sliding contact with ion-beam-deposited BN film in ultrahigh vacuum.

oxygen peak [4.6]. The friction results, as shown in Fig. 4.3, indicate that for the sputter-cleaned surfaces the coefficients of friction were low at higher percentages of d valence bond character but were generally the same at all percentages for the as-received surfaces. Only the as-received surfaces of the BN-titanium couple showed a somewhat higher coefficient of friction, perhaps because the contaminant films adsorbed on these surfaces were partially destroyed during sliding. An adsorbate layer on the surface consisted of water vapor and hydrocarbons that may have condensed from the environment and become physically adsorbed to the BN film. After the BN film surfaces had been sputter cleaned with argon ions, although there were still small carbon and oxygen contamination peaks, the boron and nitrogen peaks predominated (Fig. 4.4).

The boron nitride B_{1s} photoelectron emission lines (Fig. 4.4) peaked primarily at 190 eV, which is associated with BN. They also included a small amount of boron carbide (B_4C). The N_{1s} photoelectron emission lines peaked primarily at 397.9 eV, which again is associated with BN. The C_{1s} photoelectron lines taken from the as-received surface at 284.6 eV indicate the presence of adventitious adsorbed carbon contamination with a small amount of carbide. After sputter cleaning the adsorbed carbon contamination peak disappeared from the spectrum, and the relatively small carbide peak could be seen. The O_{1s} photoelectron lines of the as-received BN surface peaked at 531.6 eV because of adsorbed oxygen contamination and oxides. After argon sputter cleaning the adsorbed oxygen contamination peak disappeared from the spectrum, but the small oxide peak remained. Thus, the peak intensities for both boron and nitrogen associated with BN increased with argon ion sputter cleaning, but those for carbon and oxygen decreased markedly.

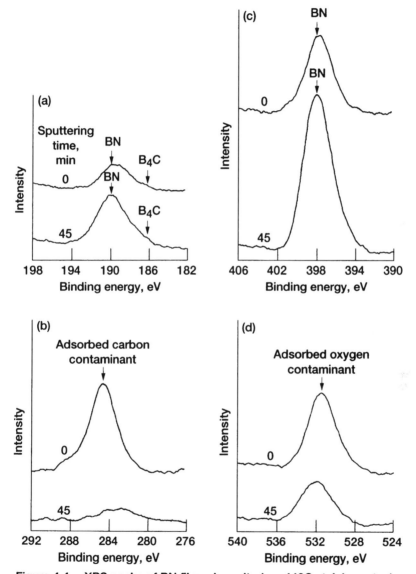

Figure 4.4.—XPS peaks of BN films deposited on 440C stainless steel. Pass energy, 25 eV; energy resolution, 0.5 eV. (a) B_{1s}. (b) N_{1s}. (c) C_{1s}. (d) O_{1s}.

The BN film deposited on a 440C stainless steel substrate was nonstoichiometric, with a boron-to-nitrogen (B/N) ratio of 1.6.

Elemental depth profiles for the ion-beam-deposited BN film on silicon as a function of sputtering time were obtained from AES analysis (Fig. 4.5) [4.6, 4.7]. The boron and nitrogen content rapidly increased in the first minute of sputtering, but the oxygen and carbon content decreased in the first 1 to 2 min and remained constant thereafter. The BN film deposited on silicon had a B/N ratio of about 2. Thus, XPS and AES analyses clearly revealed that BN was nonstoichiometric and that small amounts of oxides and carbides were present on the surface and in the bulk of the BN film. Contaminants such as carbides (e.g., B_4C) and oxides may have been introduced and absorbed into the BN film during ion beam deposition.

The BN films deposited on silicon substrates were probed by using secondary ion mass spectroscopy [4.6, 4.7]. The spectra indicated the presence of the following secondary ions: B^+, B_2^+, C^+, O^+, Si^+, Si_2^+, and SiO^+. The peak observed at 14 atomic mass units (amu) could result from N^+, CH_2^+, and SiO_2^+. Additional peaks at 24 and 25 amu were related to BN^+ and thus supported the XPS data. The SiO^+ signal was associated with the oxide present at the BN-silicon interface. AES depth profiles similar to Fig. 4.5 were obtained for a 116-nm-thick BN film deposited on indium phosphide (InP) and for a 129-nm-thick BN film deposited on gallium arsenide (GaAs) [4.7].

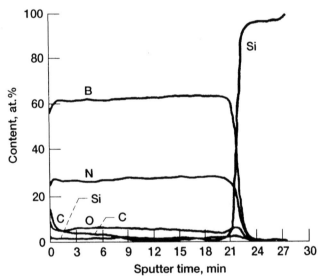

Figure 4.5.—AES elemental depth profiles as function of sputtering time for BN films ion beam deposited on silicon. Deposition temperature, 200 °C; ion beam energy, 150 eV.

Oxide ceramics.—Analogy between adhesion and friction was also found with oxide ceramic–metal couples, such as manganese-zinc (Mn-Zn) ferrite (Fig. 4.6). The contaminant layers on the as-received surfaces of Mn-Zn ferrite–metal couples reduced adhesion and friction. In this case, adhesion and friction did not depend on the material properties, such as shear modulus. Adhesion and friction were greater with the sputter-cleaned surfaces than with the as-received surfaces and, again, could be correlated with the Young's and shear moduli for the metals, as described in Chapter 3. Contaminants are weakly bound to the surfaces because the binding is physical rather than chemical. Under actual sliding conditions the contaminant surface layers may be removed by repeated sliding, making direct contact of the fresh, clean surfaces unavoidable [4.1, 4.2]. For example, when a tungsten carbide (WC)-coated ring was brought into contact with a sapphire (Al_2O_3) ring in air, the coefficient of friction increased as the number of passes increased, as shown in Fig. 4.7. Here, the coefficient of friction was not only high but erratic as well between 300 and 1600 passes. A plausible explanation is that the contaminant surface layers were removed by the repeated reciprocating sliding, allowing direct contact of the fresh, clean surfaces and increasing the coefficient of friction. Thus, the coefficient of friction of the sapphire-WC couple is also dependent on surface condition.

Debris generated by sliding action provides a useful history of the wear process. In addition to the quality and size of the wear debris particles, microscopic observation of their nature and shape yielded much more useful information [4.8, 4.9]. Scanning electron microscope (SEM) analysis of the wear scar produced on the sapphire ring by the sliding action revealed that the entire wear scar was generally smooth (Fig. 4.8). Most wear debris accumulated outside the wear scar and generally consisted of individual, fine (submicrometer and micrometer) particles. The friction behavior of an Si_3N_4 flat in contact with an Si_3N_4 pin in air was similar to the friction behavior of sapphire on tungsten dispersed with carbides.

Figure 4.9 shows another example [4.10]. When a natural diamond pin was in repeated sliding contact with a diamond film in ultrahigh vacuum, the coefficient of friction increased as the number of passes increased, reaching an equilibrium value after a certain number of passes. Again, repeatedly sliding the pin over the same track in ultrahigh vacuum had removed some contaminant surface film from the contact area. Stronger interfacial adhesion resulted between the diamond pin and the diamond film, raising the coefficient of friction.

4.2.2 Adsorbed Oxygen

Most gases, with the exception of the noble gases, adsorb readily to clean metal surfaces [4.1]; and many adsorb to nonmetals, such as silicon carbide (SiC), as well. Adhesion and friction are so sensitive to the presence of these gases, both qualitatively and quantitatively, that even hydrogen and fractions of a monolayer of other gases exert an effect. Practically all published works agree that extremely

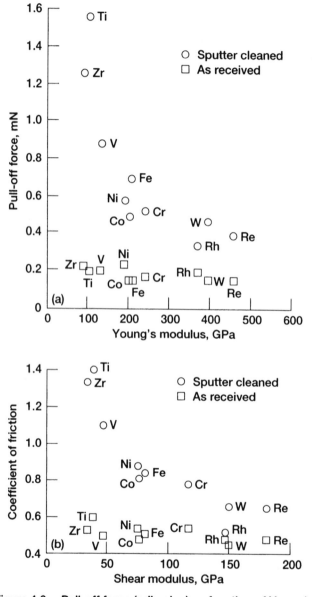

Figure 4.6.—Pull-off force (adhesion) as function of Young's modulus (a) and coefficient of friction as function of shear modulus (b) of metals in contact with polycrystalline Mn-Zn ferrite in ultrahigh vacuum.

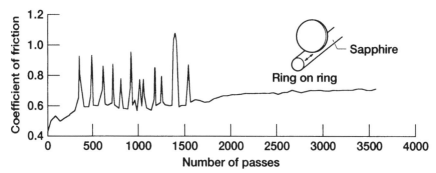

Figure 4.7.—Coefficient of friction as function of number of passes for sapphire (Al_2O_3) sliding against tungsten coating with dispersed carbides chemically vapor deposited on molybdenum in air. Reciprocating motion; load, 3 N; sliding velocity, 86 mm/min; track length, 2 mm; air environment; room temperature; relative humidity, 5.9%; cleaning agent, 1,1,1-trichloroethane.

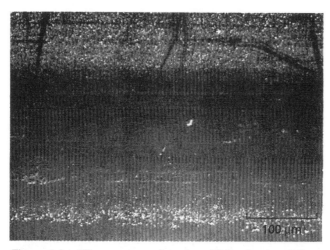

Figure 4.8.—Wear scar produced on Al_2O_3 ring by sliding contact in air. (The friction trace is shown in Fig. 4.7.)

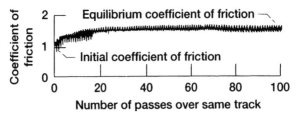

Figure 4.9.—Typical friction trace for bulk diamond
pin in sliding contact with diamond film deposited
on α-SiC substrate in vacuum.

small amounts of oxygen or other contaminant gases can greatly reduce the adhesion between metals [4.11–4.15]. However, there are situations where some amounts of oxides or a monolayer of oxygen when adsorbed can increase adhesion and friction.

On clean surface.—At room temperature and low pressures oxygen generally adsorbs on a clean SiC surface to an equilibrium layer thickness of no more than two monolayers [4.16]. For example, AES analysis has indicated that an SiC surface exposed to gaseous oxygen at 1.3 Pa and 25 °C for 1 hr will be covered with silicon oxide as well as with a simple adsorbed oxygen film [4.17]. Figure 4.10 shows the effect that such exposure has on friction behavior [4.18]. The presence of oxygen on the surfaces has reduced the coefficient of friction to two-thirds of that observed for the clean surfaces in contact. These sliding friction experiments were conducted with a single-crystal SiC pin in sliding contact with a single-crystal SiC disk in ultrahigh vacuum. The basal planes of both the pin and disk specimens were parallel to the sliding interface. The friction data presented in Fig. 4.10 were obtained as a function of the number of passes over the same track for argon-sputter-cleaned surfaces and for surfaces containing adsorbed oxygen. At a load of 0.3 N the adsorbed film gave the lower coefficient of friction after one to three passes of the pin. Friction markedly increased when the number of passes was increased from three to four, but beyond four passes the coefficients of friction were almost the same as those for sputter-cleaned SiC. This marked change in friction behavior is believed to be due to the breakdown of the adsorbed oxygen film by sliding action. At a load of 0.1 N, however, the adsorbed oxygen film did not break down for at least 20 passes.

On oxygen-ion-bombarded surface.—An argon-sputter-cleaned SiC surface was bombarded with oxygen ions at a 1000-V potential under a pressure of 1.3 Pa at 25 °C for 30 min. The AES spectrum of this surface (Fig. 4.11(a)) has three characteristic peaks: silicon peaks at 68 and 82 electron volts (eV), a carbon peak at 272 eV, and an oxygen peak at 516 eV [4.18]. The silicon peaks have two regions (I and II in the figure), which reflect the contribution to the valence band of silicon and oxygen bonding and nonbonding molecular orbitals [4.17]. The contribution from silicon-silicon bonds, which appeared at 92 eV, is not shown in this spectrum. Thus, the oxygen-ion-bombarded SiC surface seems to have been covered by

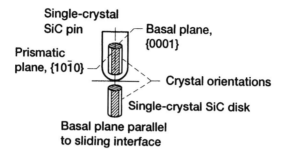

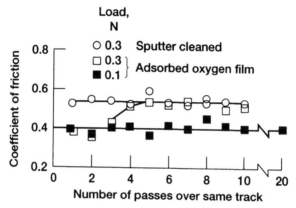

Figure 4.10.—Average coefficient of friction, obtained from maximum peak heights in friction trace, as function of number of passes of SiC pin across SiC {0001} disk. Oxygen adsorption conditions (disk and pin), 1.3 Pa and room temperature for 1 hr.

an SiO layer and, above that, an SiO_2 outer layer. The carbon peak, which was of the SiC type, is nearly undetectable in Fig. 4.11(a).

A sputter-cleaned and then oxygen-ion-bombarded SiC surface was exposed to oxygen at 1.3 Pa for 10 min. The Auger spectrum of this surface (Fig. 4.11(b)) shows the silicon peak to be at 78 eV. The position of the peak reflects the contribution to the valence band of silicon and oxygen nonbonding molecular orbitals [4.17]. The carbon peak is small, as in Fig. 4.11(a).

The coefficients of friction obtained for the oxygen-ion-bombarded SiC surfaces (Fig. 4.11(c)) were slightly higher than those obtained for sputter-cleaned SiC surfaces (Fig. 4.10). Also, at a load of 0.3 N the adsorbed film on the oxygen-ion-bombarded surface gave a fairly constant low coefficient of friction throughout the experiment without breaking down.

On reacted oxide surface.—To form a reacted oxide film on a single-crystal SiC surface, the SiC was first exposed to air at atmospheric pressure and 700 °C

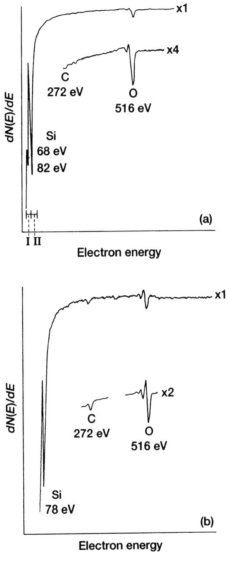

Figure 4.11.—Auger spectra and coefficients of
friction for single-crystal SiC {0001} surfaces.
(a) Auger spectrum for surface with ion-
bombarded oxygen film. (b) Auger spectrum for
surface with adsorbed oxygen film on ion-
bombarded oxygen film. (c) Coefficients of
friction for both surfaces.

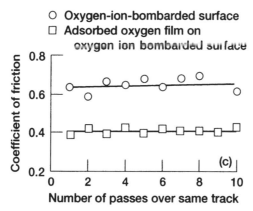

Figure 4.11.—Concluded. (c) Coefficients of friction.

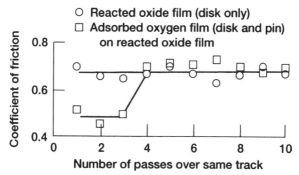

Figure 4.12.—Average coefficients of friction for single-crystal SiC {0001} surfaces with sputter-cleaned, reacted oxide film with and without adsorbed oxygen film.

for 10 min [4.18]. Then, the oxidized SiC surface was argon ion sputter cleaned in vacuum. After sputter cleaning the carbon peak was barely discernible in the AES spectrum. An oxygen peak and a chemically shifted silicon peak (at 82 eV) indicated a layer of SiO_2 on the SiC surface. After oxygen adsorption on the reacted oxide film, a slightly higher oxygen AES peak was present on the SiC surface than on the clean reacted oxide film.

Figure 4.12 presents the friction data obtained for surfaces with a reacted oxide film and with an adsorbed oxygen film on the reacted oxide. The adsorbed oxygen film gave a lower coefficient of friction during the first three passes but the same values beyond three passes as did the clean reacted film. Again, the marked increase in friction beyond three passes is believed to be due to sliding action breaking down the adsorbed oxygen film.

Figures 4.13 and 4.14 summarize the coefficients of friction measured at a load of 0.3 N with a single sliding pass for SiC-to-SiC and SiC-to-titanium interfaces with reacted, ion-bombarded, and adsorbed oxygen or nitrogen films [4.18]. The effects of oxygen interactions on the friction behavior were generally the same for both couples. The reacted oxide and oxygen-ion-bombarded surfaces interacted with the SiC surfaces to produce two effects: (1) SiC oxidized and formed a protective oxide surface layer and (2) the layer increased the coefficients of friction for both couples.

The effects of oxygen in increasing friction are related to the relative chemical thermodynamic properties of silicon, carbon, and titanium to oxygen. Table 4.1 presents free energies of formation for silicides, oxides, carbides, and nitrides [4.19, 4.20]. The greater the degree of oxidation or oxygen implantation by ion bombardment, the more chemically active the surface (Table 4.1) and the higher the coefficient of friction (Fig. 4.13). In such a situation oxygen will tend to chemically bond to the surface.

By contrast, adsorption of oxygen on argon-ion-sputter-cleaned, oxygen-ion-bombarded, and reacted oxide surfaces generally decreased the coefficient of friction (Fig. 4.14). Also, adsorption of nitrogen on argon-ion-sputter-cleaned surfaces decreased the coefficient of friction (Fig. 4.14). When oxygen and nitrogen adsorb on the surface, the forces of attraction between the adsorbing gas and the SiC surface seem to be relatively weak. Thus, the adsorption of nitrogen or oxygen generally reduces adhesion and friction.

4.2.3 Defined Exposure to Oxygen

As discussed in the previous section, the coefficient of friction is strongly affected by gas interactions with the surface, such as the adsorption of a species (physically or chemically adsorbed material) or the chemical reactions of the surface with a species.

Figure 4.15 presents the coefficients of friction as a function of the d valence bond character of various transition metals in contact with nickel-zinc (Ni-Zn) ferrite. Both the argon-ion-sputter-cleaned metal and the ferrite were exposed to 1000 langmuirs (1×10^{-6} torr-s) of oxygen gas in vacuum at 1.33×10^{-4} Pa [4.21]. At completion of the exposure the vacuum system was evacuated to 30 nPa or lower for the sliding friction experiments.

Figure 4.15 also presents comparative data for clean metals in contact with clean Ni-Zn ferrite. The adsorption of an oxygen monolayer on argon-sputter-cleaned metal and ferrite surfaces produced two effects: (1) the metal oxidized and formed an oxide surface layer and (2) the oxide layer increased the coefficients of friction for the ferrite-to-metal interfaces. Oxygen adsorption had the same effect on the adhesion and friction of Ni-Zn ferrite–metal contacts as was observed for metals in sliding contact with Mn-Zn ferrite (see Chapter 3). That is, oxygen

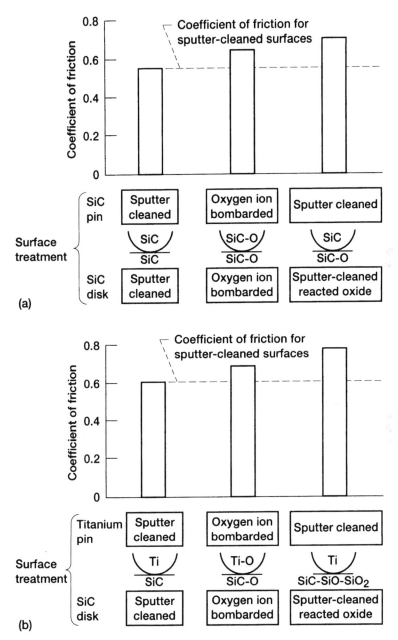

Figure 4.13.— Coefficients of friction for SiC-to-SiC (a) and SiC-to-titanium (b) contacts after various surface treatments. Single-pass sliding.

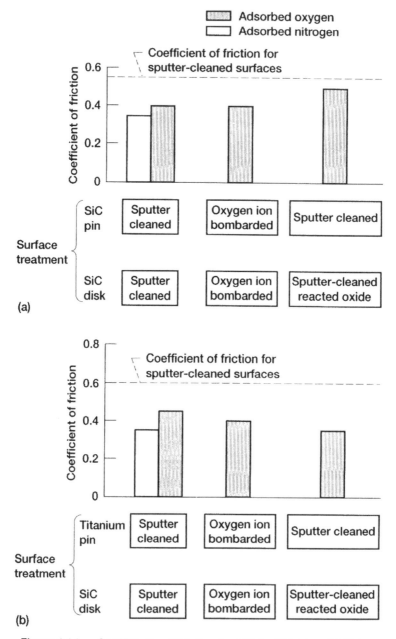

Figure 4.14.— Coefficients of friction for SiC-to-SiC (a) and SiC-to-titanium (b) contacts with adsorbed oxygen and nitrogen films after various surface treatments. Single-pass sliding.

TABLE 4.1.—FREE ENERGY OF FORMATION

Formula	Free energy of formation	
	kcal/mole	J/mole
SiC	−26.1	$−109 \times 10^3$
SiO	−32.77	−137
SiO_2	−192.4	−805
CO	−32.81	−137
CO_2	−94.26	−394
Si_3N_4	−154.7	−647
TiSi	−31.0	−130
TiC	−53	−222
TiO	−116.0	−485
TiO_2	−203.8	−853
TiN	−66.1	−277

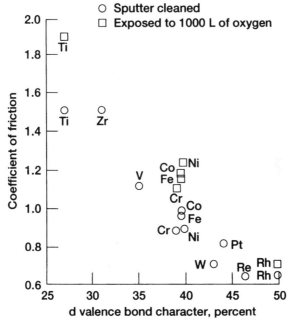

Figure 4.15.—Effect of adsorbed oxygen on friction for various metals in sliding contact with Ni-Zn ferrite. Single-pass sliding; sliding velocity, 3 mm/min; load, 0.05 to 0.2 N; vacuum, 30 nPa; room temperature.

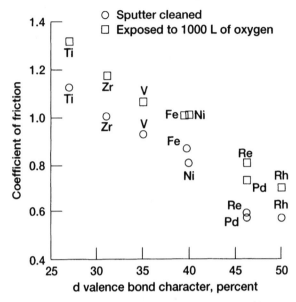

Figure 4.16.— Effect of adsorbed oxygen on friction for various metals in sliding contact with BN. Single-pass sliding; sliding velocity, 3 mm/min; load, 0.05 to 0.2 N; vacuum, 30 nPa; room temperature.

exposure strengthened metal-to-ferrite adhesion and increased friction. The enhanced bonding of the metal oxide to the ferrite may be due to a complex oxide forming on contact.

Figure 4.16 presents the coefficients of friction for clean metals in contact with BN coatings after exposure to oxygen gas [4.6]. The data reveal that the adsorption of oxygen on the surfaces of argon-sputter-cleaned metals and BN coatings produced two effects: (1) the metal and BN oxidized and formed an oxide surface layer and (2) the oxide layers increased the coefficient of friction for metal-to-BN interfaces. Thus, the oxygen exposure strengthened metal-to-BN adhesion and increased friction.

4.2.4 Humidity

Effect on adhesion.—The concentration of humidity (water vapor) in the atmosphere has a marked effect on a material's mechanical, chemical, and tribological properties. Figure 4.17 presents the pull-off forces necessary to separate a hemispherical Si_3N_4 pin in contact with an Si_3N_4 flat in dry, moist, and water-vapor-saturated nitrogen atmospheres [4.22]. Adhesion (pull-off force) remained low at relative humidities below 80% but rose rapidly above 80%. There was no change

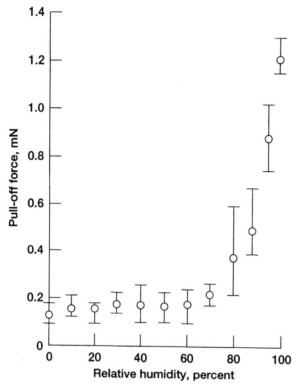

Figure 4.17.—Pull-off force (adhesion) as function of relative humidity for hemispherical Si_3N_4 pin in sliding contact with Si_3N_4 flat in nitrogen atmosphere.

in adhesion with normal load between 0.2 and 0.8 mN in dry, moist, and water-vapor-saturated nitrogen.

Probably, the greater adhesion observed in the water-vapor-saturated nitrogen atmosphere arose primarily from the surface tension effects of a thin water film adsorbed on the Si_3N_4 surfaces. The happenings at the interface may be visualized as follows:

1. An Si_3N_4 pin was essentially in elastic contact with an Si_3N_4 flat surface with a thin water film between them (Fig. 4.18(a)).
2. When the normal load was removed and the elastic stresses within the bulk of the specimens were released, the interfacial junctions were broken one by one.
3. When the pin (of radius R) detached from the flat (Fig. 4.18(b)), the applied separation force was balanced by the surface tension of a thin water film resisting surface extension.

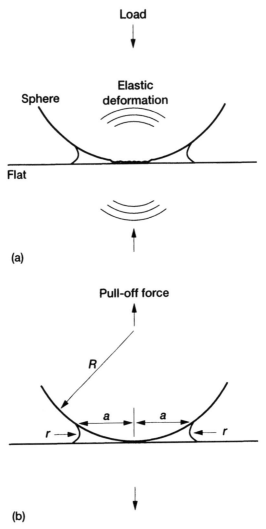

Figure 4.18.—Meniscus formed at contact area
between spherical and flat surfaces. (a) Contact.
(b) Separation.

4. Suppose the liquid collected to form a pool at the tip of the hemispherical pin.
(The radius of curvature of the meniscus profile is r.) If the meniscus
were extremely small ($r \ll R$) and the liquid completely wetted the surface
(i.e., zero contact angle), the pressure p inside the liquid would be less than
atmospheric pressure by approximately T/r, where T is the surface tension of
the liquid. This low pressure would act over an area πa^2 of the water pool,
giving a total adhesive force of $p\pi a^2$ (i.e., $\pi a^2 T/r$). To a close approximation

$a^2 = 2R \times 2r$, where R is the radius of curvature of the spherical surface. The resulting adhesive force would be [4.23, 4.24]

$$Z = 4Rr\pi \frac{T}{r} = 4\pi RT \qquad (4.1)$$

Adhesion is thus independent both of water film thickness and applied normal load. The surface tension calculated from these experimental results (Fig. 4.17) is 58×10^{-5} to 65×10^{-5} N/cm. The accepted value for water is 72.7×10^{-5} N/cm. This discrepancy may be due to the surface roughness and irregular roundness or flatness of the Si_3N_4 specimens. Surface irregularities can affect the radii of curvature of the hemispherical pin and the meniscus.

The results shown in Fig. 4.17 are of great interest. In practical fields, such as information storage systems, the role humidity plays in adhesion and friction is relevant to understanding the tribological problem. Therefore, a case study was conducted that shows the important role water vapor plays in the friction of a magnetic tape sliding against a ferrite (a magnetic head material) pin.

The effects of humidity in moist nitrogen on the friction and deformation behavior of polymeric magnetic tapes in contact with a polycrystalline Ni-Zn ferrite hemispherical pin [4.25] were examined in the case study. Experiments were conducted with loads to 1.0 N at sliding velocities to 6 mm/min in single-pass and multipass sliding at room temperature. Multipass sliding experiments were conducted in reciprocating motion.

Figure 4.19 shows the layer structure of the polymeric magnetic tape. Primary components of the magnetic layer were the magnetic oxide particles, the binder, the lubricant, the dispersant, and other minor additives. The magnetic tapes used in this investigation contained powders coated on a polyester film backing (film thickness, 23 μm; film width, 12.7 mm). The two kinds of tape used in this investigation were made with polyester-polyurethane binders, the most widely used binders for

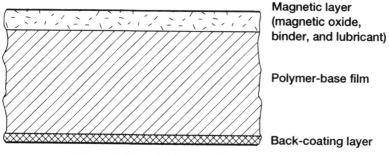

Figure 4.19.— Schematic of magnetic tape.

magnetic tape applications. The magnetic layers contained magnetic oxide (γ-Fe$_2$O$_3$) particles equal to 70% of the layer by weight and as much as 60% by volume.

Effect on friction and deformation.—Traces of friction at various loads on the tape as a function of sliding time are relatively smooth, with no evidence of stick-slip. The coefficient of friction was not constant but decreased as the load increased to 0.25 N (Fig. 4.20(a)). As the load increased above 0.25 N, however, the coefficient of friction increased. When repeated passes were made (Fig. 4.20(b)), the coefficient of friction for the tape generally exhibited small changes at loads less than 1.0 N. The data of Fig. 4.20, particularly part (a), raise the question of how the interface deforms with sliding action.

The tracks on the tape made by the ferrite pin looked different at various loads when examined by optical and scanning electron microscopy. Essentially no detectable wear track existed on the tape surface at 0.1 N. The wear track was similar to that shown in Fig. 4.21(a), the surface of as-received magnetic tape (tape 1). At 0.25 N and above the sliding action produced a visible wear track on the magnetic tape, similar to that shown in Fig. 4.21(b). This scanning electron micrograph clearly reveals a degree of plastic deformation at the asperity tips on the magnetic tape. Thus, although sliding occurred at the interface, both the tape and the Ni-Zn ferrite pin were elastically deformed at loads to 0.25 N. At 0.25 N

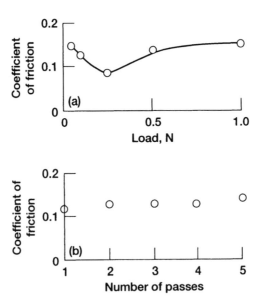

Figure 4.20.—Coefficient of friction as function
of load (a) and number of passes (b) for Ni-Zn
ferrite sliding on magnetic tape in laboratory air.

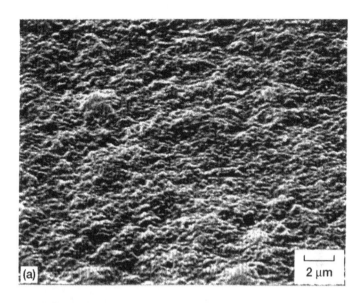

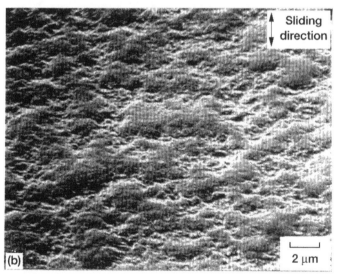

Figure 4.21.—Scanning electron micrographs of as-received surface (a) and wear track (b) on magnetic tape after five passes of Ni-Zn ferrite pin. Normal load, 1 N; sliding velocity, 1.5 mm/min; relative humidity, 40%; room temperature; laboratory air.

and above the tape deformed plastically, but the Ni-Zn ferrite pin primarily deformed elastically. Figure 4.21(b) shows the blunt appearance of the asperities on the wear track after five sliding passes at 1.0 N. This bluntness resulted primarily from plastic deformation.

From the nature of deformation at the interface the friction behavior of the hemispherical Ni-Zn ferrite pin and the magnetic tape can generally be divided into two categories (i.e., elastic and plastic contact). In the elastic contact region the friction decreases as the load increases. The relation between coefficient of friction μ and load W is given by an expression of the form $\mu = KW^{-1/3}$, where K is a constant [4.26, 4.27]. The exponent can be interpreted simply as arising from an adhesion mechanism, with the contact area being determined by elastic deformation.

For example, the coefficients of friction for the polyester tape backing in contact with the Ni-Zn ferrite pin decreased as the load increased. That is, with sliding, both the polyester backing and Ni-Zn ferrite pin surfaces deformed elastically [4.28]. By contrast, when deformation was plastic, the coefficient of friction for a hard, spherical solid pin in contact with a soft, solid tape increased as the load increased. Figure 4.20(a) presents a typical example of this at loads of 0.25 N and above [4.28]. The role of deformation in friction will be discussed in more detail in Chapter 5.

Effect on plastic deformation and softening.—The coefficient of friction measured for tape 2 was 0.14 in dry nitrogen at a load of 0.5 N (Fig. 4.22(a)). The atmosphere was then humidified to the desired relative humidity (up to 78%) by admitting humid nitrogen into the system. On humidifying (open symbols) the coefficient of friction remained low below 40% relative humidity but then increased rapidly. On dehumidifying (solid symbols) the coefficient of friction decreased rapidly between 78% and 40% relative humidities and then remained low. The friction behavior of the tape on dehumidifying was similar to that on humidifying.

At 0.25 N (Fig. 4.22(b)) the results were consistent with those at 0.5 N. At both 0.25 and 0.5 N the sliding action produced a visible wear track on the magnetic tape. The question was, Would the interface deform differently in high humidity than in dry nitrogen with sliding action (i.e., would the tape deform readily in a moist atmosphere)? To examine the deformation behavior of the tape surface, multipass sliding friction experiments were conducted with magnetic tapes in sliding contact with polycrystalline Ni-Zn ferrite pins in both dry and humid nitrogen at 78% relative humidity.

Figure 4.23 presents the coefficients of friction for magnetic tape (tape 2) in nitrogen at 78% relative humidity as a function of the number of repeated passes. After 50 passes the coefficient of friction had decreased slightly, but continuously, as the number of passes increased. However, the coefficient of friction measured in dry nitrogen was constant after 10 passes. These results suggest that a polymeric magnetic tape in contact with an Ni-Zn ferrite pin would deform plastically in humid nitrogen (78% relative humidity) more than it would in dry nitrogen.

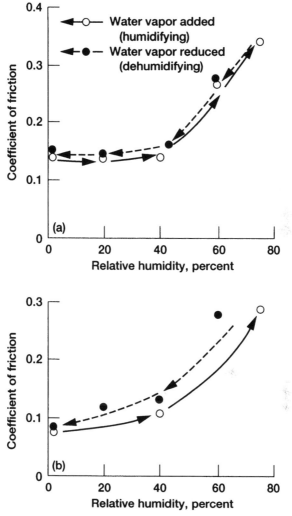

Figure 4.22.—Effect of humidifying and dehumidifying on friction of magnetic tape (tape 2) in
sliding contact with Ni-Zn ferrite pin at normal
loads of 0.5 and 0.25 N. Sliding velocity, 0.1 mm/s;
room temperature; environment, nitrogen.
(a) Normal load, 0.5 N. (b) Normal load, 0.25 N.

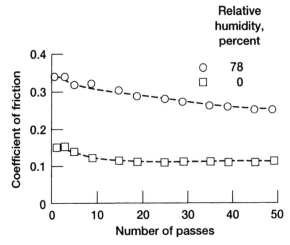

Figure 4.23.—Coefficient of friction as function of
number of passes for magnetic tape (tape 2) in
sliding contact with Ni-Zn ferrite pin. Normal load,
0.5 N; sliding velocity, 0.1 mm/s; room temperature; environments, dry and humid nitrogen.

To determine tape surface deformation with sliding action in both dry and humid
nitrogen, the wear tracks on the tape were examined by scanning electron microscopy after 50 repeated passes. Figure 4.24 presents scanning electron micrographs
of the as-received magnetic tape surface and the wear tracks generated in dry
nitrogen and in humid nitrogen (78% relative humidity). The as-received tape
surface (Fig. 4.24(a)) had the coarsest structure. Both wear tracks (Figs. 4.24(b) and
(c)) clearly reveal a degree of plastic deformation at the asperity tips. Considerable
plastic flow occurred in the tape during sliding contact with the Ni-Zn ferrite pin in
humid nitrogen, much more than in the dry nitrogen. Softening of the tape surface
due to the water vapor in the humid nitrogen caused the plastic deformation.

Effect on elastic deformation.—At loads of 0.05 and 0.1 N both the tape and the
Ni-Zn ferrite primarily deformed elastically and the sliding occurred at the interface. Figure 4.25 presents the coefficients of friction as a function of relative
humidity. On humidifying (open symbols) the coefficient of friction increased
linearly with increases in relative humidity. On dehumidifying (solid symbols) the
coefficient of friction decreased linearly with decreases in relative humidity. The
friction behavior of the tape on dehumidifying was similar to that on humidifying.
The results at 0.05 N are consistent with those at 0.1 N.

Frictional response to humidity changes.—The first experiment looked at the
change from a dry to a humid environment. A tape was preconditioned in dry
nitrogen before each of forty-nine 90-s sliding passes. On each pass after 40 s of
sliding in dry nitrogen the area around the tape–Ni-Zn ferrite pin contact (less than
100 mm^2) was flooded with nitrogen having 61% relative humidity. Figure 4.26

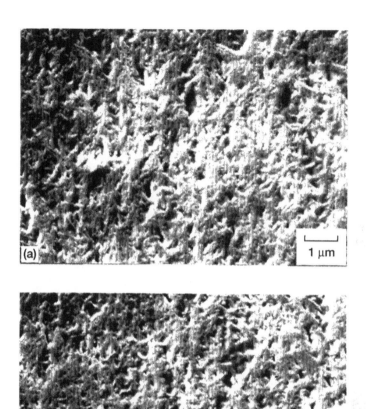

Figure 4.24.—Scanning electron micrographs of as-received
surface and wear tracks on magnetic tape (tape 2) after 50
passes against Ni-Zn ferrite pin. Normal load, 0.5 N;
sliding velocity, 0.1 mm/s; room temperature; environments, dry and humid nitrogen. (a) New tape (as-received
surface). (b) Wear track obtained in dry nitrogen.

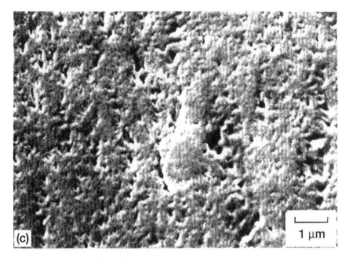

Figure 4.24.—Concluded. (c) Wear track obtained in humid
nitrogen (78% RH).

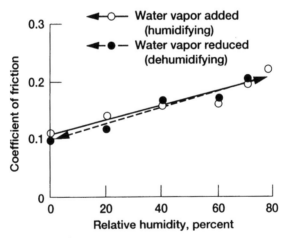

Figure 4.25.—Effect of humidifying and dehumidifying on friction of magnetic tape (tape 2) in
sliding contact with Ni-Zn ferrite pin at normal
load of 0.1 N. Sliding velocity, 0.1 mm/s; room
temperature; environment, nitrogen.

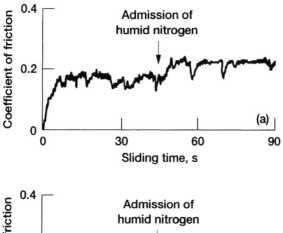

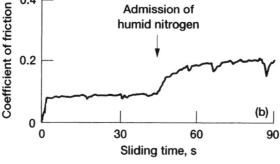

Figure 4.26.—Effect of humidity change due to admission of humid nitrogen (61% RH) on coefficient of friction for magnetic tape (tape 2) sliding against Ni-Zn ferrite pin. Normal load, 0.5 N; sliding velocity, 0.1 mm/s; room temperature; environment, nitrogen. (a) First pass. (b) Forty-ninth pass.

presents coefficient of friction as a function of sliding time under these environmental conditions. In the first pass, right after admitting humid nitrogen, the coefficient of friction increased. The coefficients of friction measured in the area flooded with humid nitrogen were 30 to 40% higher than those measured in dry nitrogen. With repeated sliding passes the tape exhibited about 2 to 2.5 times higher friction when local humidity was raised than it did in dry nitrogen.

The second experiment looked at the change from a humid to a dry environment. A tape was preconditioned in nitrogen with 63% relative humidity before each of forty-nine 90-s sliding passes. On each pass after 40 s of sliding in humid nitrogen the area around the tape–Ni-Zn ferrite pin contact was flooded with dry nitrogen. Figure 4.27 clearly shows another surprising aspect of humidity dependence. When the contact area was flooded with dry nitrogen, the coefficient of friction decreased dramatically, being 80% of the friction obtained in the humid nitrogen atmosphere in single-pass sliding and 60% in multipass sliding.

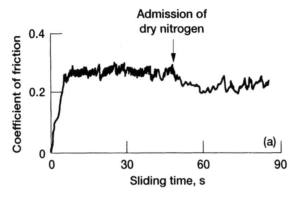

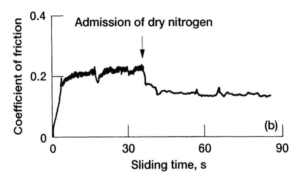

Figure 4.27.—Effect of humidity change due to
admission of dry nitrogen on coefficient of friction
for magnetic tape (tape 2) sliding against Ni-Zn
ferrite pin. Normal load, 0.5 N; sliding velocity,
0.1 mm/s; room temperature; environment,
nitrogen. (a) First pass. (b) Forty-ninth pass.

The third experiment studied two reversals in environment. The area surrounding the tape–Ni-Zn ferrite pin contact was flooded with dry nitrogen. At approximately 30 s of sliding time the supply of dry nitrogen was stopped and then humid nitrogen, at 61% relative humidity, was admitted to the contacting area for about 30 s. The supply of humid nitrogen was then stopped and dry nitrogen was again allowed to flow into the contact area. Figure 4.28 clearly indicates the marked increase in coefficient of friction as the humidity increased. Only a short transient time (~10 s) was needed for the friction to decrease or increase in relation to the humidity changes.

Mechanism of tape friction and humidity effect.—The previous sections have shown that sliding occurred primarily at the interface and that the coefficient of friction was greatly influenced by the interaction of the tape and Ni-Zn ferrite surfaces. When these surfaces are brought into elastic contact, interfacial adhesion can take place and shearing of adhesive bonds at the interface is responsible for

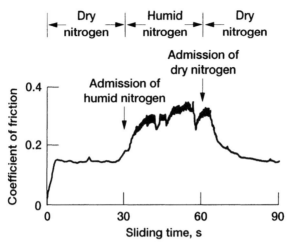

Figure 4.28.—Effect of humidity changes due to admission of humid or dry nitrogen on coefficient of friction as function of sliding time for magnetic tape (tape 2) sliding against Ni-Zn ferrite pin. Normal load, 0.5 N; sliding velocity, 0.1 mm/s; room temperature; environment, nitrogen; third pass.

friction. As Fig. 4.25 shows, the coefficient of friction increased linearly with increasing humidity and decreased when humidity was lowered. In elastic contact the changes in friction on humidifying and dehumidifying were reversible.

When an Ni-Zn ferrite pin was brought into sliding contact with an Ni-Zn ferrite flat, water adsorption on the surface did not affect the coefficient of friction, as shown in Fig. 4.29. There was no change in friction with relative humidity. The experiments with the ferrite-ferrite contact were identical to those with the tape-ferrite contact. Therefore, the effect of humidity on friction for the tape-ferrite contact, seen in Fig. 4.28, was primarily due to alteration of the tape surface. Humidity affects the friction behavior of a tape in elastic contact because it changes the chemistry and interaction of the tape. Most probably, the lubricant and binder in the tape react with the water vapor.

As Fig. 4.22 shows, when a tape was plastically deformed during sliding, the coefficient of friction remained low and constant below 40% relative humidity. There was no humidity effect on friction. Because the load applied to the contacting surfaces was sufficiently high and plastic deformation occurred, the Ni-Zn ferrite pin broke through the adsorbed water vapor film. The adhesive bonding at the contact, where fresh surfaces were continuously exposed through the water vapor film below 40% relative humidity, was similar to that in dry nitrogen.

As the relative humidity increased above 40%, however, the coefficient of friction increased rapidly (Fig. 4.22). The humidity softened the tape surface and

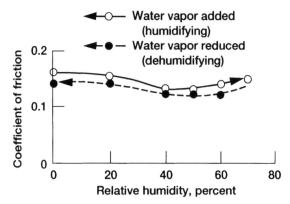

Figure 4.29.—Effect of humidifying and dehumidifying on friction of Ni-Zn ferrite pin in sliding
contact with Ni-Zn ferrite flat. Normal load, 0.5 N;
sliding velocity, 0.1 mm/s; room temperature;
environment, nitrogen.

changed the chemistry and interaction of the tape. Probably, the lubricant reacted
with the water vapor and the binder became less stable. Exposure to high humidity
and elevated temperatures for several weeks can result in hydrolytic reaction of the
tape-binder system [4.29]. The primary mechanism relevant to magnetic tape
involves the scission of the chainlike polymer structure by the action of water,
which effectively breaks the polymer into small fragments having low molecular
weights.

The experiments described in this case study were conducted at room temperature, and the tape was exposed to humid nitrogen for less than 1 hr. Figures 4.22 and
4.25 show that friction reversed on humidifying and dehumidifying. Removing the
adsorbed water vapor film from the tape and pin surfaces lowered the coefficients
of friction to those before humidifying. Thus, exposing a tape to humid nitrogen for
short periods at room temperature may result in a negligibly small amount of
hydrolytic degradation of the tape binder. The scission of the chainlike polymer
structure by the action of water vapor would also be negligible.

With the mechanical activity that takes place during sliding, however, water
vapor adsorbed on the tape surface tends to promote chemical reaction and degradation of the tape long before the surface may otherwise be ready for such deterioration. Binder degradation can lead to deterioration of the mechanical properties.

We know that the surfaces of materials in sliding contact are highly strained by
the mechanical activity that takes place. Under such conditions, on the surface and
in the surficial layers of a tape with an adsorbed water vapor film, the binder and
lubricant chemistry can be changed markedly by the strain. The higher the degree
of strain, the lower the chemical stability of the binder system and the greater the
hydrolytic reaction and degradation of the binder.

Similar phenomena have been observed with other materials. For example, the crystallinity and crystallographic orientation of a crystalline metal surface can be changed markedly by strain. The higher the degree of strain, the lower the recrystallization temperature. Consequently, high strain tends to promote recrystallization of a solid surface long before the surface may otherwise be ready for such recrystallization [4.30–4.32].

Furthermore, on the surface of an amorphous alloy (metallic glass), sliding is accompanied by a high degree of strain at room temperature, even when the sliding velocity is extremely low. This strain induces crystallization of the amorphous alloy surface long before the surface may otherwise be ready for such crystallization [4.33].

4.3 Effects of Surface Contamination by Diffusion

Surface contaminant films may form by diffusion of bulk contaminants through the solid itself. An example of this type of surface contamination and the usefulness of XPS for analyzing the surface chemical composition was demonstrated for three ferrous-based amorphous alloys (metallic glasses).

Sliding friction experiments were conducted with foil specimens of Al_2O_3 in contact with amorphous alloys in vacuum at temperatures to 750 °C [4.34, 4.35]. Friction force traces resulting from such sliding were generally characterized by fluctuating behavior with evidence of stick-slip. Figure 4.30 presents the coefficients of friction as a function of sliding temperature. The Al_2O_3 pin was sputter cleaned with argon ions at room temperature. The foil specimen was also sputter cleaned with argon ions in the vacuum system and then heated from room temperature to 750 °C. Temperature effects drastically changed the friction behavior of the amorphous alloys. There was also a considerable difference in the friction measured for the three alloy compositions.

As the temperature increased from room temperature to 350 °C, the friction increased (Fig. 4.30). Why did this appreciable increase occur? Is the increase in friction associated with changes in the chemical and microstructural states of the contacting materials? Further, at 500 °C (Fig. 4.30) the friction underwent a marked decrease from that measured at 350 °C. Does surface analysis supply the answer? To answer these questions, surface and bulk analyses of the amorphous alloys were conducted.

4.3.1 Bulk Composition and Microstructure of Amorphous Alloys

Three amorphous alloy compositions were examined in this investigation. Table 4.2 lists the compositions and some of their properties. The alloys were foils (30- to 33-μm-thick ribbon) and were in the as-cast condition. To establish the exact crystalline state of the amorphous alloy surfaces, their microstructures were

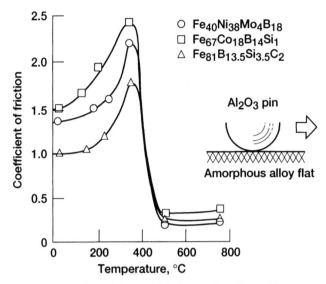

Figure 4.30.—Coefficient of friction as function of temperature for Al_2O_3 pins sliding on ferrous-base amorphous alloys (metallic glasses).

TABLE 4.2.—PROPERTIES OF AMORPHOUS ALLOYS

Nominal alloy composition, wt%	Crystallization temperature, °C	Density, g/cm³	Hardness, GPa	Ultimate tensile strength, GPa	Bend ductility[a]
$Fe_{67}Co_{18}B_{14}Si_1$	430	7.56	10	1.5	1
$Fe_{81}B_{13.5}Si_{3.5}C_2$	480	7.3	10.3	0.7	$9×10^{-3}$
$Fe_{40}Ni_{38}Mo_4B_{18}$	410	8.02	10.5	1.38	1

[a] $\varepsilon = t(d - t)$, where t is ribbon thickness and d is micrometer spacing at bend fracture.

examined by transmission electron microscopy and diffraction in a microscope operating at 100 kV.

Figure 4.31 shows a typical structure of the as-received amorphous alloy foils. Amorphous alloys lack the macroscopic structural features common in crystalline metals. In the absence of macroscopic crystallinity neither grains, grain boundaries, grain orientation, nor additional phases exist. However, black spots, believed to be nuclei and/or crystallites with a size range to 4 nm, are apparent in the photomicrograph. Diffused honeycomb-shaped structures formed by segregation and clustering of transition metals and metalloids are also apparent in the photomicrograph (dark gray bands). The electron diffraction pattern for the as-received foil (Fig. 4.31) indicates that the foil was not completely amorphous but contained extremely small nuclei and/or clusters of a few nanometers in size.

Figure 4.32 shows typical examples of an amorphous alloy foil structure after heating to 350 and 430 °C for 20 min in ultrahigh vacuum. The foil structure heated to 350 °C (Fig. 4.32(a)) was similar to the as-received foil (Fig. 4.31). After

Figure 4.31.—Typical microstructure and electron diffraction pattern of an amorphous alloy ($Fe_{67}Co_{18}B_{14}Si_1$).

heating to 430 °C (Fig. 4.32(b)), the so-called crystallization temperature for this particular $Fe_{67}Co_{18}B_{14}Si_1$ foil, the black spots were slightly larger than those shown in Figs. 4.31 and 4.32(a). The diffused honeycomb-shaped structure (dark gray bands) was enriched, suggesting that the amorphous alloy structure subdivided as a result of decomposition and separation of the amorphous phase. That is, the constituents were segregating and preferential clustering occurred at this temperature. The electron diffraction pattern also indicates the presence of an extra outer ring related to clustering or crystallinity.

Figures 4.32(c) and (d) present microstructures of the same amorphous alloy heated to 500 and 750 °C. They reveal two kinds of extremely fine-grained crystals: a dark grain and a light grain. Energy-dispersive x-ray analysis (EDXA) indicated that the light grains contained about 20 times more silicon than the dark grains. The dark grains contained more iron and FeB alloy. The transmission electron diffraction patterns for both the dark and light grains contained diffraction

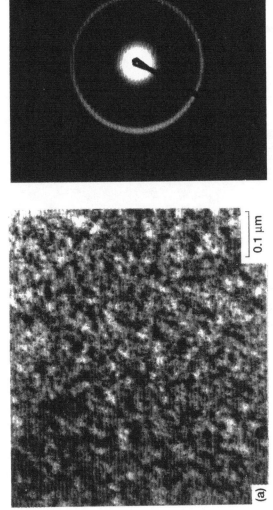

Figure 4.32.—Microstructures and electron diffraction patterns of an amorphous alloy ($Fe_{67}Co_{18}B_{14}Si_1$) heated to (a) 350 °C, (b) 430 °C, (c) 500 °C, and (d) 750 °C in vacuum (10 nPa).

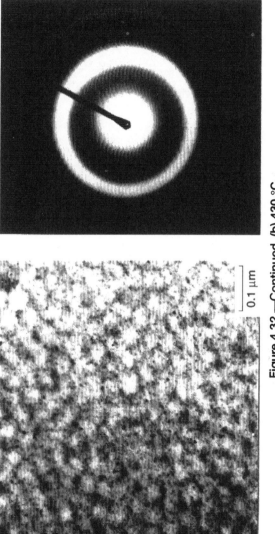

Figure 4.32.—Continued. (b) 430 °C.

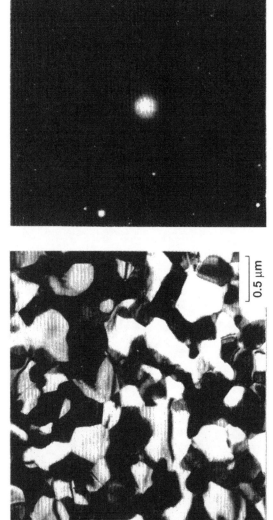

Figure 4.32.—Continued. (c) 500 °C.

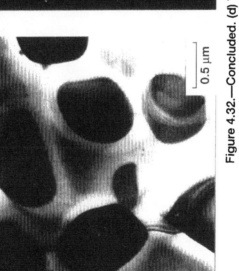

(d)

0.5 μm

Figure 4.32.—Concluded. (d) 750 °C.

spots and Kikuchi lines indicative of single-crystal structure. The crystallized grain size was 0.7 μm at 500 °C and as large as 1.4 μm at 750 °C. Thus, the microstructure of the amorphous alloy changed discontinuously and drastically during the amorphous-to-crystalline transition. The amorphous phase separated into two crystalline phases.

4.3.2 Surface Chemistry

All XPS spectra taken from the surfaces of three as-received amorphous alloys clearly revealed adsorbed oxygen and carbon contaminants in addition to the various alloying constituents of the nominal bulk composition. Beneath this adsorbate layer there was a mixture layer comprising oxides of the alloying constituents.

Figure 4.33 presents Fe_{2p} peaks for the as-received $Fe_{67}Co_{18}B_{14}Si_1$ specimen. Before cleaning, the surface clearly contained iron oxides, such as Fe_2O_3. After the amorphous alloy surface had been argon sputter cleaned for 2 and 4 min, the Fe_{2p} lines were split asymmetrically into doublet peaks, associated with iron oxides and iron. After 60 min of argon sputter cleaning the Fe_{2p} peaks associated with iron are clearly present but the iron oxide peak is extremely small. The results after 30 min of sputtering were similar to those after 60 min.

The as-received $Fe_{67}Co_{18}B_{14}Si_1$ amorphous alloy contained the oxides of cobalt, boron, and silicon as well as iron oxides [4.36]. The oxide layers are extremely important for amorphous alloys from a tribological point of view because they are hard and tenaciously bonded to the alloying constituents. Thus, the oxide layers can provide good wear resistance, as discussed in the following section.

The XPS spectra of the surface after argon ion sputtering for 60 min contained small amounts of carbide and oxide contaminants, such as SiC and SiO_2, although the $Fe_{67}Co_{18}B_{14}Si_1$ amorphous alloy was not supposed to contain carbon and oxygen. The contaminants (i.e., oxygen and carbon) may be introduced from the environment into the bulk of the alloy and form bulk contaminants, such as carbides and oxides, in the alloy during the casting process.

Table 4.3 summarizes the argon-ion-sputter-cleaned surface conditions of the three amorphous alloy foils as analyzed by XPS. The relative concentrations of the various constituents on the surfaces were different from the nominal compositions.

Temperature effects drastically changed the surface chemistry of the amorphous alloys as analyzed by XPS—and not only above the crystallization temperature but below it as well. These changes are due to segregation and diffusion of constituents, especially of metalloids such as boron and silicon.

Figure 4.34 presents the XPS spectra for one of the ferrous-based amorphous alloys of Fig. 4.30 in the as-received condition, after sputter cleaning, and when heated to 350 and 750 °C. The XPS spectra of Fe_{2p}, Co_{2p}, B_{1s}, and Si_{2p} peaks are presented as a function of binding energy. The B_{1s} photoelectron peaks of the as-received specimen (Fig. 4.34(c)) indicate the presence of boric oxide (B_2O_3) as

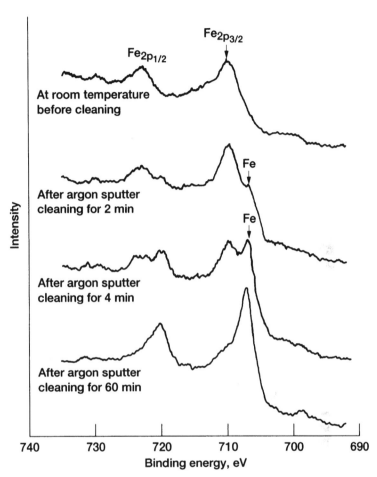

Figure 4.33.—Fe$_{2p}$ XPS peaks on Fe$_{67}$Co$_{18}$B$_{14}$Si$_1$ surface.

TABLE 4.3.—COMPOSITION OF ARGON-SPUTTER-
CLEANED SURFACE LAYER OF AMORPHOUS ALLOYS

Nominal bulk composition		Composition on surface,
wt%[a]	at.%	at.%[b]
Fe$_{67}$Co$_{18}$B$_{14}$Si$_1$	Fe$_{42}$Co$_{11}$B$_{46}$Si$_1$	Fe$_{49}$Co$_{14}$B$_{17}$Si$_6$C$_9$O$_5$
Fe$_{81}$B$_{13.5}$Si$_{3.5}$C$_2$	Fe$_{48}$B$_{42}$Si$_4$C$_6$	Fe$_{43}$B$_{15}$Si$_8$C$_{21}$O$_{14}$
Fe$_{40}$Ni$_{38}$Mo$_4$B$_{18}$	Fe$_{23}$Ni$_{21}$Mo$_1$B$_{55}$	Fe$_{18}$Ni$_{28}$Mo$_1$B$_{24}$C$_{15}$O$_{14}$

[a]Manufacturer's analysis.
[b]Relative concentrations of the various constituents were determined by using peak area sensitivity factors.

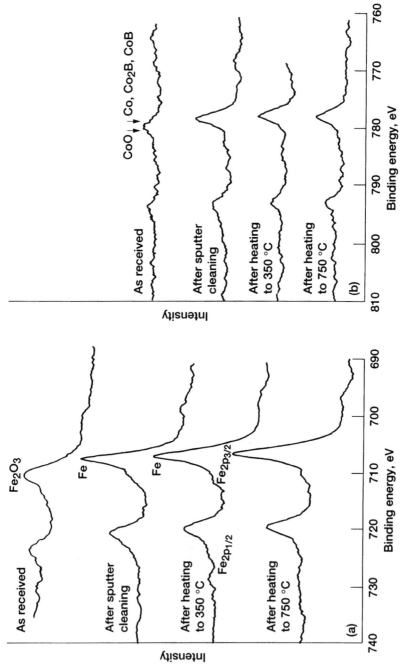

Figure 4.34.—Representative (a) Fe_{2p}, (b) Co_{2p}, (c) B_{1s}, and (d) Si_{2p} XPS peaks on $Fe_{67}Co_{18}B_{14}Si_1$ surface.

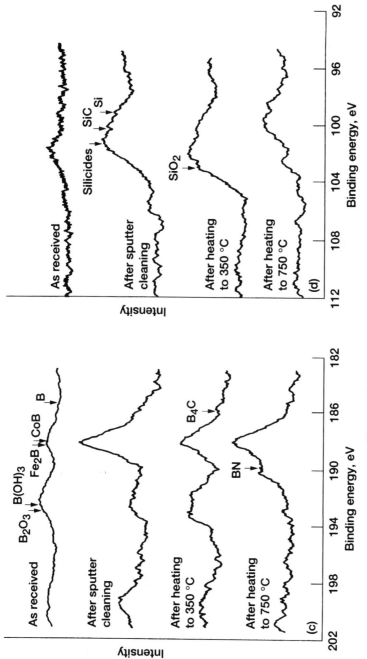

Figure 4.34.—Concluded. (c) B_{1s}. (d) Si_{2p}.

TABLE 4.4.—COMPOSITION OF SURFACE OF AMORPHOUS ALLOYS

Alloy composition	Surface			
	As received	Argon sputter cleaned	Heated to 350 °C	Heated to 750 °C
$Fe_{67}Co_{18}B_{14}Si_1$	Oxides of Fe, Co, B, Si, and C Adsorbed film of oxygen and carbon	Alloy Small amount of oxides and carbides	Alloy Boric oxides and silicon oxides migrated from bulk Small amount of carbides	Alloy Boron nitride migrated from bulk Very small amount of boric oxides and silicon oxides
$Fe_{81}B_{13.5}Si_{3.5}C_2$	Oxides of Fe, B, Si, and C Adsorbed film of oxygen and carbon	Alloy Small amount of oxides and carbides	Alloy Boric oxides and silicon oxides migrated from bulk Small amount of carbides	Alloy Boron nitride migrated from bulk Very small amount of boric oxides and silicon oxides
$Fe_{40}Ni_{38}Mo_4B_{18}$	Oxides of Fe, B, Ni, Mo, and C Adsorbed film of oxygen and carbon	Alloy Small amount of oxides and carbides	Alloy Boric oxides migrated from bulk Small amount of carbides	Alloy Boron nitride migrated from bulk Very small amount of boric oxides

well as Fe_2B and CoB. After argon sputter cleaning large peaks for boron and its alloys, as well as small B_2O_3 peaks, were present. After heating to 350 °C the alloy surface was again clearly contaminated with B_2O_3 and at 750 °C with BN that had migrated from its bulk. The surface heated to 500 °C was also contaminated with BN and was the same as the surface heated to 750 °C.

The Si_{2p} photoelectron peaks of the as-received surface (Fig. 4.34(d)) reveal silicides. Even the spectrum of the argon-sputter-cleaned surface reveals silicides as well as silicon dioxide (SiO_2) on the surface. After heating to 350 °C the alloy surface was contaminated with SiO_2 that had migrated from its bulk. After heating to 750 °C the alloy had a smaller amount of SiO_2 on its surface than at 350 °C.

The Fe_{2p} photoelectron peaks of the $Fe_{67}Co_{18}B_{14}Si_1$ heated to 350 and 750 °C (Fig. 4.34(a)) are almost the same as those for the argon-sputter-cleaned surface shown in Fig. 4.33.

The Co_{2p} photoelectron peaks of the as-received $Fe_{67}Co_{18}B_{14}Si_1$ (Fig. 4.34(b)) indicate cobalt oxide (CoO). The spectra for the surfaces that had been argon sputter cleaned and heated to 350 and 750 °C reveal cobalt and its alloy peaks. The cobalt peak is negligible.

Table 4.4 summarizes the surface conditions for the amorphous alloys as analyzed by XPS. Generally, the surfaces of the as-received alloys had a layer of oxides of the alloying elements as well as a simple adsorbed film of oxygen and carbon. The argon-ion-sputter-cleaned surfaces contained the alloying elements and small amounts of oxides and carbides. In addition to nominal element constituents the surfaces heated to 350 °C contained boric and silicon oxides (on $Fe_{67}Co_{18}B_{14}Si_1$ and $Fe_{81}B_{13.5}Si_{3.5}C_2$) and boric oxides (on $Fe_{40}Ni_{38}Mo_4B_{18}$), as well as small amounts of carbides that had segregated and migrated from the bulk. The surfaces heated to 750 °C contained BN that had also migrated from the bulk, as well as small amounts of oxides.

4.3.3 Frictional Response to Thermal and Chemical Changes

The increase in coefficient of friction with increasing temperature from room to 350 °C shown in Fig. 4.30 is associated with changes in the alloys' thermal, chemical, and microstructural states. First, the increase in friction is related to an increase in the adhesion resulting from the segregation of oxides, such as boric oxides and silicon oxides, to the alloy surface, as described earlier. The removal of adsorbed contaminants by sliding and the segregation of the oxides increase the adhesion strength and hence the friction.

Second, as the temperature is raised, the materials in contact soften and accordingly the total contact area increases. Thus, the enhanced strength of the interfacial bonds and the increased real area of contact are the main causes of the observed friction behavior at temperatures to 350 °C in Fig. 4.30.

The rapid decrease in friction in Fig. 4.30 between 350 and 500 °C can be attributed to the complete transformation of the alloys from the amorphous to the crystalline state between 410 and 480 °C. After this transformation BN, a solid lubricant, diffused to the alloy surfaces and brought about the friction reduction observed in Fig. 4.30. Thus, the marked reduction in friction at 500 and 750 °C in Fig. 4.30 is related to the presence of BN on the alloy surfaces and the reduction in the amount of surface oxides, as typically indicated at 750 °C in Fig. 4.34.

In situ examination of the surface chemistry in the heating stage gives valuable information on the behavior of surface segregation and decomposition. The segregation and migration of compounds, such as boric oxide and nitride, influence the friction behavior of amorphous alloys, as shown in Fig. 4.30.

4.4 Chemical Changes With Selective Thermal Evaporation

Increasing the surface temperature of a ceramic material or metal tends to promote surface chemical reactions and to volatilize some element, as described in Chapter 2. These chemical reactions cause products, such as graphite, to appear on the surface [4.37] and thus can alter adhesion and friction [4.38, 4.39].

4.4.1 Silicon Carbide

SiC-SiC couples.—Figure 4.35 presents the average pull-off forces for SiC {0001} surfaces in contact with a sintered polycrystalline SiC pin as a function of temperature in a vacuum [4.39]. The average pull-off force and the maximum and minimum measured pull-off forces were obtained from seven or more measurements.

Figure 4.36 presents the coefficients of static and dynamic friction for the SiC {0001} surfaces in contact with sintered polycrystalline SiC pins as a function of temperature in a vacuum. The average coefficient of friction and the maximum and minimum measured coefficients of friction were obtained from five or more measurements in this case. Comparing Figs. 4.35 and 4.36 shows that the static and dynamic friction characteristics were the same as those for adhesion. Adhesion and friction generally remained low to 300 °C but increased rapidly between 300 and 400 °C. Adhesion and friction decreased slightly at 600 °C but remained relatively high between 400 and 700 °C. Above 800 °C adhesion and friction decreased rapidly.

Figure 4.37 presents the C_{1s} and Si_{2p} XPS spectra obtained from narrow scans on the single-crystal SiC surface for preheating temperatures to 1500 °C. The as-received crystal (after bakeout) was preheated at various temperatures in a 10-nPa vacuum. Preheating temperatures given in Fig. 4.37 are the highest temperature to which each crystal had been heated. All the XPS spectra were taken at room temperature after bakeout and preheating. The Si_{2p} photoelectron peak energies associated with SiC at the various temperatures underwent a gradual

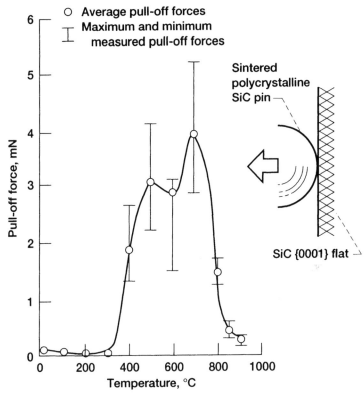

Figure 4.35.—Pull-off force (adhesion) as function of temperature for SiC {0001} flat surfaces in sliding contact with sintered polycrystalline SiC pins in vacuum.

change from 16.06 attojoules (aJ) (100.4 eV) (at 250 °C bakeout temperature) to 16.0 aJ (100.0 eV) (at 1500 °C preheating temperature). The vertical heights, peak to baseline, of the Si_{2p} peaks in the spectra were lowest at the bakeout temperature and highest at 800 °C. They were nearly the same at 400 and 600 °C. But above 800 °C the Si_{2p} for SiC decreased gradually.

The C_{1s} photoelectron emission lines for SiC were split asymmetrically into doublet peaks (Fig. 4.37), showing a significant influence of temperature on the SiC surface. Three spectral features, dependent on the chemical nature of the specimen, can be observed: (1) two kinds of doublet peaks, (2) a change in the vertical height of the peaks, and (3) a shift of peaks.

The doublet peaks were due to distinguishable kinds of carbon: (1) a carbon contamination peak and a carbide peak at room temperature and (2) graphite and carbide peaks at 400 to 1500 °C. For the XPS spectra of the as-received specimens (Fig. 4.37(a)) the carbon contamination peak was higher than the carbide peak. At 250 °C (Fig. 4.37(b)) the primary peaks were adsorbed amorphous carbon

192

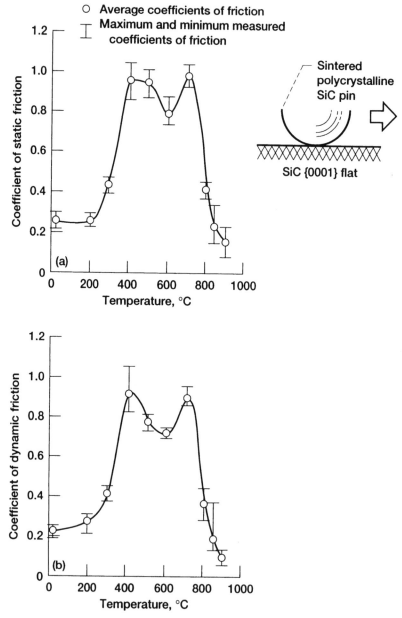

Figure 4.36.—Coefficients of static (a) and dynamic (b) friction as functions of temperature for SiC {0001} flat surfaces in sliding contact with sintered polycrystalline SiC pins in vacuum.

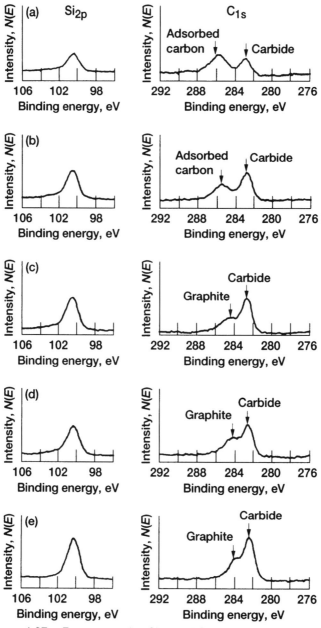

Figure 4.37.—Representative Si_{2p} and C_{1s} peaks on SiC {0001} surface preheated at various temperatures. (a) Room temperature. (b) 250 °C. (c) 400 °C. (d) 600 °C. (e) 800 °C.

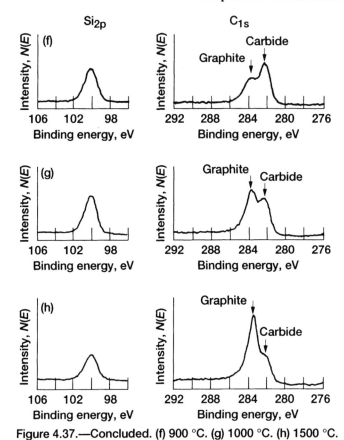

Figure 4.37.—Concluded. (f) 900 °C. (g) 1000 °C. (h) 1500 °C.

contamination and carbide, and the contaminant peak was lower than the carbide peak. At 400 °C and above (Figs. 4.37(c) to (h)) the carbon contamination peak disappeared from the spectrum and was replaced by the graphite and carbide peaks. Both the graphite and the carbide peaks increased with an increase in preheating temperature. A high carbide peak was distinguished at 800 °C (Fig. 4.37(e)).

AES analysis of an SiC surface preheated at 800 °C indicated that the AES spectrum was the same as that obtained for an argon-sputter-cleaned surface and that the spectrum included carbide-type carbon as well as a silicon peak on the surface. The AES spectrum indicated that the surface was pure SiC. But XPS analysis, which can provide more detailed chemical information than AES, clearly indicated the presence of graphite on the SiC surface preheated at 800 °C. At 900 °C (Fig. 4.37(f)) the carbide peak was lower and the graphite peak higher than at 800 °C. Nevertheless, the carbide peak remained higher than the graphite peak until 1000 °C (Fig. 4.37(g)). At 1500 °C (Fig. 4.37(h)) the carbide peak became extremely low and the graphite peak extremely high.

AES analysis of an SiC surface preheated at 1500 °C indicated that the silicon AES peak had almost disappeared and that the carbon peak was only the graphite form. However, XPS analysis of the same surface indicated evidence for silicon and carbide as well as graphite. This difference can be accounted for by the fact that XPS analyzes to greater depths and is more sensitive to the presence of silicon than is AES.

The narrow scans in Fig. 4.37 also show the shift of C_{1s} and Si_{2p} peaks on the SiC surface. All the C_{1s} of adsorbed carbon, carbide, and graphite and the Si_{2p} of SiC shifted toward lower binding energy with increasing temperature. For the carbon contamination peak C_{1s} at 45.70 aJ (285.6 eV) the shift was small (0.05 aJ to 0.06 aJ), almost the same as the instrument error (±0.05 aJ). The carbide peak shifted from 45.23 aJ (282.7 eV) at room temperature to 45.15 aJ (282.2 eV) at 1500 °C, and the graphite peak from 45.52 aJ (284.5 eV) at 400 °C to 45.38 aJ (283.6 eV) at 1500 °C. The Si_{2p} peak shifted from 16.08 aJ (100.5 eV) on the as-received surface to 16.02 aJ (100.0 eV) at 1500 °C.

Figure 4.38 presents the XPS heights of the high-resolution Si_{2p}, C_{1s}, and O_{1s} spectral peaks on single-crystal SiC surfaces as a function of a preheating temperature to 1500 °C. An adsorbed carbon contaminant on the as-received surface at room temperature is evident as well as a minute amount of oxygen contaminant. The concentration of silicon carbide Si_{2p} and C_{1s} on the single-crystal SiC surface (Fig. 4.38(a)) increased with increasing temperature from 23 to 800 °C but decreased above 800 °C. Large Si_{2p} and C_{1s} peaks associated with SiC were at maximum intensity at 800 °C. The peak heights of silicon dioxide Si_{2p} and O_{1s} (Fig. 4.38(b)) remained relatively low at 23 to 1500 °C and were almost negligible at 800 to 1500 °C. The adsorbed carbon contaminant C_{1s} peak height (Fig. 4.38(c)) decreased with increasing temperature and was negligible above 400 °C. Also, the graphite C_{1s} peak was absent from room temperature to 250 °C and remained low from 400 to 800 °C but increased rapidly from 900 to 1500 °C. Because the graphite C_{1s} peak on the SiC surface was extremely high at 1000 °C and above, the SiC surface must have been covered with a graphite layer above 1000 °C.

Comparing Fig. 4.38 with Figs. 4.35 and 4.36 shows that the low adhesion and friction between room temperature and 300 °C were caused by carbon contaminants on the SiC flat surface. The rapid increase in adhesion and friction between 300 and 400 °C can be attributed to the absence of adsorbed contaminants. The high adhesion and friction between 400 and 800 °C were primarily related to the absence of adsorbed contaminants and to a small amount of SiO_2 on the SiC surface. The somewhat low adhesion and friction at 600 °C were probably caused by the known α-SiO_2-to-β-SiO_2 transition of a small amount of SiO_2 at about 583 °C [4.40] and by changes in the amount of SiO_2 on the single-crystal SiC surface. The somewhat low adhesion and friction at 600 °C were related to the slight increase in SiO_2. The silicon dioxide Si_{2p} and O_{1s} peak heights obtained from the SiC surface at 800 °C and above were extremely small (Fig. 4.38).

The low adhesion and friction above 800 °C correlated with the graphitization of the SiC surface. Note that using both XPS and AES to study the SiC surface

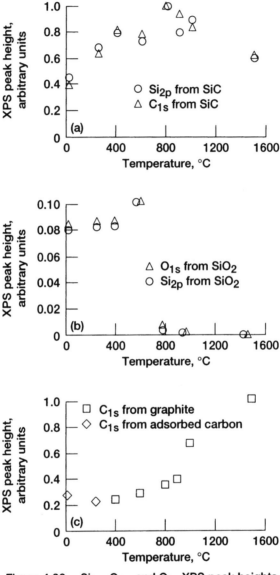

Figure 4.38.—Si_{2p}, C_{1s}, and O_{1s} XPS peak heights on single-crystal SiC surface preheated at various temperatures. (a) Si_{2p} and C_{1s} peak heights from SiC. (b) Si_{2p} and O_{1s} peak heights from SiO_2. (c) C_{1s} peak heights from graphite and adsorbed carbon.

revealed that the low adhesion and friction above 800 °C (Figs. 4.35 and 4.36) resulted from graphitization of the SiC surface by selective thermal evaporation of silicon with heating. The XPS and AES analyses complemented each other in this determination.

Iron-SiC couples.—The principal contaminants (as determined by XPS) on the as-received, sintered polycrystalline SiC surface at room temperature were adsorbed carbon and oxygen, residual graphite, and SiO_2 (Fig. 4.39). The residual graphite was generated during fabrication in an argon atmosphere. The adsorbed carbon contaminants disappeared on heating to 400 °C. Above 400 °C primarily graphite and SiO_2 appeared on the SiC surface. The amount of SiO_2 decreased rapidly from 600 to 800 °C (Figs. 4.39(a) and (c)). At 800 °C the silicon carbide Si_{2p} and C_{1s} peaks were at maximum intensity. Above 800 °C the graphite concentration increased rapidly, whereas the SiC concentration decreased rapidly, in intensity (Fig. 4.39(b)). The SiC surface graphitized predominantly at 1000 to 1200 °C.

Temperature distinctly influenced the Si_{2p}, C_{1s}, and O_{1s} concentrations on sintered, polycrystalline SiC surfaces. Figure 4.40 shows the heights of the Si_{2p} peaks (SiC and SiO_2), the C_{1s} peaks (adsorbed carbon contaminant, graphite, and SiC), and the O_{1s} peaks (adsorbed oxygen contaminant and SiO_2) as a function of preheating temperature. Over the entire temperature range (room to 1200 °C) the trend of Si_{2p} concentration from SiC was similar to that of C_{1s} from SiC (Fig. 4.40(a)). The silicon carbide Si_{2p} and C_{1s} peak heights increased linearly to 800 °C and then decreased above 800 °C. The carbon contaminant C_{1s} peak height decreased rapidly to 400 °C (Fig. 4.40(b)). Whereas the graphite C_{1s} peak was absent on as-received, single-crystal SiC, it was present on as-received, sintered, polycrystalline SiC and is thus a distinguishing feature of that surface. The graphite C_{1s} peak height (Fig. 4.40(b)) was relatively constant to 400 °C but increased above 400 °C, becoming large at 1000 and 1200 °C. In other words, the SiC surface is covered with a graphite layer.

The trends of the silicon dioxide O_{1s} and Si_{2p} peak heights were similar (Fig. 4.40(c)), increasing to 600 °C and then decreasing rapidly. Three times more SiO_2 was present on the polycrystalline SiC than on the single-crystal SiC.

Figure 4.41 presents the coefficients of friction for sintered polycrystalline SiC flat surfaces in sliding contact with an iron pin as a function of sliding temperature. The iron pin was sputter cleaned with argon ions. The as-received SiC was baked out in the vacuum system and then heated to the sliding temperature before the friction experiment began. The low coefficients of friction below 250 °C can be associated with both carbon and graphite contaminants on the as-received specimen. The rapid increase in friction at 400 °C can be attributed to the absence of carbon contaminants, increased SiO_2, and increased plastic flow causing junction growth in the contact zone. The rapid decrease in friction above 800 °C correlated with SiC surface graphitization, similar to what occurred with single-crystal SiC.

Figure 4.42 presents the coefficients of friction for SiC {0001} surfaces in contact with an iron pin as a function of sliding temperature. The iron pin was sputter

198

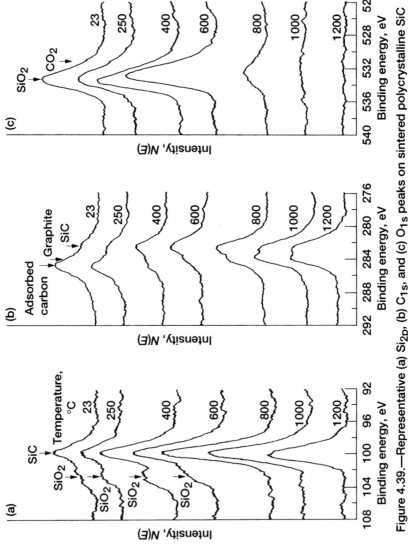

Figure 4.39.—Representative (a) Si_{2p}, (b) C_{1s}, and (c) O_{1s} peaks on sintered polycrystalline SiC surface preheated at various temperatures.

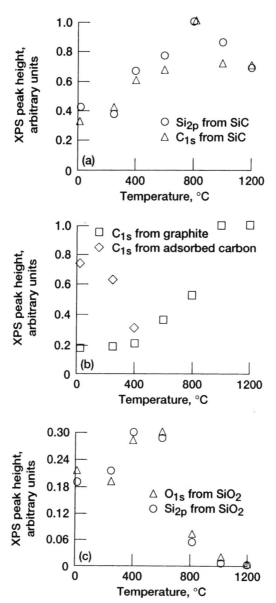

Figure 4.40.—Si_{2p}, C_{1s}, and O_{1s} XPS peak heights on sintered polycrystalline SiC surface preheated at various temperatures. (a) Si_{2p} and C_{1s} peak heights from SiC. (b) C_{1s} peak heights from graphite and adsorbed carbon. (c) Si_{2p} and O_{1s} peak heights from SiO_2.

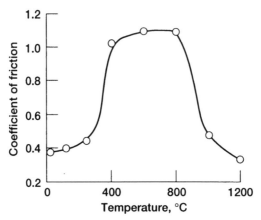

Figure 4.41.—Coefficient of friction as function
of temperature for sintered polycrystalline SiC
surface sliding against iron pin that was argon
ion sputter cleaned before experiments.
Normal load, 0.1 to 0.2 N; vacuum, 30 nPa.

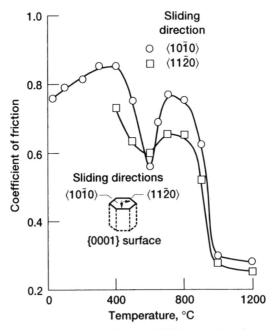

Figure 4.42.—Coefficient of friction as function
of temperature for SiC {0001} surface sliding
against iron pin that was argon ion sputter
cleaned before experiments. Normal load,
0.2 N; vacuum, 10 nPa.

cleaned with argon ions. The as-received SiC was baked out in the vacuum system and then heated to the sliding temperature before the friction experiment began. The coefficient of friction increased slightly below 400 °C and then decreased from 400 to 600 °C because carbon and oxygen contaminants were gradually removed from the surface. The coefficient of friction increased from 600 to 800 °C because both adhesion and plastic flow increased in the contact area. Above 800 °C the coefficient of friction decreased rapidly in correlation with graphitization of the single-crystal SiC surface. The coefficients of friction on single-crystal and sintered polycrystalline SiC surfaces at high temperatures were nearly the same as those on pyrolytic graphite in sliding contact with iron in a vacuum [4.41].

Heating single-crystal SiC from 800 to 1500 °C graphitized the SiC surface. XPS analysis demonstrated that silicon segregated and evaporated in the SiC surficial region and that the remaining carbon reconstructed at the SiC surface. After the SiC specimen had cooled to room temperature, sliding friction experiments from room temperature to 1200 °C in vacuum were conducted on the graphitized surface. Figure 4.43 presents the friction properties. The SiC specimens preheated to 1500 °C produced coefficients of friction lower, by one-half, than those for the surfaces cleaned by argon ion sputtering. The marked difference in friction between graphitized and nongraphitized surfaces was observed even in air [4.42]. Thus, with heating the SiC generates its own solid lubricant surface films in the form of graphite.

Mechanisms of SiC graphitization.—Meyer and Loyen [4.43] conducted ellipsometric measurements above 1200 °C with two different SiC crystal {0001} faces, one consisting of silicon atoms {0001} and the other consisting of carbon atoms {0001}. In 1 hr of heating at 1300 °C the carbon (graphite) layer on the

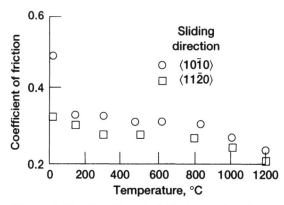

Figure 4.43.—Coefficient of friction as function of temperature and crystallographic orientation for single-crystal SiC {0001} surface preheated to 1500 °C sliding against iron pin. Normal load, 0.2 N; vacuum, 10 nPa.

carbon face grew to about 100 nm, whereas the layer on the silicon face did not grow thicker than 10 nm even with longer heating. Figure 4.44 presents a typically complete elemental depth profile for an SiC surface preheated to 1500 °C as a function of sputtering time. The graphite C_{1s} peak decreased rapidly in the first 30 min of sputtering and then decreased gradually to about 18 hr. After 18 hr the graphite peak did not change much with sputtering time. On the other hand, the Si_{2p} and carbide C_{1s} peaks increased gradually to 20 hr. The SiC {0001} surfaces consisted of both silicon and carbon atoms because etching the SiC surface in molten salt ($1NaF + 2KCO_3$) gave both a smooth surface for the silicon face and a rough one for the carbon face. Figure 4.45 shows that the apparent thickness of the intermediate layer (a mixture of graphite and SiC) produced by heating above 1200 °C for 1 hr was about 100 nm (1000 Å), equivalent to the depth of a layer sputter etched for about 18 hr.

The graphitization behavior in the outermost surficial layer can be as follows. The AES analysis depth is 1 nm or less, and an elemental concentration as low as 0.1% of a monolayer can be detected and identified. The XPS analysis depth is 2 nm or less, and the ultimate sensitivity is sufficient to allow fractions of a monolayer to be detected and identified. Therefore, Meyer and Loyen [4.43] concluded the outermost SiC surficial layer, consisting mostly of graphite with very little silicon, to be 2 nm. This estimation is consistent with the proposition of Van Bommel et al. [4.44]; that is, the collapse of the carbon in two or three successive SiC layers after

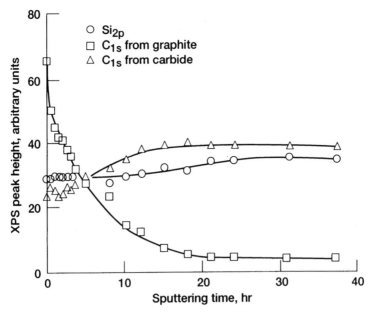

Figure 4.44.—Elemental depth profiles of SiC {0001} surface pre-heated to 1500 °C for 1 hr.

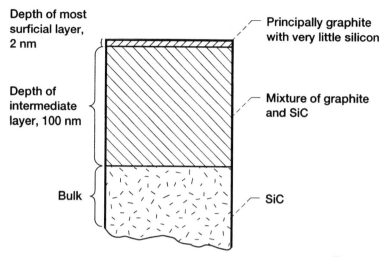

Figure 4.45.—Surface condition of SiC after heating above 1200 °C.

the silicon has evaporated from two or three successive SiC layers (Fig. 4.46) is the most probable mechanism in the initial stages of SiC basal plane graphitization.

4.4.2 Diamondlike Carbon Coatings

Diamondlike carbons (DLC's), such as amorphous hydrogenated carbon (a-C:H), can be considered as metastable carbons produced as thin coatings with a broad range of structures (primarily amorphous with variable sp^2/sp^3 carbon bonding ratio) and compositions (variable hydrogen concentration). DLC converts to graphite or other nondiamond carbon at lower temperatures than diamond [4.45]. An initial transformation has been observed as low as 250 °C. The transformation usually occurs at 400 °C and proceeds by loss of hydrogen and subsequent graphitization. Maximum long-term use temperature for DLC ranges from 250 to 300 °C.

Figure 4.47 presents the hydrogen concentration in as-deposited and rapid thermally annealed DLC films [4.46]. The DLC films were deposited by a direct ion beam deposition technique using methane ions. Thermally annealed DLC films were prepared by rapidly heating the as-deposited DLC films at 300 deg C/s in a nitrogen atmosphere for 2 min at various temperatures to 1000 °C.

The as-deposited DLC film contained 30% hydrogen, as did the rapid thermally annealed DLC films when the annealing temperature was below 500 °C. Therefore, the chemical composition of DLC does not change with temperature below 500 °C. The hydrogen concentration decreased linearly with annealing temperature between 500 and 900 °C to 5% at 900 °C. Several researchers have reported that heat treating hydrocarbon films, which leads to hydrogen evolution, causes structural changes between 400 and 600 °C [4.47, 4.48].

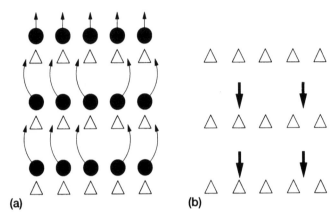

Figure 4.46.—Graphitization of SiC surface. (a) Evaporation of silicon. (b) Collapse of carbon.

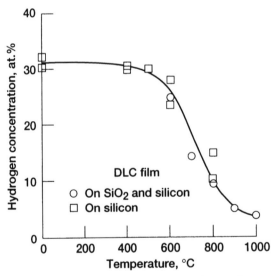

Figure 4.47.—Hydrogen concentration as function of rapid thermal annealing temperature for DLC films.

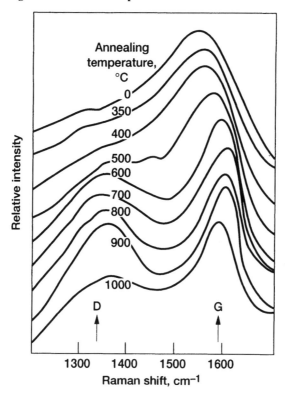

Figure 4.48.—Raman spectra from as-radio-
frequency-plasma-deposited and rapid
thermally annealed DLC films.

Figure 4.48 reveals that at 350 and 400 °C the Raman spectra of the as-
radiofrequency-plasma-deposited DLC film and the rapid thermally annealed DLC
films are quite similar. No phase change in the films was observed for annealing
temperatures below 400 °C, consistent with Rutherford backscattering (RBS) and
hydrogen forward-scattering (proton recoil detection) data. However, at 500 to
1000 °C the Raman spectra for the annealed DLC films were significantly different
from the as-deposited DLC spectrum. As the rapid thermal annealing temperature
increased, the full width at half-maximum (FWHM) of the "G" peak narrowed,
and the peak shifted toward the graphitic peak at 1590 cm^{-1}, indicating a transition
from less tetragonal bonding to more trigonal bonding. At the same time the
intensity and FWHM of the "D" peak at 1350 cm^{-1} increased. The Raman spectra
of the as-deposited DLC films heated above 400 °C show the appearance of the "D"
peak at 1350 cm^{-1}, which is characteristic of disordered graphite with crystalline
size less than 20 nm (200 Å) [4.49].

Experiments were performed in humid air (relative humidity (RH), ~40%)
and in dry nitrogen (RH, <1%) with a 1.6-mm-radius Si_3N_4 pin contacting an

as-deposited DLC film. Figure 4.49 presents the coefficients of friction obtained as a function of the number of repeated passes in reciprocating sliding. Much higher coefficients of friction were measured in humid air (above 0.15) than in dry nitrogen (~0.01). In the presence of moisture the coefficient of friction remained constant with increasing number of passes. This relatively higher, constant friction was the result of tribochemical interactions between the Si_3N_4 pin and moisture at the sliding interface [4.50–4.52]. The tribochemical interactions produced reaction products, such as SiO_2, at the sliding interface. These reaction products maintained the relatively higher constant friction in this case.

In dry nitrogen the initial coefficients of friction obtained were relatively high (0.12 to 0.15 in Fig. 4.49) due to adsorbed contaminants, primarily moisture, at the sliding interface. As sliding progressed, the coefficient of friction decreased to near 0.01 after approximately 2000 passes. This decrease in friction suggests that some contaminants, such as SiO_2 and moisture, were removed from the sliding surface of the Si_3N_4 pin during repeated passes in dry nitrogen.

The rapid thermally annealed DLC films exhibited similar friction behaviors as the as-deposited DLC film (see Fig. 4.49) in their respective environments. Figure 4.50 summarizes the equilibrium coefficients of friction for the as-deposited and rapid thermally annealed DLC films sliding against Si_3N_4 pins. These data indicate that annealing temperature had almost no effect on the equilibrium coefficients of friction in their respective environments.

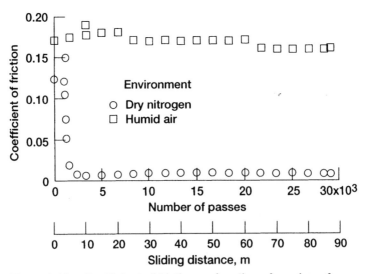

Figure 4.49.—Coefficient of friction as function of number of passes for Si_3N_4 pin (1.6 mm in radius) in sliding contact with as-deposited DLC films in humid air and dry nitrogen at room temperature.

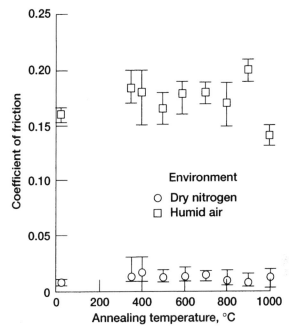

Figure 4.50.—Coefficients of friction in equilibrium
state for as-deposited and rapid thermally
annealed DLC films in sliding contact with Si_3N_4
pin in humid air and dry nitrogen at room
temperature.

Figure 4.51 presents the average wear rates of as-deposited and rapid thermally
annealed DLC films contacting Si_3N_4 pins in humid air and dry nitrogen at room
temperature. The effect of rapid thermal annealing on the wear rate in humid air and
dry nitrogen was moderate except at annealing temperatures between 600 and
800 °C. The wear rates for the DLC films annealed between 600 and 800 °C
drastically increased in humid air. As shown in Fig. 4.47 the hydrogen concentration
of DLC films drastically decreased with increasing annealing temperature in this
temperature range. The rapid evolution of hydrogen concentration in DLC films
during annealing can be attributed to the creation of micropores and the roughing
of the DLC surfaces. The high wear rates in humid air for DLC films annealed at 600
to 800 °C may be due to the high friction in humid air and the transformation of the
DLC structure.

Although the rapid thermal annealing of DLC films does not affect their friction
properties in humid air and dry nitrogen, thermal annealing significantly affects
their friction properties in vacuum. For example, Fig. 4.52 shows an abrupt
increase in coefficient of friction for DLC films annealed at temperatures between
500 and 600 °C in vacuum. In this investigation in situ single-pass sliding friction

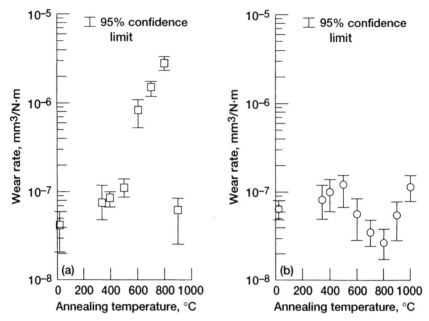

Figure 4.51.—Wear rates for as-deposited and rapid thermally annealed DLC films in sliding contact with Si_3N_4 pin (a) in humid air and (b) in dry nitrogen.

experiments were conducted in an ultrahigh vacuum (10^{-7} Pa) with plasma-deposited DLC (amorphous hydrogenated carbon, a-C:H) films on an Si_3N_4 flat in contact with an ion-sputter-cleaned, hemispherical Si_3N_4 pin (1.6 mm in radius). Loads were to 1.7 N (average Hertzian contact pressure, 1.5 GPa), and sliding velocity was 3 mm/min at temperatures to 700 °C. For the plasma-deposited DLC (a-C:H) films the coefficient of friction remained low to 500 °C and rapidly increased at 600 °C and above, remaining high from 600 to 700 °C. The mechanisms involved in this rapid increase in friction at 600 to 700 °C should be related to the two-step process—namely, carbonization and polymerization [4.53]. The carbonization stage includes loss of volatile matter, identified with hydrogen loss in this case [4.54]. This stage generally occurs in DLC films between 500 and 600 °C. The polymerization stage includes the formation of graphitic crystallites or sheets. The two-stage pyrolysis process of carbonization and polymerization occurs simultaneously in DLC films such as a-C:H. Under such conditions the sliding action produces failure in the film and at the interfacial adhesive bonds between the film and the substrate and causes film breakthrough in the contact area.

When the DLC film was ruptured, fresh surfaces of the Si_3N_4 couples came into direct contact, and the coefficient of friction rapidly increased to 0.7, almost the same as that (0.75) for a bare Si_3N_4 flat in contact with a clean Si_3N_4 pin in ultrahigh vacuum, as shown in Fig. 4.52.

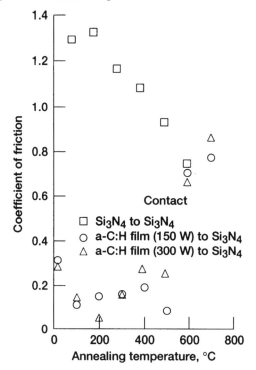

Figure 4.52.—Coefficients of friction for
a-C:H-to-Si$_3$N$_4$ and Si$_3$N$_4$-to-Si$_3$N$_4$
contacts in vacuum.

4.4.3 Silicon Nitride Coatings

Like DLC (amorphous hydrogenated carbon, a-C:H) films, amorphous hydro-genated silicon nitride (a-SiN$_x$:H) thin films transform at high temperatures (say above 500 °C). In as-deposited Si$_3$N$_4$ films hydrogen largely passivates nitrogen dangling bonds [4.55]. The hydrogen-nitrogen bonds can dissociate by high-temperature postdeposition heating or annealing, leaving behind charged nitrogen sites. As a matter of course, temperature could significantly influence the adhesion, friction, and wear of the Si$_3$N$_4$ films.

Adhesion and sliding friction experiments were conducted with Si$_3$N$_4$ films in contact with a hemispherical, monolithic, hot-pressed, polycrystalline, magnesia-doped Si$_3$N$_4$ pin at temperatures to 700 °C in vacuum [4.56]. Thin films (80 nm thick) containing Si$_3$N$_4$ were deposited by low- and high-frequency plasmas (30 kHz and 13.56 MHz) on silicon, gallium arsenide (GaAs), and indium phos-phide (InP) substrates by using silane (SiH$_4$), ammonia (NH$_3$), and nitrogen gases in a parallel-plate plasma reactor.

Silicon nitride films.—AES analyses provided complete elemental depth pro-files for the Si$_3$N$_4$ films on GaAs as a function of sputtering time. Figure 4.53

presents typical examples. The Si/N ratio was higher for the films deposited at 13.56 MHz than for those deposited at 30 kHz. XPS analyses (Table 4.5) also showed that Si/N ratios were much higher for the films deposited at 13.56 MHz and thus supported the AES data.

Table 4.6 gives film thicknesses determined with a rotating ellipsometer. The error margins are the 90% confidence limits. The results were obtained by using a model that assumed a single film on a substrate. The small error margins show that this model is an excellent description of the sample.

Figure 4.54 presents representative results for the refractive index η and the absorption coefficient α (where $\alpha = 4\pi K/\lambda$, K is the extinction coefficient, and λ is the wavelength) for two Si_3N_4 films on GaAs. The refractive index for the films deposited at 13.56 MHz was higher than that of pure amorphous Si_3N_4 [4.57], indicating that a small, but not insignificant, amount of amorphous silicon with its higher refractive index was present. In addition, pure Si_3N_4 does not absorb at all above 300 nm, whereas amorphous silicon does show absorption in the wavelength range used here. This fact and the results shown in Fig. 4.54(b), indicating absorption in the films deposited at 13.56 MHz, reinforced the conclusion that amorphous silicon was present in these films.

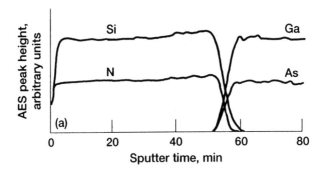

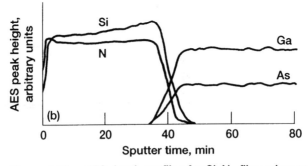

Figure 4.53.—AES depth profiles for Si3N4 films plasma deposited on GaAs (a) at high frequency (13.56 MHz) and (b) at low frequency (30 kHz).

TABLE 4.5.—Si/N RATIO IN
Si₃N₄ FILMS

Substrate	Plasma deposition frequency	
	30 kHz	13.56 MHz
	Si/N ratio	
Si	1.1	1.4
GaAs	1.2	1.4
InP	1.1	1.7

TABLE 4.6.—FILM THICKNESS OF
Si₃N₄ FILMS

Substrate	Plasma deposition frequency	
	30 kHz	13.56 MHz
	Film thickness, nm	
Si	77.1±0.2	85.6±0.2
GaAs	77.0±0.1	88.6±0.7
InP	76.1±0.1	76.6±0.1

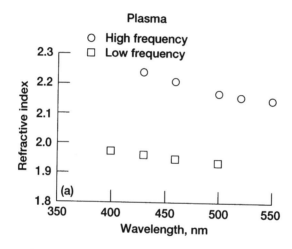

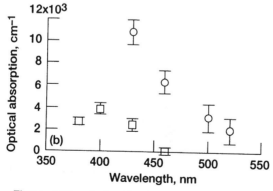

Figure 4.54.—Refractive index (a) and optical absorption (b) as function of wavelength for Si₃N₄ films plasma deposited on GaAs.

The films deposited at 30 kHz had lower refractive indexes than pure Si_3N_4 and almost vanishing absorption. Therefore, these films contained either a small number of voids or a small amount of oxygen and a negligibly small amount of amorphous silicon. (Oxygen usually exists in the form of silicon oxynitride.)

Measurements of Si_3N_4 films on silicon and InP were similar to those obtained on GaAs. Thus, the difference in the properties of high- and low-frequency deposition is due only to the plasma, since there is no indication of substrate interaction.

Adhesion.—Adhesion experiments were conducted with plasma-deposited Si_3N_4 films in contact with monolithic, magnesia-doped Si_3N_4 at temperatures to 700 °C in ultrahigh vacuum. The adhesive strength is again defined as the force necessary to pull the surfaces apart. The monolithic Si_3N_4 pins were sputter cleaned with argon ions before the adhesion experiments. The plasma-deposited Si_3N_4 films were in the as-received state after they had been baked out in the vacuum system. At room temperature adsorbates from the environment were present on the plasma-deposited Si_3N_4 films. The results indicated no significant change in the pull-off force with respect to load over the load range 1 to 6 mN. The data, however, clearly suggest that adhesion depends on the temperature and the plasma deposition frequency.

Figure 4.55 presents the average pull-off forces for as-received Si_3N_4 films deposited by high- and low-frequency plasmas as a function of temperature. Although the pull-off force (adhesion) for the high-frequency films increased slightly from 0 to 400 °C, it generally remained low at these temperatures (Fig. 4.55(a)). The pull-off force increased significantly at 500 °C and remained high between 500 and 700 °C. Extremely strong adhesive bonding can take place at the contacting interface at temperatures in this range. However, the adsorbed carbon contaminant (determined by XPS) decreased rapidly with increasing temperature to 400 °C. Above 400 °C it had disappeared from the surface of the Si_3N_4 film. The rapid increase in adhesion at 500 to 700 °C correlated with this contaminant removal, resulting in a strong surface chemical reaction between the monolithic Si_3N_4 and the plasma-deposited Si_3N_4 film. In contrast, the low adhesion below 400 °C correlated with the presence of adsorbates. Also, the increase in adhesion at 500 °C correlated with the dissociation of the hydrogen-nitrogen bonds in the Si_3N_4 films. The hydrogen-nitrogen bond density does decrease during high-temperature heating [4.55].

For the Si_3N_4 films deposited by low-frequency plasma (30 kHz), the pull-off force increased with increasing temperature (Fig. 4.55(b)). When compared with the results of Fig. 4.55(a), however, the pull-off forces were generally lower in the high-temperature range (500 to 700 °C).

Friction and wear.—Figure 4.56 presents the coefficients of friction for Si_3N_4 films deposited by high-frequency plasma as a function of sliding temperature. The static friction characteristics (Fig. 4.56(a)) are the same as the adhesion characteristics (Fig. 4.55(a)). The coefficient of static friction increased slightly to 400 °C and rapidly above 500 °C, remaining high in the range 500 to 700 °C. The

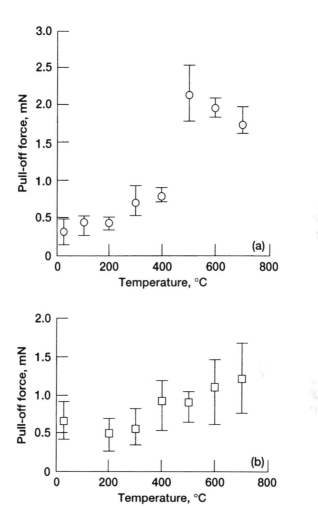

Figure 4.55.—Pull-off force (adhesion) as function of
temperature for (a) high-frequency (13.56 MHz)
and (b) low-frequency (30 kHz), plasma-deposited
Si₃N₄ films in sliding contact with hemispherical
monolithic Si₃N₄ pins in vacuum.

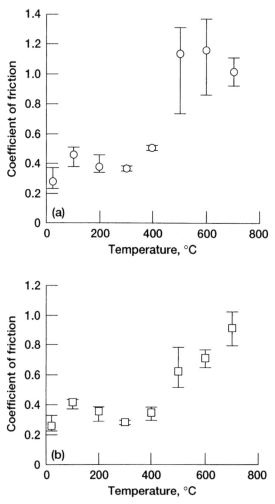

Figure 4.56.—Coefficient of friction as function of
temperature for high-frequency-plasma-deposited
Si_3N_4 films in sliding contact with hemispherical
monolithic Si_3N_4 pins in vacuum. (a) Static friction.
(b) Dynamic friction.

trend for the coefficient of dynamic friction (Fig. 4.56(b)) is also quite similar to that for adhesion. When compared with static friction, the coefficient of dynamic friction was generally lower below 700 °C.

Figure 4.57 shows the coefficients of friction for Si_3N_4 films deposited by low-frequency plasma as a function of sliding temperature. The static friction character-istics (Fig. 4.57(a)) are similar to the adhesion characteristics (Fig. 4.55(a)). The coefficient of static friction increased slightly to 500 °C and quite rapidly after 500 °C, remaining high between 600 and 700 °C. The trend for the coefficient of dynamic friction (Fig. 4.57(b)) was also similar to that for adhesion. When

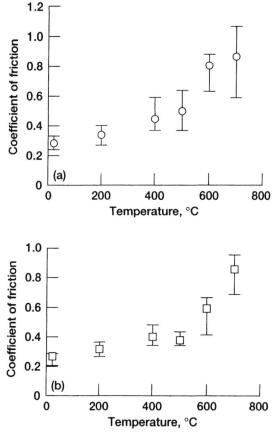

Figure 4.57.—Coefficient of friction as function of temperature for low-frequency-plasma-deposited Si_3N_4 films in sliding contact with hemispherical monolithic Si_3N_4 pins in vacuum. (a) Static friction. (b) Dynamic friction.

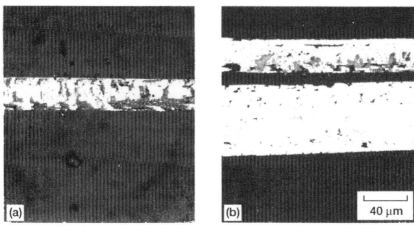

Figure 4.58.—Optical micrographs of wear tracks generated by hemispherical monolithic Si_3N_4 pins sliding at high temperatures in vacuum on high-frequency-plasma-deposited (13.56 MHz) Si_3N_4 film surfaces. (a) At 500 °C. (b) At 700 °C.

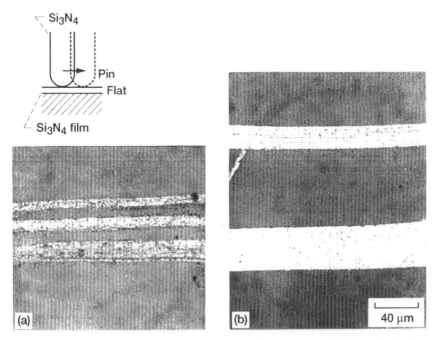

Figure 4.59.—Optical micrographs of wear tracks generated by hemispherical monolithic Si_3N_4 pins sliding at high temperatures in vacuum on low-frequency-plasma-deposited (30 kHz) Si_3N_4 film surfaces. (a) At 500 °C. (b) At 700 °C.

compared with static friction, the coefficient of dynamic friction was generally lower below 700 °C.

When the sliding temperature was increased to 500 to 700 °C, the sliding action produced failure in the films and at the interfacial adhesive bonds between the film and the substrate and caused film breakthrough in the contact area (Figs. 4.58 and 4.59). Also, wear debris particles of Si_3N_4 film were observed. Thus, the increase in friction at higher loads was caused by the gross failure of the Si_3N_4 film. A large amount of frictional energy was dissipated as the film failed during sliding.

Thus, as anticipated, there was a significant temperature influence on the adhesion, friction, and wear of Si_3N_4 films. The changes in tribological properties at high temperatures are primarily related to the dissociation of hydrogen-nitrogen bonds.

The adhesion behavior of the plasma-deposited Si_3N_4 films at temperatures to 700 °C was similar to that of the coefficient of friction. Although adhesion and friction remained low to 400 °C and the temperature effect was small, they increased greatly above 500 to 600 °C.

Source wear of the Si_3N_4 film occurred in the contact area at high temperatures. The wear correlated with the increase in adhesion and friction for the low- and high-frequency-plasma-deposited films above 600 and 500 °C, respectively.

References

4.1 D.H. Buckley, *Surface Effects in Adhesion, Friction, Wear and Lubrication, Tribology Series,* Elsevier Scientific Publishing Co., New York, Vol. 5, 1981.

4.2 D. Tabor, Status and direction of tribology as a science in the 80's—understanding and prediction, *New Directions in Lubrication, Materials, Wear, and Surface Interactions* (W.R. Loomis, ed.), Noyes Publications, Park Ridge, NJ, Vol. 1, 1984, pp. 1–17. (Also NASA CP-2300.)

4.3 K. Miyoshi, Uses of Auger and x-ray photoelectron spectroscopy in the study of adhesion and friction, *Advances in Engineering Tribology* (Y.-W. Chung and H.S. Cheng, eds.), STLE SP–31, Society of Tribologists and Lubrication Engineers, Park Ridge, IL, 1991, pp. 3–12.

4.4 K. Miyoshi and Y.W. Chung, *Surface Diagnostics in Tribology,* World Scientific Publishing Co., Inc., River Edge, NJ, 1993.

4.5 L. Pauling, A resonating-valence-bond theory of metals and intermetallic compounds, *Proc. R. Soc. London A196*: 343–362 (1949).

4.6 K. Miyoshi, J.J. Pouch, and S.A. Alterovitz, Plasma-deposited amorphous hydrogenated carbon films and their tribological properties, *Mater. Sci. Forum 52–53*: 645–656 (1989).

4.7 J.J. Pouch, S.A. Alterovitz, K. Miyoshi, and J.D. Warner, Boron nitride: composition, optical properties, and mechanical behavior, *Materials Modification and Growth Using Ion Beams* (U.J. Gibson, A.E. White, and P.P. Pronko, eds.), Materials Research Society, Pittsburgh, PA, Vol. 93, 1987, pp. 323–328.

4.8 F.T. Barwell, Metallic wear, *CRC Handbook of Lubrication* (E.R. Booser, ed.), CRC Press, Inc., Boca Raton, FL, Vol. 2, 1984, pp. 163–184.

4.9 E. Rabinowicz, Wear coefficients, *CRC Handbook of Lubrication* (E.R. Booser, ed.), CRC Press, Inc., Boca Raton, FL, Vol. 2, 1984, pp. 201–208.

4.10 K. Miyoshi et al., Friction and wear of plasma-deposited diamond films, *J. Appl. Phys. 74, 7*: 4446–4454 (1993).

4.11 I. Langmuir, The constitution and fundamental properties of solids and liquids, Part I–Solids, *J. Am. Chem. Soc. 38, 11*: 2221–2295 (1916).

4.12 J.E. Lennard-Jones, Processes of adsorption and diffusion on solid surfaces, *Trans. Faraday Soc. 28*: 333–359 (1932).

4.13 H.S. Taylor, The activation energy of adsorption processes, *J. Am. Chem. Soc. 53, 2*: 578–597 (1931).

4.14 J.K. Roberts, The adsorption of hydrogen on tungsten, *Proc. R. Soc. London A 152, 876*: 445–477 (1935).

4.15 C.W. Oatley, The adsorption of oxygen and hydrogen on platinum and the removal of these gases by positive ion bombardment, *Proc. Phys. Soc. London 51, 2*: 318–328 (1939).

4.16 J.A. Dillon, Jr., The Interaction of oxygen with silicon carbide surfaces, *Silicon Carbide, a High Temperature Semiconductor* (J.R. O'Connor, and J. Smitems, eds.), Pergamon Press, Oxford, 1960, pp. 235–240.

4.17 J.S. Johannessen, W.E. Spicer, and Y.E. Strausser, An Auger analysis of the SiO_2-Si interface, *J. Appl. Phys. 47, 7*: 3028–3037 (1976).

4.18 K. Miyoshi and D.H. Buckley, Effect of oxygen and nitrogen interactions on friction of single-crystal silicon carbide, NASA TP–1265 (1978).

4.19 L. Brewer, L.A. Bromley, P.W. Gilles, and N.L. Lofgren, Thermodynamic and physical properties of nitrides, carbides, sulfides, silicides, and phosphides, *The Chemistry and Metallurgy of Miscellaneous Materials: Thermodynamics* (L.L. Quill, ed.), McGraw-Hill, New York, 1950, pp. 40–59.

4.20 F.D. Rossin et al., Selected values of chemical thermodynamic properties, *Circular of the National Bureau of Standards 500*, U.S. Government Printing Office, Washington, DC, 1952.

4.21 K. Miyoshi and D.H. Buckley, X-ray photoelectron spectroscopy and friction studies of nickel-zinc and manganese-zinc ferrites in contact with metals, NASA TP–2163 (1983).

4.22 K. Miyoshi, C. Maeda, and R. Masuo, Development of a torsion balance for adhesion measurements, *IMEKO XI: Instrumentation for the 21st Century*, Proceedings of the 11th Triennial World Congress of the International Measurement Confederation (IMEKO), Instrument Society of America, Triangle Park, NC, 1988, pp. 233–248.

4.23 D.D. Eley, *Adhesion*, Oxford University Press, London, 1961.

4.24 F.P. Bowden and D. Tabor, Action of extreme pressure lubricants, *The Friction and Lubrication of Solids—Part XI*, Clarendon Press, Oxford, UK, 1958, pp. 228–314.

4.25 K. Miyoshi and D.H. Buckley, Effects of water vapor on friction and deformation of polymeric magnetic media in contact with a ceramic oxide, *Tribology and Mechanics of Magnetic Storage Systems* (B. Bushan et al., eds.), STLE SP-16, Society of Tribologists and Lubrication Engineers, Park Ridge, IL, 1984, pp. 27–34.

4.26 K. Miyoshi and D.H. Buckley, Friction and deformation behavior of single-crystal silicon carbide, NASA TP–1053 (1977).

4.27 F.P. Bowden and D. Tabor, *The Friction and Lubrication of Solids—Part II*, Clarendon Press, Oxford, UK, 1964.

4.28 K. Miyoshi, D.H. Buckley, and B. Bhushan, Friction and morphology of magnetic tapes in sliding contact with nickel-zinc ferrite, NASA TP–2267 (1984).

4.29 R.L. Bradshaw and B. Bhushan, Friction in magnetic tapes III: role of chemical properties, *ASLE Trans. 27, 3*: 207–219 (1984).

4.30 V.D. Scott and H. Wilman, Surface reorientation caused on metals by abrasion—its nature, origin, and relation to friction and wear, *Proc. Roy. Soc. London A 247, 1250*: 353–368 (1958).

4.31 J. Goddard, H.J. Harker, and H. Wilman, The surface reorientation caused by unidirectional abrasion on face-centered cubic metals, *Proc. Phys. Soc. London 80*: 771–782 (1962).

4.32 D.H. Buckley, Recrystallization and preferred orientation in single-crystal and polycrystalline copper in friction studies, NASA TN D–3794 (1967).

4.33 K. Miyoshi and D.H. Buckley, Sliding-induced crystallization of metallic glass, NASA TP–2140 (1983).

4.34 K. Miyoshi and D.H. Buckley, Surface chemistry, microstructure, and friction properties of some ferrous-base metallic glasses at temperatures to 750 °C, NASA TP-2006 (1982).

4.35 K. Miyoshi and D.H. Buckley, Friction and wear of some ferrous-base metallic glasses, *ASLE Trans. 27, 4*: 295–304 (1984). (Also NASA TM-83067.)

4.36 K. Miyoshi and D.H. Buckley, Friction and surface chemistry of some ferrous-base metallic glasses, NASA TP-1991 (1982).

4.37 D.V. Badami, X-ray studies of graphite formed by decomposing silicon carbide, *Carbon 3, 1*: 53–57 (1965).

4.38 K. Miyoshi and D.H. Buckley, Surface chemistry and wear behavior of single-crystal silicon carbide sliding against iron at temperatures to 1500 °C in vacuum, NASA TP-1947 (1982).

4.39 K. Miyoshi, D.H. Buckley, and M. Srinivasan, Tribological properties of sintered polycrystalline and single-crystal silicon carbide, *Am. Ceram. Soc. Bull. 62, 4*: 494–500 (1983).

4.40 R.C. Weast, ed., *CRC Handbook of Chemistry and Physics: A Ready Reference Book of Chemical and Physical Data*, 68th ed., CRC Press Inc., Boca Raton, FL, 1987.

4.41 D.H. Buckley and W.A. Brainard, Friction and wear of metals in contact with pyrolytic graphite, *Carbon 13, 6*: 501–508 (1975).

4.42 K. Miyoshi and D.H. Buckley, Considerations in friction and wear, *New Directions in Lubrication, Materials, Wear, and Surface Interactions* (W.R. Loomis, ed.), Noyes Publications, Park Ridge, NJ, Vol. I, 1984, pp. 282–319. (Also NASA CP-2300.)

4.43 F. Meyer and G.J. Loyen, Ellipsometry applied to surface problems—optical layer thickness measurement, *Acta Electron 18*: 33–38 (1975).

4.44 A.J. Van Bommel, J.E. Crombeen, and A. Van Tooren, LEED and Auger electron observations of the SiC (0001) surface, *Surf. Sci. 48, 2*: 463–472 (1975).

4.45 H.O. Pierson, Structure and properties of diamond and diamond polytypes, *Handbook of Carbon, Graphite, Diamond, and Fullerenes*, Noyes Publications, Park Ridge, NJ, 1993, pp. 244–277.

4.46 R.L.C. Wu, K. Miyoshi, R. Vuppuladhadium, and H.E. Jackson, Physical and tribological properties of rapid thermal annealed diamond-like carbon films, *Surf. Coat. Tech., 54/55*: 576–580 (1992).

4.47 D.I. Jones and A.D. Stewart, Properties of hydrogenated amorphous carbon films and the effects of doping, *Phil. Mag. B 46*: 423–434 (1982).

4.48 B. Dischler, A. Bubenzer, and P. Koidl, Bonding in hydrogenated hard carbon studied by optical spectroscopy, *Solid State Commun. 48, 2*: 105–108 (1983).

4.49 R.O. Dillon, J.A. Woollam, and V. Katzanant, Use of Raman scattering to investigate disorder and crystallite formation in as-deposited and annealed carbon films, *Phys. Rev. B 29*: 3482–3489 (1984).

4.50 T. Sugita, K. Ueda, and Y. Kanemura, Material removal mechanism of silicon nitride during rubbing in water, *Wear 97*: 1–8 (1984).

4.51 T.E. Fischer and H. Tomizawa, Interaction of tribochemistry and microfracture in the friction and wear of silicon nitride, *Wear of Materials 1985* (K.C. Ludema, ed.), American Society of Mechanical Engineers, New York, 1985, pp. 22–32.

4.52 H. Ishigaki and K. Miyoshi, Tribological properties of ceramics, *Proceedings of the 6th International Conference on Production Engineering, Osaka 1987*, Japan Society of Precision Engineering, Tokyo, 1987, pp. 661–666.

4.53 J. Robertson, Amorphous carbon, *Adv. Phys. 35, 4*: 317–374 (1986).

4.54 J.C. Angus, P. Koidl, and S. Domitz, Carbon thin films, *Plasma-Deposited Thin Films* (J. Mort and F. Jansen, eds.), CRC Press, Inc., Boca Raton, FL, 1986, pp. 89–127.

4.55 W.L. Warren et al., Creation and properties of nitrogen dangling bond defects in silicon nitride thin films, *J. Electrochem. Soc. 143, 11*: 3685–3691 (1996).

4.56 K. Miyoshi et al., Adhesion, friction, and wear of plasma-deposited thin silicon nitride films at temperatures to 700 °C, *Wear 133*: 107–123 (1989).

4.57 E.D. Palik, *Handbook of Optical Constants of Solids*, Academic Press, Orlando, FL, 1985.

Chapter 5
Abrasion: Plowing and Cutting

5.1 Introduction

Abrasion causes friction and wear by displacing material from one of two surfaces in relative motion. It may be caused by hard protuberances (asperities) on the second contact surface (two-body conditions) or by hard particles between the surfaces (three-body conditions) or embedded in one of them [5.1, 5.2]. The hard asperities and hard particles plow (groove) or cut one of the rubbing surfaces.

Abrasion is involved in the finishing of many surfaces. Filing, sanding, grinding, and lapping of engineering surfaces all involve abrasion. However, in many mechanical engineering systems abrasion is a common wear phenomenon of great economical importance. Approximately 50% of the wear encountered in industry is caused by abrasion mechanisms [5.3, 5.4].

A metal's resistance to abrasive wear is related to its static hardness under two-body conditions [5.5]. That is, the abrasive wear rate is inversely proportional to Vickers hardness for many annealed pure metals. Theory and experiments indicate that a metal's resistance to abrasive wear is proportional to the Vickers hardness of the fully work-hardened surface region [5.6]. Similar results have been obtained under three-body conditions [5.7].

Abrasion produces a worked or deformed layer and promotes a chemical interaction, as discussed in Chapter 2. For example, abrasion in a moist environment quickly admits hydrogen into aluminum, not only near the surface but in the bulk [5.8]. The solubility of hydrogen in aluminum is known to be low, only 0.6 atomic part per million (0.00006%) at the atmospheric pressure of hydrogen gas, even just below the melting point [5.9]. Aluminum is also highly reactive with water or oxygen and forms a chemically stable oxide surface, which is a strong chemical barrier against hydrogen permeation. Although these features of aluminum make the admission of hydrogen into aluminum difficult, the concentration of hydrogen admitted into aluminum by abrasion using an abrasive paper (silicon carbide (SiC) 800-mesh powder) is approximately two orders of magnitude greater than its solubility in aluminum at atmospheric pressure near the melting point. Hydrogen could lead to embrittlement and corrosion of materials.

5.2 Nature of Abrasion

Figures 5.1(a) and (b) are a scanning electron micrograph and an x-ray dispersive analysis map of a nearly spherical SiC particle and a groove, produced plastically by the plowing action of the particle, which slid and/or rolled on the softer metal surface [5.10]. Abrasion can arise when a hard particle or a hard asperity plows or cuts a series of grooves (Fig. 5.2, [5.11]). In plowing, the material first moves upward ahead of the particle's rake face and then moves around it into side ridges. In cutting, a ribbon of material is separated from the surface and moves upward past the rake face. How the material moves strongly depends on the nature of the surface; the tribological and bulk properties of the material, hard particle, or asperity; and the environment. Figures 5.3(a) and (b) are scanning electron micrographs of the wear tracks created on a manganese-zinc (Mn-Zn) ferrite ({100} surfaces) by sliding contact with a hemispherical (100-μm radius) diamond pin under dry and lubricated conditions. The lubricant was a high-purity olive oil, which is commonly used in fine lapping of ceramic materials and semiconductors. As shown, the tangential force introduced by sliding generated surface cracks much more easily in dry sliding than with the lubricant. The critical loads to fracture were 4 N in lubricated sliding and 1.5 N in dry sliding. Therefore, liquid lubrication of ceramics not only reduces friction and wear but also arrests crack formation.

Abrasion encountered in industrial situations can be broken down into two mechanisms: plowing and cutting. However, there are situations where one type changes to another or where the two mechanisms operate together.

5.3 Abrasion by Single Asperity

Experimental modeling and measurements can clarify the complex abrasion process and help in predicting friction and wear. The experiments can be conducted relatively simply by using a pin-on-flat sliding contact configuration. When a hard, spherical pin or wedge pin (e.g., Fig. 5.4) is brought into sliding contact with a softer flat surface, the materials behave both elastically and plastically. With rubber-like materials the elastic properties play an important role in contact deformation. With ductile and semiductile materials, such as metals and intermetallic compounds, however, the behavior is different. Although the elastic moduli are large, the range over which metals deform elastically is relatively small. Consequently, when metals are deformed or indented, the deformation is predominantly outside the elastic range. Often, a plastically deformed groove (Fig. 5.5) is generated in plowing, or a ribbon-like metal chip (Fig. 5.6) moves upward past the wedge's rake face in cutting. The sliding involves plastic flow and generates a considerable amount of metal wear debris.

Brittle materials, such as ceramics and some intermetallic compounds, behave micromechanically in a ductile fashion up to a certain contact stress when in contact with themselves or other solids. Even at room temperature ceramics, such as

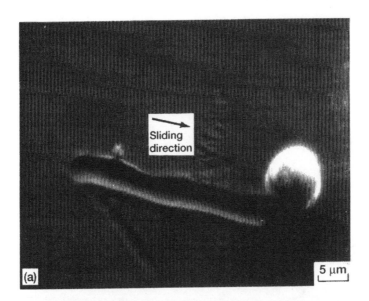

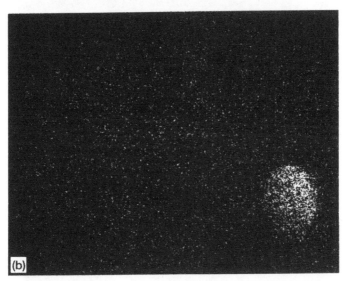

Figure 5.1.—Scanning electron micrograph and x-ray dispersive analysis of wear scar produced by spherical SiC particle on alloy in vacuum (10^{-8} Pa). Sliding velocity, 3 mm/min; load, 0.2 N; room temperature. (a) Spherical SiC particle and groove. (b) Silicon K_α x-ray map on 1.02 at.% Ti-Fe alloy.

224

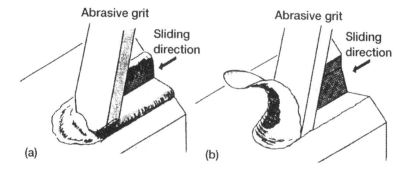

Figure 5.2.—Abrasion models. (a) Plowing. (b) Cutting.

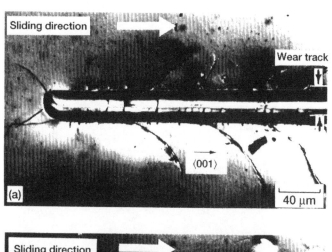

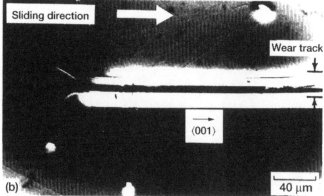

Figure 5.3.—Diamond pin sliding on single-crystal Mn-Zn
ferrite {100} surface. (a) In air. (b) Lubricated with olive oil.

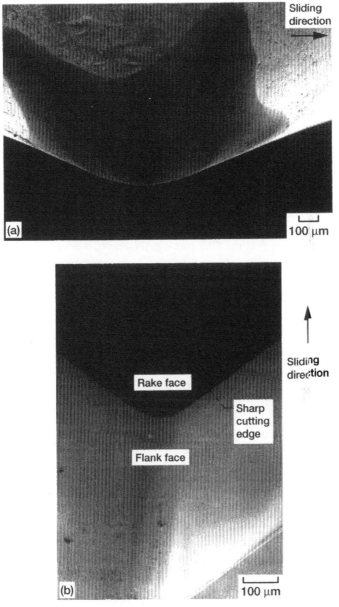

Figure 5.4.—Idealized diamond abrasive particles. (a) Single-crystal spherical pin (0.2-mm radius). (b) Single-crystal wedge (sliding direction perpendicular to rake face).

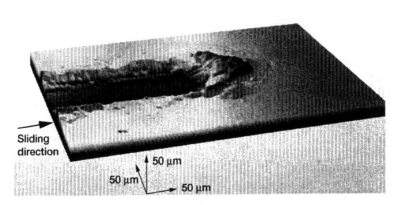

Sliding direction

50 μm

50 μm

50 μm

Figure 5.5.—Plastically deformed groove on γ-TiAl surface produced
by single-pass sliding of 0.2-mm-radius diamond pin at 30-N load.

aluminum oxide (Al_2O_3) and SiC, behave both elastically and plastically at low
stresses under relatively modest rubbing contact, but they microfracture under more
highly concentrated contact stresses [5.12–5.19]. This microfracture, known as
brittle fracture, is detrimental to the reliability of ceramics and must be considered
in designing for structural and tribological applications.

Thus, the deformation and fracture behavior of materials, which plays an
important role in assessing abrasive friction and wear, falls into three main
categories: elastic contact, plastic contact, and brittle contact.

5.3.1 Plowing by Spherical Pin

In elastic contact (Fig. 5.7(a)) the coefficient of friction decreases as the load
increases. The relation between coefficient of friction μ and load W is given by
[5.20]:

$$\mu = KW^{-1/3} \tag{5.1}$$

The minus 1/3 power can be interpreted simply as arising from an adhesion
mechanism, with the contact area being determined by elastic deformation. In
elastic contact frictional energy is dissipated during sliding by shearing at the
interface. By contrast, in plastic contact the coefficient of friction increases as the
load increases (Fig. 5.7(b)), a complete reversal in friction behavior from that in
elastic contact. In plastic contact frictional energy is dissipated during sliding by
shearing at the interface and by plastic deformation of the soft material (i.e., plowing
of the soft flat by the hard pin).

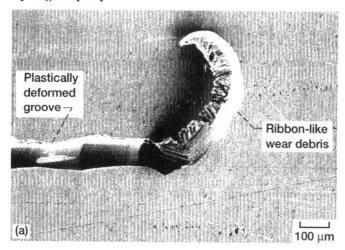

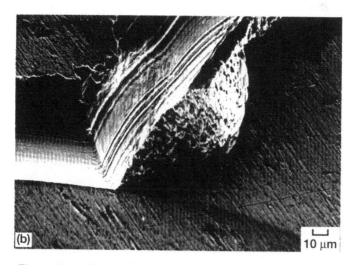

Figure 5.6.—Ribbon-like wear debris and plastically deformed groove on aluminum surface produced by single-
pass sliding of diamond wedge. (a) Overview of groove
and ribbon. (b) High magnification.

With a brittle solid, such as a ceramic, under high contact pressure, sliding action
causes gross surface and subsurface cracking as well as plastic deformation [5.17].
Cracking produces wear debris particles and large fracture pits. The area of a
fracture pit is a few times larger than the area of the plastically deformed groove. The
coefficient of friction is also much higher in brittle contact (Fig. 5.7(c)) than in
elastic and plastic contacts. In brittle contact frictional energy is dissipated during
sliding by shearing at the interface, by plastic deformation of the soft material, and
by cracking in the brittle material [5.21, 5.22].

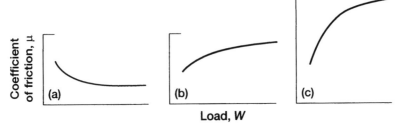

Figure 5.7.—Schematic representation of load effect on coefficient of friction for spherical, hard pin sliding on flat, soft solid. (a) Elastic contact. (b) Plastic contact. (c) Brittle contact.

Elastic contact.—Figure 5.8(a) presents coefficients of friction measured for a smooth polyester film (23 μm thick and 12.7 mm wide) in sliding contact with a smooth, polycrystalline nickel-zinc (Ni-Zn) ferrite, spherical pin (2-mm radius) in humid air at various loads. The data clearly indicate that the coefficient of friction decreased as the load increased. With sliding, elastic deformation occurred in the surfaces of both the polyester film and the polycrystalline Ni-Zn ferrite pin [5.23].

Another example of friction behavior in elastic contact is shown in Fig. 5.8(b). When boron nitride (BN) deposited on an AISI 440C stainless steel flat was placed in sliding contact with BN deposited on a 440C stainless steel, hemispherical pin at various loads in ultrahigh vacuum, the coefficient of friction decreased as the load increased. To a first approximation for the load range investigated, the relation between coefficient of friction μ and load W on logarithmic coordinates is given by $\mu = 0.29W^{-1/3}$ for sputter-cleaned specimen couples and $\mu = 0.17W^{-1/3}$ for as-received specimen couples. The friction was a function of the shear strength of this elastic contact area [5.18]. Thus, ceramics behave elastically up to a certain load under sliding contact.

Transition from elastic to plastic contact.—Single-pass and multipass sliding friction experiments were also conducted with the three tapes in Table 5.1 in contact with a polycrystalline Ni-Zn ferrite, spherical pin (2-mm radius) in laboratory air at various loads. Friction traces were relatively smooth, with no evidence of stick-slip. As the load increased, the coefficient of friction decreased, up to 0.25 N, and then increased (Fig. 5.9). When repeated passes were made, the coefficient of friction generally exhibited only small changes with the number of passes at any load up to 1.0 N (data not shown). The data of Fig. 5.9 raise the question of how the interface deforms with sliding action.

The wear tracks on the tape, which the ferrite pin was made to traverse, varied with different loads up to 1.0 N when examined by optical and scanning electron microscopes. At 0.1 N essentially no detectable wear track existed on the tape surface, which looked like the surface of the as-received magnetic tape (as shown in Chapter 4, Fig. 4.21(a)). At loads up to 0.25 N, although sliding occurred at the interface, both the tape and the Ni-Zn ferrite pin deformed elastically. At 0.25 N and

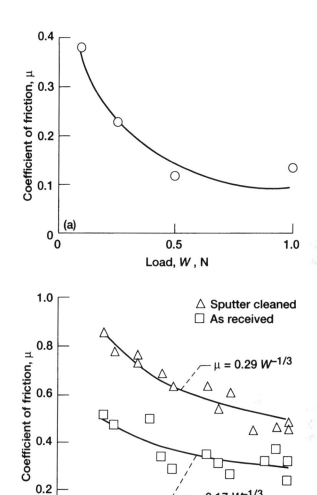

Figure 5.8.—Coefficient of friction as function of load
in elastic contact. (a) Ni-Zn ferrite pin (2-mm radius)
sliding on polyester film. Sliding velocity, 1.5 mm/min;
relative humidity, 40%; temperature, 23 °C. (b) BN
film deposited on hemispherical pin in contact with
BN film deposited on flat in vacuum.

TABLE 5.1.—COMPOSITION AND PROPERTIES OF MAGNETIC TAPES

[Polyester base film 23 μm thick and 12.7 mm wide. From [5.23].]

Tape	Magnetic particle	Binder	Lubricant	Surface roughness,[a] nm	Knoop hardness at 23 °C,[b] MPa
1	CrO_2	Polyester-polyurethane and phenoxy	Butyl stearate and butyl palmitate	14.4	225
2	γ-Fe_2O_3	Polyester-polyurethane and epoxy	Butyl myristate	8.5	157
3	Cobalt-doped γ-Fe_2O_3	Polyester-polyurethane, polyvinyl chloride, and polyvinyl acetate	Butyl ethyl stearate	13.6	118

[a]Root-mean-square roughness.
[b]Measuring load of hardness, 0.0013 N.

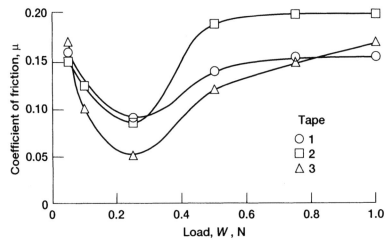

Figure 5.9.—Coefficient of friction as function of load for Ni-Zn ferrite pin (2-mm radius) sliding on magnetic tapes (tapes 1 to 3 in Table 5.1). Sliding velocity, 1.5 mm/min; relative humidity, 40%; temperature, 23 °C.

above the Ni-Zn ferrite pin primarily deformed elastically, but the tape deformed plastically. The sliding action produced a visible wear track on the magnetic tape. Figure 4.21(b) shows the blunt appearance, resulting primarily from plastic deformation, of the asperities on the wear track after five sliding passes at 1.0 N.

Plastic contact.—When a monolithic SiC surface was placed in sliding contact with a spherical diamond pin under relatively low load (up to 0.5 N), elastic deformation occurred in both surfaces (Fig. 5.10(a)). With the initiation of tangential motion, sliding occurred at the interface. Under these low loads neither groove formation due to plastic flow nor SiC cracking during sliding was observed. Friction was a function of the shear strength of the elastic contact area (i.e., the relation

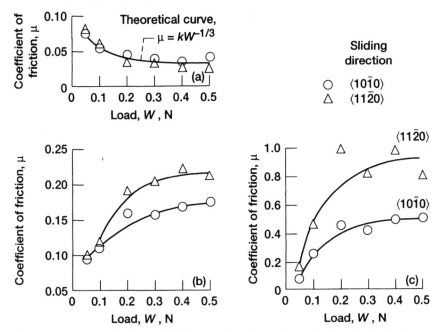

Figure 5.10.—Coefficient of friction as function of load for hemispherical diamond pins of different radii and conical diamond pin sliding on {0001} surface of single-crystal SiC in argon at atmospheric pressure. (a) Elastic contact (diamond radius, 0.3 mm). (b) Plastic contact (diamond radius, 0.02 mm). (c) Brittle contact (conical diamond pin with apical angle of 117°).

between coefficient of friction μ and load W is given by Eq. (5.1)). Over the entire load range the mean contact pressure ranged from 1.5 to 3.5 GPa. The maximum pressure at the center of the contact area calculated from a Hertzian stress distribution was 2.3 to 4.9 GPa.

Silicon carbide, like most ceramics, both in monolithic and in coating form, deforms in a ductile fashion as the contact pressure is increased. The increase in applied load, however, results in a complete reversal in friction characteristic from that in elastic contact. Figure 5.10(b) reveals an entirely different mode of deformation and energy dissipation with an estimated maximum Hertzian contact pressure ranging from 14 to 30 GPa. Plastic deformation occurred in the SiC, causing permanent grooves during sliding, but there was little or no evidence of small cracks being generated in the SiC. The diamond pin indented the SiC without suffering any permanent deformation itself. The frictional energy dissipated during sliding following solid-state contact was caused by shearing at the interface and by plastic deformation of the SiC (i.e., plowing by the diamond). The relation between coefficient of friction μ and load W now took the form

$$\mu = kW^{m} \tag{5.2}$$

where $m = 0.3$ to 0.4. The exponent m depends on the crystallographic orientation, plastic deformation, and hardness of the single-crystal SiC.

With a conical diamond pin ($117°$ apical angle with radius of curvature at apex less than 5 μm), ceramics behave in a brittle fashion (Fig. 5.10(c)). This subject is discussed in the next section.

Similar contact and friction characteristics also occurred for diamond pins (0.2-mm radius) on BN films [5.18, 5.19]. At certain loads the sliding action of the diamond pin permanently grooved the BN films deposited on both metallic [5.18] and nonmetallic [5.19] substrates. For example, Fig. 5.11 presents the widths of plastically deformed grooves in BN films on 440C stainless steel substrates. Comparative data for uncoated 440C stainless steel are also presented. When the widths D of the resulting grooves in the BN films are plotted against the load W on logarithmic coordinates, the data can be expressed as Meyer's law:

$$W = kD^n \qquad\qquad (5.3)$$

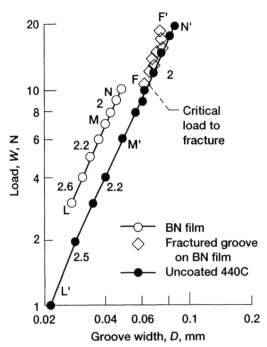

Figure 5.11.—Groove width as function of load for BN film deposited on 440C stainless steel in contact with hemispherical diamond pin (2-mm radius) in laboratory air.

The portions of the curves LM for the BN film and L'M' for the uncoated 440C stainless steel are considered to be approximately straight with transitional slopes of 2.6, 2.5, and 2.2. The portions MN for the BN film and M'N' for the uncoated 440C stainless steel are straight lines with a slope of 2. The portions MN and M'N' are the ranges over which Meyer's law is valid for BN film and for uncoated 440C stainless steel. Here the Meyer index n is constant and equal to 2. Thus, BN films on metallic [5.18] and nonmetallic [5.19] substrates behave plastically much like metals when they are brought into contact with hard solids such as diamond [5.18, 5.24]. The contact pressure gradually increases until deformation passes to a fully plastic state. The mean contact pressure P_m at a fully plastic state increases by a factor of 2 with the presence of BN film. When the load exceeds a certain critical value, the sliding action of diamond on the monolithic SiC and on the BN film causes fracture in both specimens.

The width D and height H of a groove are defined in Fig. 5.12. The mean contact pressure P_m during sliding may be defined by $P_m = W/A_s$, where W is the applied load and A_s is the projected contact area given as $A_s = \pi D^2/8$. Only the front half of the pin is in contact with the flat. The relation between the groove width generated by the pin and the load is expressed by $W = kD^n$, which is known as Meyer's law [5.25].

Brittle contact.—With a conical diamond pin (causing highly concentrated stress in the contact area between diamond and SiC), sliding action produces gross surface and subsurface cracking as well as plastic deformation [5.17]. Under such conditions wear debris particles and large fracture pits caused by cracking are observed. The SEM observation indicated that the area of a fracture pit is a few times larger than the area of the plastically deformed groove. The coefficient of friction in brittle contact (Fig. 5.10(c)) is also much higher (four times or more) than in elastic and plastic contacts. Although fracture and plastic deformation in SiC are responsible for the friction behavior observed, most of the frictional energy dissipation during sliding is caused by the fracturing of the SiC. Therefore, the coefficient of friction

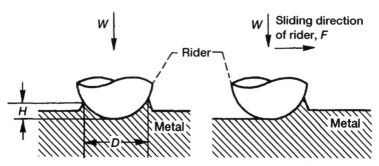

Figure 5.12.—Deformation of metal. (a) End view (lateral cross section). (b) Side view (longitudinal cross section).

is commonly influenced by the bulk properties of the ceramic, such as fracture toughness and crystallographic orientation (Fig. 5.10(c), [5.17]).

For BN films deposited onto metallic or nonmetallic substrates [5.18, 5.19], failure occurs primarily in the film or at the interface between film and substrate (or both) when the film is critically loaded. For example, in Fig. 5.11 the portion FF', representing the condition of fracture where the load exceeds the critical load, is also roughly expressed by $W = kD^n$. The fractured groove in the BN film is almost as wide as the groove in the uncoated 440C. The evidence confirms that cracks are generated from the contact area rather than from the free surface of the film. It suggests that the substrate is responsible not only for controlling the critical load, which will fracture the BN film, but also for the extent of fracture. Furthermore, the critical load required to fracture a ceramic film on a substrate can be determined by measuring the groove width.

Effects of shear strength of metals.—Abrasion both causes plastic deformation and generates metal wear debris. The metal removal mechanisms are plastic deformation, surface fatigue, and adhesive wear. Although experimental modeling by using spherical pins indeed represents some typical abrasive wear phenomena, it does not represent the principal mechanisms of metal removal in abrasion. The shear strength of the metal must also be taken into account. Abrasion occurs, for example, when SiC grit slides on metals and alloys, causing plastic flow and generating metal wear debris. A spherical SiC pin sliding on iron (Fig. 5.13) resulted in a permanent groove in the iron surface, with a considerable amount of deformed metal piled up along the groove sides, and metal wear debris. The wear debris particles were primarily on the sides of the wear track or transferred to the SiC pin at its tip. Much less metal debris was removed, however, than the volume of the groove plowed out by the SiC pin.

When relatively ductile metals, such as magnesium, aluminum, copper, zirconium, and iron (Table 5.2, [5.11]), are worn in abrasion, the principal mechanism of metal removal is chip formation in front of the cutting abrasive, which often is angular in shape [5.26–5.28]. The few exceptions to this mechanism include abrasion by abrasives that are soft relative to the metal and abrasion by lightly loaded, round abrasives. In these cases metal is removed at a relatively low rate (discussion by Jorn Larsen-Basse in [5.11]). For example, abrasion on dry emery leads to an amount of wear (metal removal) corresponding to only about 10% of the volume of the grooves plowed out by the abrasives [5.27].

A spherical pin will remove chips from a metal surface when the ratio between the groove height and the sphere radius $H/R > 0.1$. For $H/R < 0.01$ only elastic deformation takes place, and for $0.01 < H/R < 0.1$ material flows by plastic deformation to ridges along the groove sides [5.29]. For example, when $R = 25$ μm, the two values of H are 2.5 μm and 0.25 μm, respectively.

In consideration of $0.01 < H/R < 0.1$ and to obtain geometrically similar grooves, experiments with the relatively ductile metals shown in Table 5.2 were conducted at a load of 0.049 N; and experiments with relatively harder metals, such as titanium,

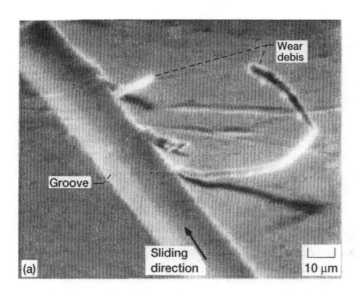

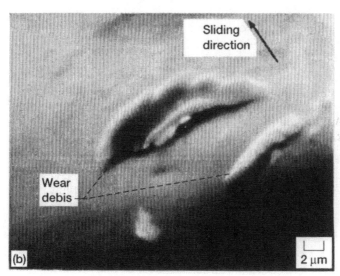

Figure 5.13.—Iron wear debris and groove. Single-pass sliding of 0.04-mm-radius SiC pin; sliding velocity, 3 mm/min; load, 0.25 N; temperature, 25 °C; environment, argon; pressure, atmospheric. (a) Iron wear debris and groove on iron surface. (b) Iron wear debris transferred to SiC.

Abrasion: Plowing and Cutting

TABLE 5.2.—CRYSTALLINE, PHYSICAL, AND CHEMICAL PROPERTIES OF METALS

Metal	Purity,[a] percent	Crystal structure at 25 °C[b]	Lattice constant,[c] Å (10⁻¹⁰ m)	Cohesive energy[b]		Shear modulus,[b] Pa
				J/g·atom	kcal/g·atom	
Iron	99.99	(c)	a = 2.8610	416.0×10^3	99.4	8.15×10^{10}
Chromium			a = 2.8786	395	94.5	11.7
Molybdenum			a = 3.1403	657.3	157.1	11.6
Tungsten			a = 3.1586	835.5	199.7	15.3
Aluminum		(d)	a = 4.0414	322	76.9	2.66
Copper	99.999		a = 3.6080	338	80.8	4.51
Nickel	99.99		a = 3.5169	428.0	102.3	7.50
Rhodium			a = 3.7956	556.5	133.0	14.7
Magnesium		(e)	a = 3.2022; c = 5.1991	148	35.3	1.74
Zirconium			a = 3.223; c = 5.123	609.6	145.7	3.41
Cobalt			a = 2.507; c = 4.072	425.5	101.7	7.64
Titanium	99.97		a = 2.923; c = 4.729	469.4	112.2	3.93
Rhenium	99.99		a = 2.7553; c = 4.4493	779.1	186.2	17.9

[a]Manufacturer's analysis. [d]Face-centered cubic.
[b]From [5.40]. [e]Hexagonal close packed.
[c]Body-centered cubic [f]From [5.41].

nickel, rhodium, molybdenum, and tungsten (Table 5.2), were conducted at a load of 0.2 N. Mineral oil was used to minimize adhesion. When a spherical, 25-μm-radius SiC pin was put in sliding contact with the various metal disks, the coefficient of friction and the groove height linearly decreased and the contact pressure linearly increased as the shear strength of the metal increased (Fig. 5.14), for both the ductile and harder metals without exception. The coefficient of friction, particularly its plowing term, and the groove height corresponding to the groove volume may be governed by two factors: the shear strength at the interface and the shear strength of the bulk metal. Thus, the data presented in Fig. 5.14 fall within the range $0.01 < H/R < 0.1$, where only elastic and plastic surface deformation are expected.

Alloying element effects.—There is no doubt that a solid's structure plays an important role in its tribological behavior, such as adhesion and friction behavior, as discussed in Chapter 3. For example, solid-solution alloying, a major mode of metal strengthening, influences friction and deformation behavior under abrasive conditions. Figures 5.15 and 5.16 present the coefficients of friction and the groove heights (corresponding to the groove volume) as a function of solute concentration (in atomic percent) for a number of binary alloys in sliding contact with 25-μm-radius, spherical, single-crystal SiC pins at loads of 0.05 and 0.1 N in mineral oil. The iron-base binary alloy systems were alloyed with manganese, nickel, chromium, rhodium, tungsten, or titanium. The coefficient of friction and the groove height decreased as the solute concentration increased. The coefficient of friction

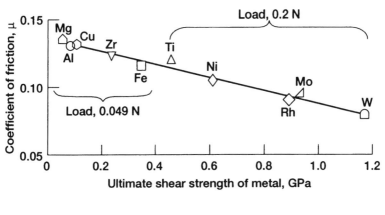

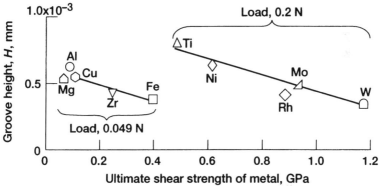

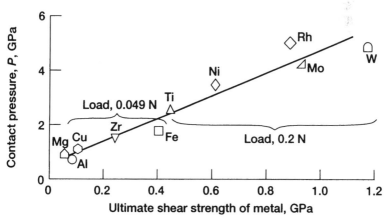

Figure 5.14.—Coefficient of friction, groove height, and contact pressure as function of shear strength for various metals. Single-pass sliding of 25-μm-radius SiC pin in mineral oil; sliding velocity, 3 mm/min; load, 0.049 or 0.2 N; temperature, 25 °C.

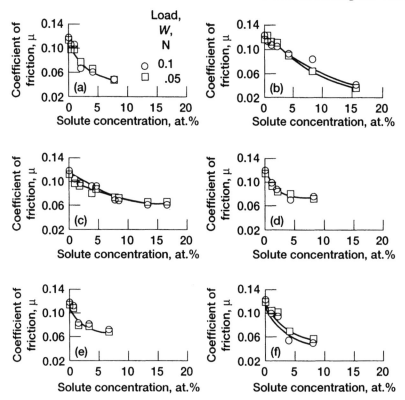

Figure 5.15.—Coefficient of friction as function of solute concentration for various iron-base alloys and pure iron. Single-pass sliding of 25-μm-radius SiC pin in mineral oil; sliding velocity, 3 mm/min; temperature, 25 °C. (a) Iron-manganese alloy.(b) Iron-nickel alloy. (c) Iron-chromium alloy. (d) Iron-rhodium alloy. (e) Iron-tungsten alloy. (f) Iron-titanium alloy.

did not change significantly with load (Fig. 5.15), but there were obvious differences in the groove height with load (Fig. 5.16). The average rates of decrease in the coefficient of friction and the groove height strongly depended on the alloying element.

Figure 5.17 presents the contact pressures (groove microhardnesses) during sliding under the same conditions as in Figs. 5.15 and 5.16 for the same binary alloys as a function of solute concentration (in atomic percent). The contact pressure was calculated from the groove width. It increased as the solute concentration increased, with the average rates of increase depending on the alloying element. The contact pressure did not change significantly with load. Controlling mechanisms for the friction and deformation of iron-base binary alloys are discussed next.

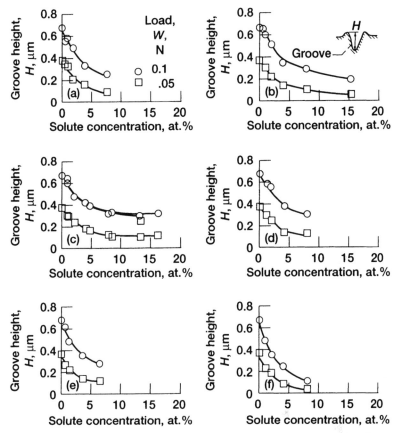

Figure 5.16.—Groove height as function of solute concentration for various iron-base alloys and pure iron. Same conditions as for Fig. 5.15. (a) Iron-manganese alloy. (b) Iron-nickel alloy. (c) Iron-chromium alloy. (d) Iron-rhodium alloy. (e) Iron-tungsten alloy. (f) Iron-titanium alloy.

The grooves (wear tracks) in this investigation were formed in the alloys primarily by plastic deformation under hydrostatic contact pressure and the plowing stress associated with sliding. In addition, there was occasional material removal (wear debris fragments) [5.11]. The formation of grooves may be similar to the formation of hardness test indentations. Therefore, the manner in which the friction and wear properties correlate with the solute-to-iron atomic radius ratio or atomic size misfit is of interest. Alloy hardening at high temperatures (300 and 411 K) and alloy softening at lower temperatures (77 and 188 K) have been observed in several iron-base binary alloys [5.30–5.34]. These investigations concluded that for many alloy systems both alloy softening and alloy hardening were controlled by atomic

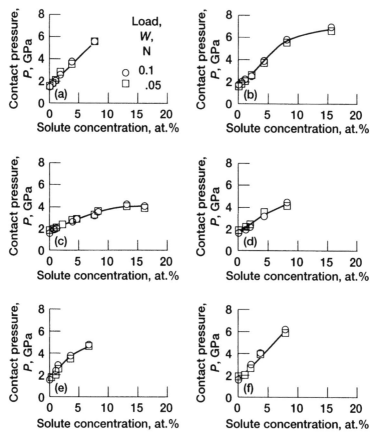

Figure 5.17.—Contact pressure (calculated) as function of solute concentration for various iron-base alloys and pure iron. Same conditions as for Fig. 5.15. (a) Iron-manganese alloy. (b) Iron-nickel alloy. (c) Iron-chromium alloy. (d) Iron-rhodium alloy. (e) Iron-tungsten alloy. (f) Iron-titanium alloy.

size misfit or solute-to-iron atomic radius ratio. The atomic size misfit parameter was concluded to be a reasonably good indicator that α-iron was strengthened by adding a low concentration of substitutional solutes [5.30].

Figure 5.18 presents the decreasing rates of coefficient of friction $-d\mu/dC$ and groove height $-dH/dC$ and the increasing rate of contact pressure dP/dC with increasing solute concentration C as a function of solute-to-iron atomic radius ratio K at loads of 0.1 and 0.05 N. The rates were estimated from the data in Figs. 5.15 to 5.17. That is, the rate is the difference between the maxima and minima of coefficient of friction, groove height, or contact pressure divided by the maximum

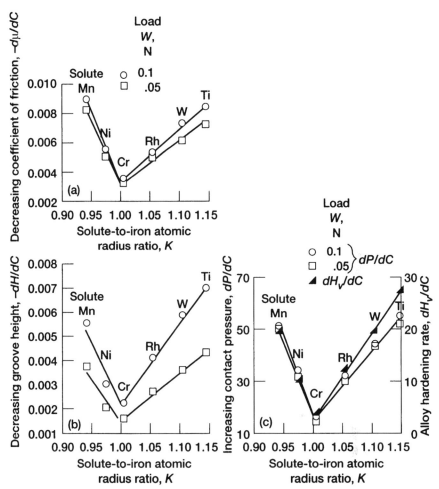

Figure 5.18.—Rates of change of coefficient of friction, groove height, and contact pressure as function of solute-to-iron atomic radius ratio.

solute concentration in each alloy. Agreement is good between the friction and deformation (wear) properties and the solute-to-iron atomic radius ratio. The correlation for each rate is separated into two cases: first, the case for alloying with manganese and nickel, which have smaller atomic radii than iron; and second, the case for chromium, rhodium, tungsten, and titanium, which have larger atomic radii than iron. The $-d\mu/dC$, $-dH/dC$, and dP/dC increase as the solute-to-iron atomic radius ratio increases or decreases from unity. The rates of change are a minimum at unity. Thus, the correlations indicate that the atomic size of the solute, as well as alloy hardening (Fig. 5.18(c), [5.34]), is an important parameter in controlling abrasive wear and friction in iron-base, binary alloys.

Deformation effects of mechanical polishing.—The effect of mechanical polishing on abrasive wear was examined. Spherical, 25-μm-radius SiC pins were slid on mechanically polished, iron-titanium binary alloy disk surfaces in both the annealed and unannealed (polished) states in mineral oil at loads of 0.05, 0.1, and 0.2 N [5.35]. The coefficient of friction and the groove height generally decreased and the contact pressure (corresponding to the microhardness) increased as the titanium concentration increased (Fig. 5.19). These changes were larger for the annealed surfaces than for the unannealed surfaces, but the trends for both surfaces were similar.

Crystallographic orientation (anisotropy) effects.—Metals and ceramics exhibit anisotropic behavior in many of their mechanical, physical, and chemical properties. Under abrasion conditions (just as under adhesive conditions described in Chapter 3), the differences in the coefficients of friction with respect to the crystallographic plane and direction are also significant, as discussed for single-crystal SiC in Section 2.2.1 (Fig. 2.10). The friction, deformation, fracture, and wear behaviors of single-crystal ceramics are anisotropic. Both oxide and nonoxide ceramics exhibit anisotropic abrasion and friction behavior.

Figure 5.20 presents the coefficient of friction as a function of the crystallographic sliding direction on the {100}, {110}, {111}, and {211} planes of Mn-Zn ferrite in sliding contact with a 20-μm-radius, spherical diamond pin at a 1-N load in laboratory air. As a reference, the Knoop hardness is also presented as a function of the crystallographic direction on the four planes of Mn-Zn ferrite. The sliding caused primarily plastic flow as well as surface cracking in the ferrite. The coefficient of friction was influenced by the crystallographic orientation. The anisotropic friction and plastic deformation of Mn-Zn ferrite may be controlled by the slip systems {110}⟨110⟩. In general, the softer the crystallographic direction, the greater the coefficient of friction.

5.3.2 Cutting by Wedge Pin

When a diamond wedge pin (Fig. 5.4(b)) slid on a metal flat surface lubricated with a drop of perfluoropolyether oil at a load of 0.2 N, the coefficient of friction, the volume of metal displaced in unit load and unit distance (wear rate), and the metal's wear resistance (the inverse of wear rate) correlated with the Vickers hardness of the metal (Fig. 5.21). Perfluoropolyether oil was used to minimize adhesion. The wedge's longitudinal, 50-μm-radius, rounded edge was oriented parallel to the sliding direction. Therefore, the sharp edges of the wedge's rake face played a major role in cutting the metal. The principal mechanism of metal removal was cutting action and subsequent chip formation in front of the wedge's rake face (Fig. 5.6). The coefficient of friction and the volume of metal displaced linearly decreased as the Vickers hardness increased at slopes of –0.77 and –1.61, respectively,

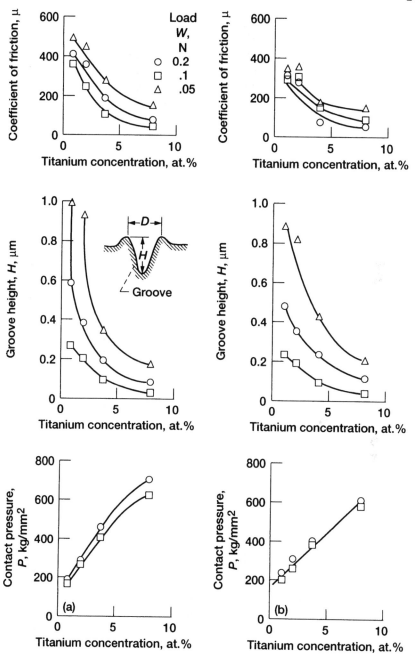

Figure 5.19.—Coefficient of friction, groove height, and contact pressure (calculated) as function of titanium concentration for annealed and mechanically polished Fe-Ti alloys. Same conditions as for Fig. 5.15. (a) Annealed surface. (b) Polished surface.

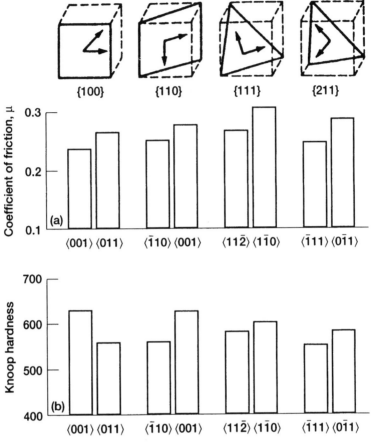

Figure 5.20.—Anisotropies on {100}, {110}, {111}, and {211} surfaces of Mn-Zn ferrite in laboratory air at room temperature. (a) Coefficient of friction; single-pass sliding of diamond pin (20-μm radius); sliding velocity, 3 mm/min; load, 1 N. (b) Knoop hardness; load, 3 N.

on logarithmic coordinates. The metal's wear resistance linearly increased as the Vickers hardness increased at a slope of 1.61 on logarithmic coordinates.

Figure 5.22 presents comparative data using a 50-μm-radius, spherical diamond pin (Fig. 5.4(a)). Again, all the variables correlated with the Vickers hardness. The coefficient of friction and the volume of metal displaced linearly decreased as the Vickers hardness increased at slopes of −0.35 and −1.33, respectively, on logarithmic coordinates. The metal's wear resistance linearly increased as the Vickers hardness increased at a slope of 1.33 on logarithmic coordinates.

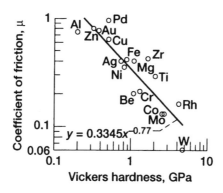

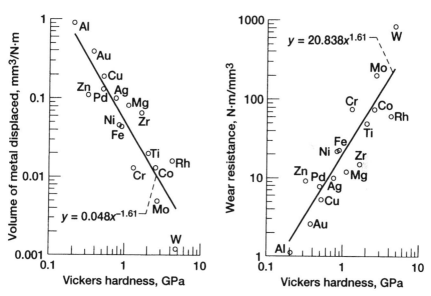

Figure 5.21.—Coefficient of friction, volume of metal displaced in unit load and unit distance, and wear resistance as function of Vickers hardness for various metals as a result of single-pass sliding of diamond wedge pin. Load, 0.2 N.

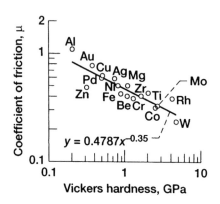

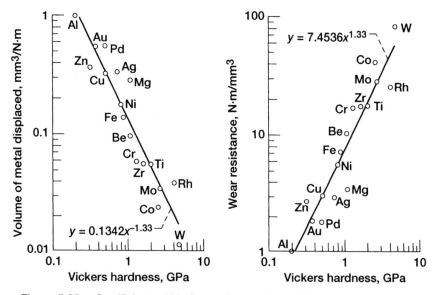

Figure 5.22.—Coefficient of friction, volume of metal displaced in unit load and unit distance, and wear resistance as function of Vickers hardness for various metals as a result of single-pass sliding of 50-μm-radius, spherical diamond pin. Load, 1 N.

5.4 Two- and Three-Body Abrasion by Hard Particles

When a number of grits of hard abrasive material (e.g., SiC and Al_2O_3) embedded in a resin matrix (binder) come into contact with a softer material (the two-body condition), the abrasive grits begin to cut or skive grooves in the softer material surfaces [5.36]. Abrasive wear can also occur when a third particle harder than one or both of the surfaces in contact becomes trapped at the interface. It can then remove material from one or both surfaces. This mode of wear is called three-body abrasion [5.36].

Many investigators have found that the abrasive wear of metals increases as the grit size increases up to a critical value and, thereafter, is independent of grit size [5.36]. The critical grit diameter is in the range 30 to 150 μm but strongly depends on abrasives, materials, and conditions, such as load, velocity, and environment. The grit size effect has also been found to hold for ceramic materials [5.37, 5.38]. For example, as shown in Fig. 5.23 the specific wear rate of polycrystalline Mn-Zn ferrite strongly depended on grit size in both two- and three-body abrasion, decreasing rapidly as grit size decreased. Detailed information on these results is provided in the following sections.

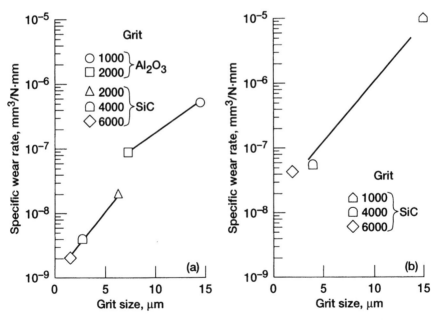

Figure 5.23.—Specific wear rate as function of grit size for polycrystalline Mn-Zn ferrite. (a) Two-body conditions. Abrasive-impregnated tapes; sliding velocity, 0.52 m/s; laboratory air; room temperature. (b) Three-body conditions. Lapping disk, cast iron; lapping fluid, olive oil; sliding velocity, 0.5 m/s; abrasive/fluid ratio, 27 wt.%; apparent contact pressure, 4 N/cm^2.

5.4.1 Two-Body Abrasion

The two-body abrasion study (Fig. 5.24(a), [5.37]) used abrasive-impregnated tapes, commonly called lapping tapes. The lapping tapes consisted of SiC or Al_2O_3 powder of various grit sizes in a polymeric binder on polyester films. The abrasive layer resembled emery paper, a familiar abrasive. The lapping tape sliding on an Mn-Zn ferrite generated abrasion (as shown in Chapter 2, Figs. 2.11 and 2.13). The ferrite wear surface revealed many plastically deformed grooves in the sliding direction formed primarily by the plowing action of abrasive particles in the lapping tape. The groove width increased as the grit size increased. Again, the wear rates of the polycrystalline Mn-Zn ferrite shown in Fig. 5.23(a) strongly depended on the kind of abrasive grit and its size. The wear rate for grits of the same kind increased rapidly as grit size increased. However, the coefficient of friction depended only on the kind of abrasive grit and was independent of grit size. The coefficient of friction was approximately 0.4 with Al_2O_3-grit lapping tapes but only 0.2 with SiC-grit lapping tapes.

Another interesting point to be observed from Fig. 5.23(a) is that, in spite of the nearly same grit size of 7.1 μm for Al_2O_3 and 6.3 μm for SiC, Al_2O_3 was approximately five times more abrasive than SiC. This difference may be related to the property, shape, and size distribution of abrasive particles on the tape as well as the degree to which the particles were enclosed by the binder.

Figure 5.24 presents the deformed-layer thickness for the Mn-Zn ferrite {110} plane sliding against lapping tapes as a function of grit size (see also Section 2.2.2). The deformed-layer thickness, as determined with reflection electron diffraction by depth profiling, increased as the grit size increased [5.39]. With 2.7-μm-SiC-grit lapping tape the deformed layer was 0.5 to 0.6 μm thick; with 14-μm-Al_2O_3-grit lapping tape the deformed layer was 0.8 to 0.9 μm thick. The deformed-layer

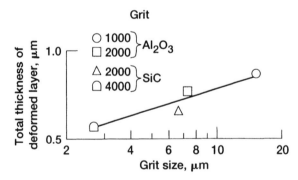

Figure 5.24.—Total thickness of deformed layer as function of grit size for Mn-Zn ferrite {110} plane in single-pass sliding contact with various lapping tapes. Initial tape tension, 0.3 N; head displacement, 120 ± 5 μm.

thickness, as determined with chemical-etching-rate depth profiling, also increased as the grit size increased.

Figure 5.25 presents the wear volume for four crystallographic planes of single-crystal Mn-Zn ferrite, {100}, {110}, {111}, and {211}, sliding against a grit-impregnated lapping tape as a function of the Vickers hardness of the wear surfaces [5.39]. Wear volume was influenced by crystallographic orientation, {110} < {100} < {111} < {211}, and by Vickers hardness. The slip (most closely packed) planes of Mn-Zn ferrite {110} exhibited the highest resistance to the two-body abrasion. Wear volume was inversely proportional to the Vickers hardness of the abraded ferrite wear surface. It, therefore, appears that the relationship of wear volume to Vickers hardness is similar for ceramics and metals.

5.4.2 Three-Body Abrasion

The three-body abrasion study (Fig. 5.23(b)) used a polishing machine capable of measuring friction during three-body abrasion. The apparatus (Fig. 5.26) contains a 160-mm-diameter, gray cast iron disk (lapping plate), a gray cast iron

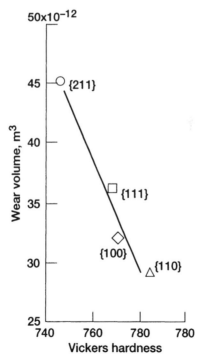

Figure 5.25.—Anisotropy of wear as function of Vickers hardness for {100}, {110}, {111}, and {211} planes of Mn-Zn ferrite. Sliding direction, ⟨110⟩; lapping tape, 2000-grit Al_2O_3; sliding velocity, 11 m/s; laboratory air; room temperature; measuring load, 0.25 N.

conditioning ring (72-mm outer diameter; 44-mm inner diameter), and an arm. The abrasive grits used (1000-, 4000-, and 6000-mesh SiC) have a cubelike shape with many sharp edges (Fig. 5.27). The grit shape does not depend on the grit size.

Figure 5.28 presents the grit size distribution of the SiC abrasives used in the study (i.e., the relative amounts of the various grits sizes in the powders). Because the grits were irregularly shaped, two dimensions (the largest diameter and the smallest diameter of each particle) were measured from the transmission electron micrographs. The bell-shaped curves reveal a generally strong correlation with the normal distribution. Therefore, they are expressed in terms of the normal density function. The average grit diameters and standard deviations are listed in Table 5.3. The 1000-, 4000-, and 6000-mesh SiC grits had average diameters of 15, 4, and 2 μm (with narrow tolerances), respectively. The crystalline shape of the SiC grits, particularly the shape along their largest diameter, makes them very sharp. Figure 5.29 presents the angle distribution of cutting edges of the abrasive grits along their largest diameter. The angle distributions were nearly the same for the 1000-, 4000-, and 6000-mesh SiC grits. Many edges had angles ranging from 90° to 100°.

The SiC grits abraded the single-crystal Mn-Zn ferrite surfaces in sliding contact under the three-body condition (as shown in Chapter 2, Fig. 2.12). The Mn-Zn ferrite wear was linearly proportional to the sliding distance, regardless of grit size,

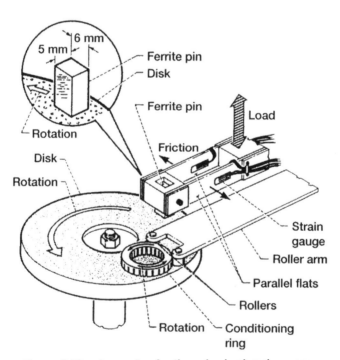

Figure 5.26.—Apparatus for three-body abrasive wear.

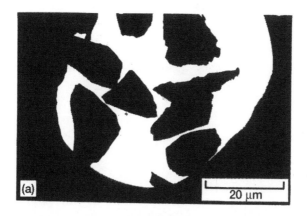

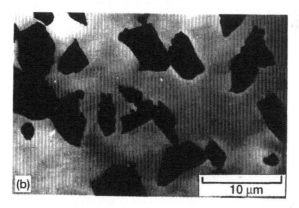

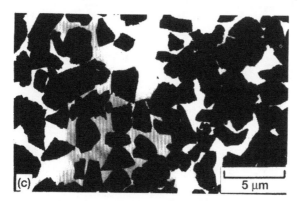

Figure 5.27.—Transmission electron micrographs
of SiC abrasive grits. (a) 1000-mesh SiC.
(b) 4000-mesh SiC. (c) 6000-mesh SiC.

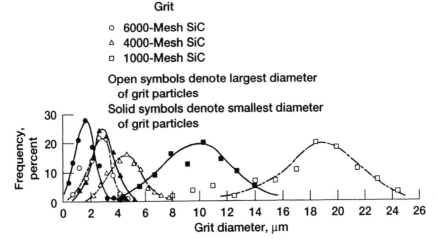

Figure 5.28.—Grit size distribution of 1000-, 4000-, and 6000-mesh SiC abrasives.

TABLE 5.3.—ABRASIVE GRIT SIZE IN THREE-BODY ABRASION STUDY

	1000-Mesh SiC	4000-Mesh SiC	6000-Mesh SiC
Average of smallest diameters of grit particles, μm	11.2	3.0	1.8
Average of largest diameters of grit particles, μm	17.9	4.5	2.7
Average grit diameter, μm	14.6	3.8	2.3
Ratio of largest to smallest grit diameter	1.7	1.6	1.7
Standard deviation of smallest grit diameters, μm	0.4	0.4	0.4
Standard deviation of largest grit diameters, μm	0.6	0.7	0.5
Standard deviation of average grit diameters, μm	0.4	0.5	0.3
Standard deviation of ratio of largest to smallest grit diameters	0.1	0.1	0.1

and was highly reproducible. However, the abrasion mechanism changed drastically with grit size. With 15-μm (1000-mesh) SiC grits, ferrite abrasion was caused principally by brittle fracture (Fig. 2.12(a)); but with the 4- and 2-μm (4000- and 6000-mesh) SiC grits, abrasion was caused by plastic deformation and fracture (Fig. 2.12(b)). Considerable microcracking (with 15-μm SiC grits) or plastic flow and fracture (with 4- and 2-μm SiC grits) drastically changed the crystalline state of the single-crystal ferrite surfaces to polycrystalline.

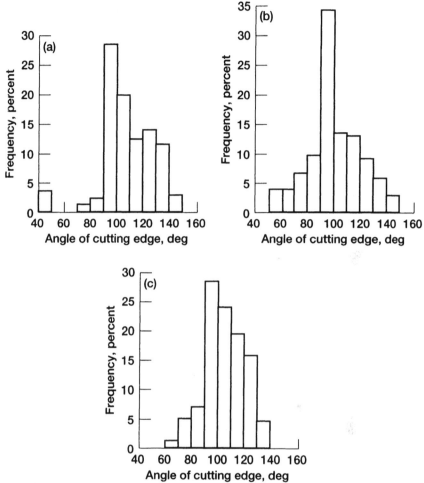

Figure 5.29.—Angle distribution of cutting edges of SiC abrasives.
(a) 1000-mesh SiC. (b) 4000-mesh SiC. (c) 6000-mesh SiC.

The specific wear rate, wear surface roughness, and total deformed-layer thickness strongly depended on grit size. The specific wear rate (Fig. 5.23(b)) increased rapidly as grit size increased. The wear surface roughness (R_{max}, maximum height of irregularities) for single-crystal {100} Mn-Zn ferrites abraded in the $\langle 110 \rangle$ direction increased as grit size increased (not shown). It was 10 to 15 times greater ($R_{max} = 2.2$ µm) when abraded by 15-µm SiC grits than when abraded by 4- and 2-µm SiC grits ($R_{max} = 0.21$ and 0.17 µm, respectively). Figure 5.30 presents schematic structures for the deformed layers of the abraded single-crystal {100} Mn-Zn ferrites as a function of etching depth (depth from the wear surface). The total deformed-layer thickness, as determined with reflection electron diffraction

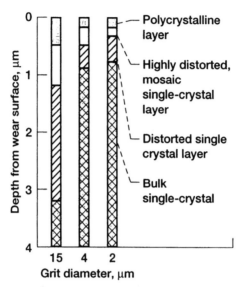

Figure 5.30.—Deformed layer produced on single-crystal {100} surface of Mn-Zn ferrite as function of grit size under three-body abrasion. Apparent contact pressure, 4 N/cm^2; sliding direction, $\langle 0\bar{1}1 \rangle$.

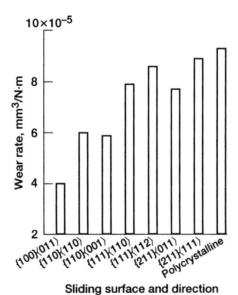

Figure 5.31.—Anisotropies on {100}, {110}, {111}, and {211} surfaces of Mn-Zn ferrite for 15-μm SiC grits under three-body abrasion.

by depth profiling, increased markedly as grit size increased. The deformed layers were 3 μm for the ferrite abraded by 15-μm SiC, 0.9 μm for the ferrite abraded by 4-μm SiC, and 0.8 μm for the ferrite abraded by 2-μm SiC.

Figure 5.31 presents specific wear rates under three-body conditions for four crystallographic planes of single-crystal Mn-Zn ferrite, {100}, {110}, {111}, and {211}, abraded with 15-μm SiC grits in different directions and for a polycrystalline Mn-Zn ferrite also abraded with 15-μm SiC grits. The wear rates of the single-crystal ferrite surfaces varied with crystallographic plane and direction. The polycrystalline ferrite exhibited the weakest wear resistance to abrasion because of grain boundary weakness.

References

5.1 *Glossary of Terms and Definitions in the Field of Friction, Wear, and Lubrication*, Research Group on Wear of Engineering Materials, Organization for Economic Cooperation and Development (OECD), Paris, 1969.

5.2 P.A. Engel, *Impact Wear of Materials*, Vol. 2, Elsevier Scientific Publishing Co., Amsterdam, 1978.

5.3 H. Czichos, *Tribology—A Systems Approach to the Science and Technology of Friction, Lubrication, and Wear*, Vol. 1, Elsevier Scientific Publishing Co., New York, 1978.

5.4 T.S. Eyre, Wear characteristics of metals, *Trib. Int. 9, 5*: 203–212 (1976).

5.5 M.M. Krushchov and M.A. Babichev, Resistance to abrasive wear and elasticity modulus of metals and alloys, *Soviet Physics-Doklady 5*: 410–412 (1960–61).

5.6 B.W.E. Avient, J. Goddard, and H. Wilman, An experimental study of friction and wear during abrasion of metals, *Proc. R. Soc. London A 258, 1293*: 159–180 (1960).

5.7 E. Rabinowicz, L.A. Dunn, and P.G. Russell, A study of abrasive wear under three-body conditions, *Wear 4*: 345–355 (1961).

5.8 S.I. Hayashi, Hydrogen introduction by abrasion in aluminum, *Japanese J. Appl. Phys. 35, 12A*: 6191–6199 (1996).

5.9 R.A.H. Edwards and W. Eichenauer, Reversible hydrogen trapping at grain-boundaries in superpure aluminum, *Scr. Metall. 14, 9*: 971–973 (1980).

5.10 K. Miyoshi and D.H. Buckley, Occurrence of spherical ceramic debris in indentation and sliding contact, NASA TP–2048 (1982).

5.11 K. Miyoshi and D.H. Buckley, The friction and wear of metals and binary alloys in contact with an abrasive grit of single-crystal silicon carbide, *ASLE Trans. 23, 4*: 460–472 (1980).

5.12 R.P. Steijn, On the wear of sapphire, *J. Appl. Phys. 32, 10*: 1951–1958 (1961).

5.13 R.P. Steijn, Sliding and wear in ionic crystals, *J. Appl. Phys. 34, 2*: 419–428 (1963).

5.14 R.P. Steijn, Friction and wear of single crystals, *Wear 7*: 48–66 (1964).

5.15 R.P. Steijn, Friction and wear of rutile single crystals, *ASLE Trans. 12, 1*: 21–33 (1969).

5.16 K.F. DuFrane and W.A. Glaeser, Study of rolling-contact phenomena in magnesium oxide, NASA CR–72295 (1967).

5.17 K. Miyoshi and D.H. Buckley, Friction, deformation, and fracture of single-crystal silicon carbide, *ASLE Trans. 22, 1*: 79–90 (1979).

5.18 K. Miyoshi, D.H. Buckley, S.A. Alterovitz, J.J. Pouch, and D.C. Liu, Adhesion, friction, and deformation of ion-beam-deposited boron-nitride films, *International Conference on Tribology—Friction, Lubrication, and Wear: Fifty Years On*, Vol. 11, Mechanical Engineering Publications, London, 1987, pp. 621–628.

5.19 K. Miyoshi, D.H. Buckley, J.J. Pouch, S.A. Alterovitz, and H.E. Sliney, Mechanical strength and tribological behavior of ion-beam-deposited BN films on nonmetallic substrates, *Surf. Coat. Technol. 33*: 221–233 (1987). (Also NASA TM–89818.)

5.20 F.P. Bowden and D. Tabor, *The Friction and Lubrication of Solids*, Part II, Clarendon Press, Oxford, UK, 1964, pp. 158–185.

5.21 K. Miyoshi and D.H. Buckley, Anisotropic tribological properties of SiC, *Wear 75, 2*: 253–268 (1982).

5.22 D.H. Buckley and K. Miyoshi, Friction and wear of ceramics, *Wear 100, 1–3*: 333–353 (1984).

5.23 K. Miyoshi, D.H. Buckley, and B. Bhushan, Friction and morphology of magnetic tapes in sliding contact with nickel-zinc ferrite, NASA TP–2267 (1984).

5.24 K. Miyoshi, Adhesion, friction, and micromechanical properties of ceramics, *Surf. Coat. Technol. 36, 12*: 487–501 (1988).

5.25 D. Tabor, *The Hardness of Metals*, Clarendon Press, Oxford, UK, 1951, pp. 6–18.

5.26 M.F. Stroud and H. Wilman, The proportion of the groove volume removed as wear in abrasion of metals, *Brit J. Appl. Phys. 13*: 173–178 (1962).

5.27 T.O. Mulhearn and L.E. Samuels, The abrasion of metals: a model of the process, *Wear 5*: 478–498 (1962).

5.28 J. Larsen-Basse, Influence of grit size on the groove formation during sliding abrasion, *Wear 11*: 213–222 (1968).

5.29 I.V. Kragelskii, *Friction and Wear*, Butterworths, Washington, 1965, pp. 20–26, 88–110.

5.30 W.C. Leslie, Iron and its dilute substitutional solid solutions, *Metall. Trans. 3*: 5–26 (1972).

5.31 W.A. Spitzig and W.C. Leslie, Solid-solution softening and thermally activated flow in alloys of Fe with three atomic % Co, Ni, or Si, *Acta Metall. 19, 11*: 1143–1152 (1971).

5.32 E. Pink, Legierungsentfestigung in kubisch-raumzentrierten Metallen (Alloy softening in body-centered cubic metals), *Zeitschrift fuer Metallkunde 64, 12*: 871–881 (1973).

5.33 D. Leemans and M.E. Fine, Solid solution softening and strain-rate sensitivity of Fe-Re and Fe-Mo alloys, *Metall. Trans. 5, 6*: 1331–1336 (1974).

5.34 J.R. Stephens and W.R. Witzke, Alloy softening in binary iron solid solutions, *J. Less-Common Metals 48, 2*: 285–308 (1976).

5.35 K. Miyoshi and D.H. Buckley, Friction and wear with a single-crystal abrasive grit of silicon carbide in contact with iron-base binary alloys in oil—effects of alloying element and its content, NASA TP–1394 (1979).

5.36 E. Rabinowicz, *Friction and Wear of Materials*, John Wiley & Sons, New York, 1964.

5.37 K. Miyoshi, Lapping of manganese-zinc ferrite by abrasive tape, *Lubr. Eng., 38, 3*: 165–172 (1982).

5.38 K. Miyoshi, Effect of abrasive grit size on wear of manganese-zinc ferrite under three-body abrasion, *Tribology and Mechanics of Magnetic Storage Systems* (B. Bhushan and N.S. Eiss, Jr., eds.), STLE SP–22, Vol. IV, 1987, pp. 123–132.

5.39 K. Miyoshi, D.H. Buckley, and K. Tanaka, Effect of crystallographical and geometrical changes of a ferrite head on magnetic signals during the sliding process with magnetic tape, *Tribology and Mechanics of Magnetic Storage Systems*, ASLE SP–21, Vol III, 1986, pp. 42–49.

5.40 K.A. Gschneider, Jr., Physical properties and interrelationships of metallic and semimetallic elements, *Solid State Physics*, Chapt. 16, F. Seitz and D. Turnbull, eds., Academic Press, 1965, pp. 275–426.

5.41 C.S. Barrett, *Structure of Metals, Crystallographic Methods, Principles, and Data*, McGraw-Hill, New York, 1943, pp. 552–554.

Chapter 6
Friction and Wear Properties of Selected Solid Lubricating Films: Case Studies

6.1 Introduction

Once the initial shortcomings relating to friction, wear, and lubrication in design and application had been dealt with, it became increasingly clear that materials science and technology ranked equal with design in reducing the friction and wear of machinery and mechanical components [6.1]. This conclusion applied particularly in the field of solid lubrication of aerospace mechanisms.

In modern technology the coefficient of friction and the wear rate are regarded as widely variable, depending on lubricants, operational variables, substrate properties, and surface films. Therefore, testing is central and of great importance to tribologists, lubrication specialists, designers, and engineers. Particularly, standard performance testing of solid lubricant systems is important for the lubricant manufacturer during development of new lubricating materials, for quality control of lubricant manufacture, and for providing quantitative criteria for manufacturing specifications.

Compiling manufacturers' standard test results for a number of lubricant formulations can aid mechanism design engineers (i.e., users) in selecting the best lubricant for an application. However, such data can at best only narrow the field to a specific class of lubricants. Deciding on the optimum lubricant formulation for a specific application requires more custom-design, element, component, and full-scale testing. However, after the optimum lubricant has been chosen, such standard tests can also be useful for quality control. This is especially important in a long space program where satellites or launch vehicles are built over a period of years during which lubricant formulations and film application procedures might undergo change. For solid lubricant films the end user should request that standard test coupons be coated along with the actual parts. Testing each lubricant batch will ensure that the manufacturing quality remains constant throughout the life of the program.

The technology of solid lubrication has advanced rapidly in the past four decades, responding primarily to the needs of the automobile and aerospace industries. Solid lubrication can be accomplished in several modes: lubricating powders, bonded films, lubricant composites (metal and plastic based), and lubricating coatings and films. This chapter, however, primarily describes bonded films and lubricating coatings and films.

6.2 Commercially Developed Dry Solid Film Lubricants

This section is limited to discussing the tribological properties, particularly friction and wear, of the solid lubricating films selected from commercially developed, affordable dry solid film lubricants (Table 6.1):

1. Bonded molybdenum disulfide (MoS_2), the most widely used mode
2. Magnetron-sputtered MoS_2
3. Ion-plated silver
4. Ion-plated lead
5. Magnetron-sputtered diamondlike carbon (MS DLC)
6. Plasma-assisted, chemical-vapor-deposited diamondlike carbon (PACVD DLC)

The friction and wear properties of the selected solid lubricating films were examined in ultrahigh vacuum, in humid air at a relative humidity of ~20%, and in dry nitrogen at a relative humidity of <1%. Unidirectional pin-on-disk sliding

TABLE 6.1.—CHARACTERISTICS OF SELECTED SOLID
LUBRICATING FILMS
[Substrate material, 440C stainless steel.]

Film	Film material	Film thickness, μm	Film surface roughness,[a] R_a, nm	
			Mean	Standard deviation
Bonded MoS_2	MoS_2, polyimide-imide, others (proprietary blend)	10±4	1.2×10^3	2.4×10^2
MS MoS_2	MoS_2	1.0±0.2	32	4.0
Ion-plated silver	Silver	0.5±0.2	30	3.2
Ion-plated lead	Lead	0.55	98	15
MS DLC	Carbon and WC	2 to 3	43	5.1
PACVD DLC	Carbon and Si	3 to 5	29	3.2

[a]The centerline-average roughness R_a was measured by using a cutoff of 1 mm.

friction experiments were conducted with AISI 440C stainless steel balls in sliding contact with the solid lubricating films at room temperature in the three environments. The resultant solid lubricating films and their wear surfaces were characterized by scanning electron microscopy (SEM), energy-dispersive x-ray spectroscopy (EDX), and surface profilometry. SEM and EDX were used to determine the morphology and elemental composition of wear surfaces and wear debris. The sampling depth of EDX for elemental information ranged between 0.5 and 1 μm. Surface profilometry was used to determine the surface morphology, roughness, and wear of the films.

6.2.1 Selected Solid Lubricating Films

Molybdenum disulfide films.—The technical use of molybdenum disulfide as a lubricating solid started in the 1940's. It is now used in more applications than any other lubricating solid. MoS_2 differs from graphite mainly in that its low friction is an inherent property and does not depend on the presence of adsorbed vapors [6.2]. Therefore, it can be used satisfactorily in both high vacuum and dry environmental conditions and has been used in many spacecraft applications. For example, the extendible legs on the Apollo lunar module were lubricated with bonded MoS_2. It is usable at low temperatures and to 350 °C in air or 1000 °C in high vacuum with high load-carrying capacity. At 350 °C in air MoS_2 begins to oxidize. Apart from oxidation it is stable with most chemicals but is attacked by strong oxidizing acids and by alkalis. MoS_2 is a poor conductor of heat and electricity and stable in vacuum.

A useful MoS_2 film can be obtained by

1. Simply rubbing or burnishing MoS_2 powder onto a substrate material
2. Dipping, brushing, or spraying with a dispersion of MoS_2 in a volatile solvent or water and allowing the liquid to evaporate
3. Using a binder material (resin, silicate, phosphate, or ceramic)
4. Vacuum sputtering

Bonded films are much thicker than sputtered films and much more readily available. Sputtered films are easier to incorporate in precision bearing systems. On the whole, MoS_2 is a versatile and useful solid lubricant [6.2–6.4].

The bonded MoS_2 films studied were relatively rough with centerline-average roughness R_a of 1.2 μm, and the magnetron-sputtered MoS_2 films studied were relatively smooth with R_a in the range of 32 nm [6.5]. The bonded films were 10 times thicker than the sputtered films.

Silver and lead films.—Soft metals like silver and lead can be sheared easily and have a number of other properties that make them attractive as solid lubricants for certain circumstances or special situations. For example, in addition to their low shear strengths and ability to be applied as continuous films over harder substrates, soft metals are good conductors of heat and electricity and are stable in vacuum or when exposed to nuclear radiation [6.6].

Soft metal films can be deposited as lubricating films on harder substrates by conventional electroplating or by physical vapor deposition methods such as evaporation, vacuum sputtering, and ion plating. Ion plating and vacuum sputtering permit close control of film deposition and thickness and can provide good adhesion to the substrate [6.7]. Soft metal films are usually about 0.25 to 1.0 μm thick and must be very smooth and uniform. Friction increases with either thicker or thinner films [6.8].

Thin-metal-film lubrication is most relevant at high temperatures or in applications where sliding is limited (e.g., rolling-element bearings). Silver and barium films have been used successfully on lightly loaded ball bearings in high-vacuum x-ray tubes. Silver and gold films have been tested successfully for high-vacuum use in spacecraft applications. Perhaps the most successful applications so far have used thin lead films as long-term, rolling-element (ball) bearing lubrication in spacecraft [6.2, 6.9]. One advantage of lead films over silver or indium is the unavoidable presence of lead oxide (PbO), a reputedly good solid lubricant, within the films [6.1, 6.10]. Ion-plated lead films are used in such mechanisms as solar array drives in European satellites [6.11]. Lead films have been used for many years to lubricate rolling-element bearings on rotating anode x-ray tubes, on satellite parts operating in space, and on other equipment exposed to high temperatures and nuclear radiation. Gold-, silver-, and lead-coated, rolling-element bearings are commercially available at reasonable prices.

The ion-plated silver and ion-plated lead films studied [6.5, 6.12] were relatively smooth with R_a of 30 and 98 nm, respectively, and ~0.5 μm thick and uniform.

Diamondlike carbon films.—A new category of solid lubricants and lubricating films, diamond and related hard materials, is continuing to grow. Particularly, mechanical parts and components coated with diamondlike carbon (DLC) are of continuously expanding commercial interest.

DLC can be divided into two closely related categories: amorphous, non-hydrogenated DLC (a-DLC or a-C) and amorphous, hydrogenated DLC (H–DLC or a-C:H) [6.8]. H–DLC contains a variable and appreciable amount of hydrogen. DLC can be considered a metastable carbon produced as a thin film with a broad range of structures (primarily amorphous with variable sp^2/sp^3 bonding ratio) and compositions (variable hydrogen concentration). A DLC's properties can vary considerably as its structure and composition vary [6.13–6.17]. Although it is a complex engineering job, it is often possible to control and tailor the properties of a DLC to fit a specific application and thus ensure its success as a tribological product. However, such control demands a fundamental understanding of the tribological properties of DLC films. The absence of this understanding can act as a brake in applying DLC to a new product and in developing the product.

Commercial applications of DLC, as protecting, self-lubricating coatings and films, are already well established in a number of fast-growing industries [6.18–6.20] making

1. Magnetic recording media and high-density magnetic recording disks and sliders (heads)
2. Process equipment (e.g., digital video camcorders and copy machines)
3. Abrasion-resistant optical products, rubbers, and plastics
4. Implant components including hip and knee joints, blood pumps, and other medical products
5. Packaging materials and electronic devices
6. Forming dies (e.g., plastic molds and stamping devices)
7. Blades (e.g., razor blades and scalpels)
8. Engine parts (e.g., cam followers, pistons, gudgeon pins, and gear pumps)
9. Mechanical elements (e.g., washers, such as grease-free ceramic faucet valve seats; seals; valves; gears; bearings; bushing tools; and wear parts)

The cost is affordable and generally similar to that of carbide or nitride films deposited by chemical or physical vapor deposition (CVD or PVD) techniques [6.13]. The surface smoothness, high hardness, low coefficient of friction, low wear rate, and chemical inertness of DLC coatings and films, along with little restriction on geometry and size, make them well suited as solid lubricants combating wear and friction.

The magnetron-sputtered (MS) DLC films and plasma-assisted (PA) CVD DLC films studied were relatively smooth with R_a of 43 and 29 nm, respectively, and ~2 to 5 μm thick and uniform [6.21]. The MS DLC films had a multilayer structure and were prepared by using two chromium targets, six tungsten carbide (WC) targets, and methane (CH_4) gas. The multilayer film had alternating 20- to 50-nm-thick WC and carbon layers. The Vickers hardness number was ~1000. The PACVD DLC films were prepared by using radiofrequency plasma and consisted of two layers, an ~2-μm-thick DLC layer on an ~2-μm-thick silicon-DLC underlayer. The DLC top layer was deposited by using CH_4 gas at a total pressure of 8 Pa with a power of 1800 to 2000 W at –750 to –850 V for 120 min. The silicon-containing DLC underlayer was deposited by using a mixture of CH_4 and tetramethylsilane ($C_4H_{12}Si$) gases. The ratio of the concentration of CH_4 and $C_4H_{12}Si$ used was 90:18 (std cm^3/min) at a total pressure of 10 Pa with a power of 1800 to 2000 W at –850 to –880 V for 60 min. The Vickers hardness number was 1600 to 1800.

The 6-mm-diameter 440C stainless steel balls (grade 10) used were smooth, having R_a of 6.8 nm with a standard deviation of 1.8 nm.

6.2.2 Comparison of Steady-State Coefficients of Friction and Wear Rates

Table 6.2 and Figs. 6.1 to 6.3 summarize the steady-state coefficients of friction, the wear rates (dimensional wear coefficients) for the solid lubricating films, and the wear rates for the 440C stainless steel balls after sliding contact in all three environments: ultrahigh vacuum, humid air, and dry nitrogen. The data presented

TABLE 6.2.—STEADY-STATE COEFFICIENT OF FRICTION, WEAR LIFE, AND
WEAR RATE FOR SELECTED SOLID LUBRICATING FILMS IN SLIDING
CONTACT WITH 440C STAINLESS STEEL BALLS

Film	Environment	Steady-state coefficient of friction	Film wear (endurance) life[a]	Film wear rate, $mm^3/N \cdot m$	Ball wear rate, $mm^3/N \cdot m$
Bonded MoS_2	Vacuum	0.045	>1 million	6.0×10^{-8}	1.3×10^{-9}
	Air	0.14	113 570	2.4×10^{-6}	8.1×10^{-8}
	Nitrogen	0.04	>1 million	4.4×10^{-8}	6.9×10^{-10}
MS MoS_2	Vacuum	0.070	274 130	9.0×10^{-8}	2.5×10^{-9}
	Air	0.10	277 377	2.4×10^{-7}	1.5×10^{-7}
	Nitrogen	0.015	>1 million	1.6×10^{-8}	9.9×10^{-10}
Ion-plated silver	Vacuum	0.20	364 793	8.8×10^{-8}	2.4×10^{-8}
	Air	0.43	8	5.5×10^{-5}	1.2×10^{-5}
	Nitrogen	0.23	1040	1.6×10^{-5}	1.6×10^{-5}
Ion-plated lead	Vacuum	0.15	30 294	1.5×10^{-6}	7.6×10^{-7}
	Air	0.39	82	3.7×10^{-6}	3.6×10^{-7}
	Nitrogen	0.48	1530	9.1×10^{-6}	3.4×10^{-6}
MS DLC	Vacuum	0.70	<10	5.7×10^{-5}	3.2×10^{-4}
	Air	0.12	>1 million	1.7×10^{-7}	4.1×10^{-8}
	Nitrogen	0.12	23 965	4.2×10^{-7}	1.1×10^{-7}
PACVD DLC	Vacuum	0.54	<10	1.1×10^{-5}	1.8×10^{-4}
	Air	0.07	>1 million	1.0×10^{-7}	2.3×10^{-8}
	Nitrogen	0.06	>1 million	1.1×10^{-8}	6.4×10^{-9}

[a]Film wear life was determined to be the number of passes at which the coefficient of
friction rose to 0.30.

in the table reveal the marked differences in friction and wear resulting from the
environmental conditions and the solid lubricating film materials.

Ultrahigh vacuum.—In sliding contact with 440C stainless steel balls in ultrahigh vacuum, the bonded MoS_2 films had the lowest coefficient of friction, lowest
film wear rate, and lowest ball wear rate (Fig. 6.1). The MS MoS_2 films also had low
coefficient of friction, low film wear rate, and low ball wear rate, similar to those
for the bonded MoS_2. The wear rates of the solid lubricating films and the
counterpart steel balls in ultrahigh vacuum were, in ascending order, bonded MoS_2
< MS MoS_2 < ion-plated silver < ion-plated lead < PACVD DLC < MS DLC. The
coefficients of friction were in a similar ascending order. The MS DLC films had
the highest coefficient of friction, the highest film wear rate, and the highest ball
wear rate in ultrahigh vacuum.

Humid air.—In sliding contact with 440C stainless steel balls in humid air, the
PACVD DLC films had the lowest coefficient of friction, lowest film wear rate, and
lowest ball wear rate (Fig. 6.2). The bonded MoS_2, MS MoS_2, and MS DLC films
generally had low coefficients of friction, low film wear rates, and low ball wear
rates, similar to those for the PACVD DLC films. The wear rates of the solid

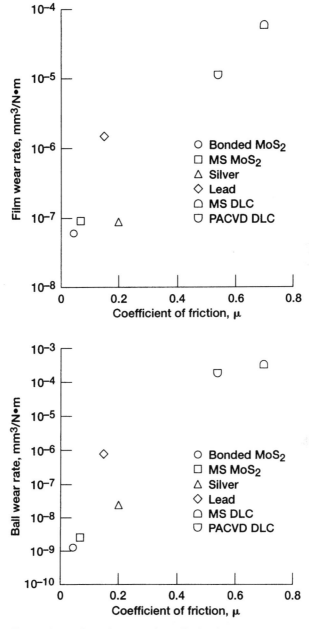

Figure 6.1.—Steady-state (equilibrium) coefficients of friction and wear rates (dimensional wear coefficients) for solid lubricating films in sliding contact with 440C stainless steel balls in ultrahigh vacuum.

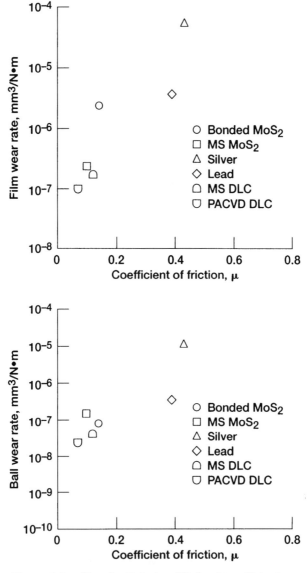

Figure 6.2.—Steady-state (equilibrium) coefficients of friction and wear rates (dimensional wear coefficients) for solid lubricating films in sliding contact with 440C stainless steel balls in humid air.

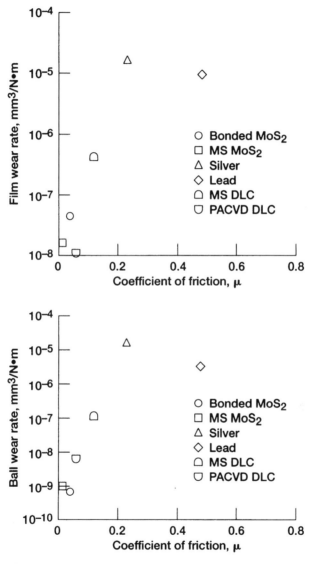

Figure 6.3.—Steady-state (equilibrium) coefficients of friction and wear rates (dimensional wear coefficients) for solid lubricating films in sliding contact with 440C stainless steel balls in dry nitrogen.

lubricating films in humid air were, in ascending order, PACVD DLC < MS DLC < MS MoS_2 < bonded MoS_2 < ion-plated lead < ion-plated silver. The coefficients of friction and the wear rates of the counterpart steel balls studied were in a similar ascending order. The ion-plated silver films had the highest coefficient of friction, highest film wear rate, and highest ball wear rate in humid air.

Dry nitrogen.—In sliding contact with 440C stainless steel balls in dry nitrogen, the MS MoS_2 films had the lowest coefficient of friction (Fig. 6.3). The bonded MoS_2, MS MoS_2, and PACVD DLC films had low coefficients of friction, low film wear rates, and low ball wear rates in dry nitrogen. However, the ion-plated silver and ion-plated lead films had high friction and high wear. The wear rates of the solid lubricating films in dry nitrogen were, in ascending order, PACVD DLC < MS MoS_2 < bonded MoS_2 < MS DLC < ion-plated lead < ion-plated silver. The coefficients of friction and wear rates of the counterpart steel balls studied were in a similar ascending order.

6.2.3 Wear Life (Endurance Life)

The sliding wear (endurance) life of the solid lubricating films deposited on 440C stainless steel disks was determined to be the number of passes at which the coefficient of friction rose to 0.30 in a given environment. As shown in Figs. 6.4 to 6.6 and Table 6.2, the sliding wear life varied with the environment and the type of solid lubricating film.

Ultrahigh vacuum.—In sliding contact with 440C stainless steel balls in ultrahigh vacuum, the bonded MoS_2 films had the longest wear life, over 1 million passes (Fig. 6.4). The wear lives of the solid lubricating films in ultrahigh vacuum were, in descending order, bonded MoS_2 > ion-plated silver > MS MoS_2 > ion-plated lead > PACVD DLC > MS DLC.

Humid air.—In sliding contact with 440C stainless steel balls in humid air, the PACVD DLC and MS DLC films had the longest wear lives, over 1 million passes (Fig. 6.5). The wear lives of the solid lubricating films in humid air were, in descending order, PACVD DLC > MS DLC > MS MoS_2 > bonded MoS_2 > ion-plated lead > ion-plated silver.

Dry nitrogen.—In sliding contact with 440C stainless steel balls in dry nitrogen, the bonded MoS_2, MS MoS_2, and PACVD DLC films had the longest wear lives, over 1 million passes (Fig. 6.6). The wear lives of the solid lubricating films in dry nitrogen were, in descending order, bonded MoS_2 > MS MoS_2 > PACVD DLC > MS DLC > ion-plated lead > ion-plated silver. The bonded MoS_2 films had wear lives of over 1 million passes in ultrahigh vacuum and dry nitrogen but only 1 13 570 passes in humid air. The MS MoS_2 films also had wear lives of over 1 million passes in dry nitrogen, much greater than in either ultrahigh vacuum or humid air. The wear lives of the ion-plated silver films were relatively greater in ultrahigh vacuum than in either dry nitrogen or humid air.

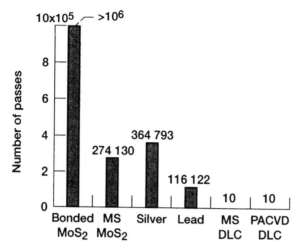

Figure 6.4.—Sliding wear lives for solid lubricating films in sliding contact with 440C stainless steel balls in ultrahigh vacuum.

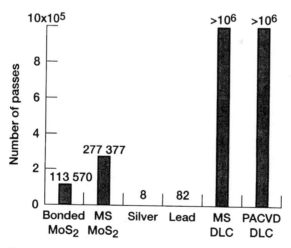

Figure 6.5.—Sliding wear lives for solid lubricating films in sliding contact with 440C stainless steel balls in humid air.

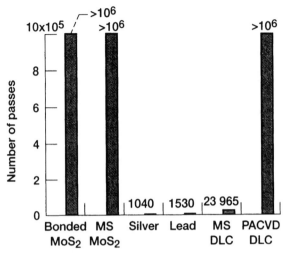

Figure 6.6.—Sliding wear lives for solid lubricating films in sliding contact with 440C stainless steel balls in dry nitrogen.

In ultrahigh vacuum the bonded MoS_2 films had longer wear lives than the MS MoS_2 and ion-plated silver films. In humid air the DLC films had much longer wear lives than the other solid lubricating films. In dry nitrogen both the bonded MoS_2 and MS MoS_2 films had longer wear lives than the ion-plated silver films.

6.2.4 Sliding Wear Behavior, Wear Debris, and Transferred Wear Fragments

Adhesion and plastic deformation played important roles in the friction and sliding wear of the selected solid lubricating films in contact with the 440C stainless steel balls in all three environments [6.5, 6.12, 6.21]. The worn surfaces of both films and balls contained wear debris particles. Examination of the surface morphology and compositions of the worn surfaces by SEM and EDX provided detailed information about the plastic deformation of the films, wear debris, and transferred wear fragments produced during sliding. Marked plastic deformation occurred in the six solid lubricating films. Smeared, agglomerated wear debris accumulated around the contact borders, particularly on the rear ends of ball wear scars. All sliding involved adhesive transfer of materials. SEM micrographs, EDX spectra, and detailed descriptions were reported in the references [6.5, 6.12, 6.21].

Bonded MoS_2.—The 440C stainless steel balls left transferred steel fragments in the wear tracks on the bonded MoS_2 films in all three environments [6.5]. During sliding the relatively coarse asperities of the bonded MoS_2 films were deformed plastically, and the tips of the asperities were flattened under load. The ball wear scars contained transferred MoS_2 fragments. Fragments of MoS_2 and steel usually adhered to the counterpart surface or came off in loose form. Another form of adhesive MoS_2 transfer was found in sliding wear. SEM and EDX showed that a thin MoS_2 layer (or sheet) was generated over the entire ball wear scars in all three environments.

Magnetron-sputtered MoS_2.—The 440C stainless steel balls left transferred steel fragments in the wear tracks on the MS MoS_2 films in all three environments [6.5]. The fine asperities of the sputtered MoS_2 films were flattened and elongated in the sliding direction by plastic deformation, revealing a burnished appearance. The ball wear scars contained transferred MoS_2 fragments and were entirely covered by a thin MoS_2 layer.

According to the elemental concentrations, in ultrahigh vacuum, humid air, and dry nitrogen much less transfer occurred between the films and the balls and vice versa with MS MoS_2 than with bonded MoS_2. A thin MoS_2 layer was generated over the entire ball wear scars in humid air and dry nitrogen.

Ion-plated silver.—The 440C stainless steel balls left a small amount of transferred steel fragments in the wear tracks on the ion-plated silver films in all three environments [6.5]. The fine asperities of the ion-plated silver films were flattened and elongated in the sliding direction by plastic deformation, revealing a burnished appearance. Severe plastic deformation and shearing occurred in the silver films during sliding.

According to the elemental concentrations, after sliding in ultrahigh vacuum the entire ball wear scar contained thick transferred layers (or sheets) of silver, and plate-like silver particles were deposited at the edges of the film wear track. In contrast, after sliding in humid air the ball wear scar contained an extremely small amount of transferred silver particles. This result suggests that oxidation of silver during sliding in humid air may prevent large silver transfer. However, plate-like silver debris was deposited at the edges of the film wear track. After sliding in dry nitrogen the ball wear scar contained transferred silver plates and particles. Plate-like silver debris was deposited at the edges of the film wear track. Severe plastic deformation and shearing occurred in the silver films during sliding in dry nitrogen.

Ion-plated lead.—The 440C stainless steel balls left a small amount of transferred steel fragments in the wear tracks on the ion-plated lead films in all three environments [6.12]. The fine asperities of the ion-plated lead films were flattened and elongated in the sliding direction by plastic deformation, revealing a burnished appearance. Severe plastic deformation and shearing occurred in the lead films during sliding.

According to the elemental concentrations, after sliding in ultrahigh vacuum the entire ball wear scar contained thick transferred layers (or sheets) of lead. Plate-like lead debris was found at the edges of the film wear track. In contrast, after sliding in humid air the ball wear scar contained an extremely small amount of transferred lead particles. This result suggests that oxidation of lead, like silver oxidation, during sliding in humid air may prevent large lead transfer. However, plate-like lead debris was deposited at the edges of the film wear track in humid air. After sliding in dry nitrogen the ball wear scar contained transferred lead plates and particles, and plate-like lead debris was deposited at the edges of the film wear track.

Magnetron-sputtered DLC.—With MS DLC films sliding involved generation of fine wear debris particles and agglomerated wear debris and transfer of the worn materials in all three environments [6.21].

According to the elemental concentrations, after sliding in ultrahigh vacuum the 440C stainless steel ball left a roughened worn surface and a small amount of transferred steel fragments in the wear track on the MS DLC film. The ball wear scar contained fine steel particles and a small amount of transferred DLC fragments. The wear mechanism was that of small fragments chipping off the DLC surface.

After sliding in humid air the 440C stainless steel ball left a small amount of transferred steel fragments in the wear track on the MS DLC film. The fine asperities of the MS DLC film were flattened and elongated in the sliding direction by plastic deformation, revealing a burnished appearance. The entire ball wear scar contained transferred patches and thick transferred layers (or sheets) of MS DLC. Plate-like DLC debris was also deposited at the edges of the wear scar. Severe plastic deformation and shearing occurred in the DLC films during sliding in humid air.

After sliding in dry nitrogen the 440C stainless steel ball left an extremely small amount of transferred steel debris in the wear track on the MS DLC film. In addition, smeared, agglomerated DLC debris was deposited on the film. The fine asperities of the MS DLC film were flattened and elongated in the sliding direction by plastic deformation, revealing a burnished appearance. The ball wear scar contained transferred DLC wear debris.

Plasma-assisted, chemical-vapor-deposited DLC.—With PACVD DLC films, like MS DLC films, sliding involved generation of fine wear debris particles and agglomerated wear debris and transfer of the worn materials in all three environments [6.21].

According to the elemental concentrations, after sliding in ultrahigh vacuum the 440C stainless steel ball left smeared, agglomerated DLC debris and a small amount of transferred steel fragments in the film wear track. The ball wear scar contained fine steel particles and large smeared, agglomerated patches containing transferred DLC fragments. The wear mechanism was adhesive, and plastic deformation played a role in the burnished appearance of the smeared, agglomerated wear debris.

After sliding in humid air the 440C stainless steel ball left a small amount of transferred steel fragments in the film wear track. The fine asperities of the PACVD DLC film were flattened and elongated in the sliding direction by plastic deformation, revealing a burnished appearance. The smooth ball wear scar contained an extremely small amount of transferred DLC debris.

After sliding in dry nitrogen the 440C stainless steel ball left DLC debris, micro-pits, and an extremely small amount of transferred steel debris in the wear track on the PACVD DLC film. The fine asperities of the film were flattened and elongated in the sliding direction by plastic deformation, revealing a burnished appearance. The ball wear scar contained fine grooves in the sliding direction, steel debris, and a small amount of transferred DLC debris.

6.2.5 Summary of Remarks

Recently developed, commercially available, dry solid film lubricants for solid lubrication applications were evaluated in unidirectional sliding friction experiments with bonded molybdenum disulfide (MoS$_2$) films, magnetron-sputtered (MS) MoS$_2$ films, ion-plated silver films, ion-plated lead films, MS diamondlike carbon (DLC) films, and plasma-assisted, chemical-vapor-deposited DLC films in contact with AISI 440C stainless steel balls in ultrahigh vacuum, in humid air, and in dry nitrogen. The main criteria for judging the performance of the dry solid lubricating films were coefficient of friction and wear rate, which had to be <0.3 and 10^{-6} mm^3/N·m, respectively. The following remarks can be made:

1. Bonded MoS$_2$ and MS MoS$_2$ films met both criteria in all three environments. Also, the wear rates of the counterpart 440C stainless steel balls met that criterion in the three environments.

2. In ultrahigh vacuum the coefficient of friction and endurance (wear) life of bonded MoS$_2$ films were superior to those of all the other dry solid film lubricants.

3. Ion-plated silver films met both criteria only in ultrahigh vacuum, failing in humid air and in dry nitrogen, where the film and ball wear rates were higher than the criterion.

4. Ion-plated lead films met both criteria only in ultrahigh vacuum, failing in humid air and in dry nitrogen, where the coefficients of friction were greater than the criterion. Both the lead film and ball wear rates met that criterion in all three environments.

5. MS DLC and PACVD DLC films met the criteria in humid air and in dry nitrogen, failing in ultrahigh vacuum, where the coefficients of friction were greater than the criterion.

6. Adhesion and plastic deformation played important roles in the friction and wear of all the solid lubricating films in contact with 440C stainless steel balls in the three environments. All sliding involved adhesive transfer of materials: transfer of solid lubricant wear debris to the counterpart 440C stainless steel, and transfer of 440C stainless steel wear debris to the counterpart solid lubricant.

6.3 Characteristics of Magnetron-Sputtered Molybdenum Disulfide Films

Solid lubricants must not only display low coefficients of friction but also maintain good durability and environmental stability [6.2–6.7]. The ability to allow rubbing surfaces to operate under load without scoring, seizing, welding, or other manifestations of material destruction in hostile environments is an important lubricant property. For solid lubricating films to be durable under sliding conditions, they must have low wear rates and high interfacial adhesion strength between the films and the substrates. The actual wear rates, wear modes, and interfacial adhesion strengths of solid lubricating films (e.g., MS MoS_2 films), however, are widely variable, depending on operating variables and substrate preparation. This section describes the coefficient of friction, wear rate, and endurance life for MS MoS_2 films deposited on 440C stainless steel disk substrates and slid against a 440C stainless steel bearing ball [6.22, 6.23]. The MoS_2 films produced by using a standardized process and condition were characterized and qualified, typically with a number of coupon specimens and test conditions. Aspects of practical engineering decision-making were simulated to illustrate wear mechanisms as well as surface engineering principles. Eventually, this approach will become necessary for evaluating and recommending solid lubricants and operating parameters for optimum performance.

6.3.1 Experimental Procedure

Dry, solid MoS_2 films were deposited by using a magnetron radiofrequency sputtering system and a commercially available, high-purity molybdenum disulfide target on the 440C stainless steel disks [6.22, 6.23]. Table 6.3 presents the deposition conditions. Because sputtered MoS_2 films are often nonstoichiometric, films examined in this case study were named MoS_x. Table 6.4 lists the value of x, thickness, density, surface roughness, and Vickers hardness of the standard MoS_x specimens. Some details on the MoS_x films were reported in the references [6.22, 6.23].

In a vacuum environment sputtering with rare gas ions, such as argon ions, can remove contaminants adsorbed on the surface of materials and etch the surface. The

TABLE 6.3.—MAGNETRON RADIOFREQUENCY SPUTTERING CONDITIONS

Substrate material ....440C stainless steel	
Substrate centerline-average	
surface roughness, R_a, nm ....9.0 with a standard deviation of 0.9	
Substrate average Vickers microhardness	
at loads from 0.49 to 4.9 N, GPa...................6.8 with a standard deviation of 0.08	
Ion etching of substrate before deposition:	
Power, W 550	
Argon pressure for 5 min, Pa (mtorr) 2.7 (20)	
Target material ....Molybdenum disulfide	
Target cleaning:	
Power, W 900	
Argon pressure for 5 min, Pa (mtorr) 2.7 (20)	
Target-to-substrate distance, mm 90	
Deposition conditions:	
Power, W 900	
Argon pressure, Pa (mtorr) 2.7 (20)	
Deposition rate, nm/min 110	
Power density, W/m^2 ....4.9×10^4	
Deposition temperature ....Room temperature	

ion-sputtered surface consisted of sulfur, molybdenum, and small amounts of carbon and oxygen (see Fig. 2.33 in Chapter 2).

AES analysis provided elemental depth profiles for the MoS$_x$ films deposited on the 440C stainless steel substrates. For example, Fig. 2.34 in Chapter 2 presented a typical example of an AES depth profile, with concentration shown as a function of the sputtering distance from the MoS$_x$ film surface. The concentrations of sulfur and molybdenum at first rapidly increased with an increase in sputtering distance, and the concentrations of carbon and oxygen contaminant decreased. All elements remained constant thereafter. The deposited films contained small amounts of carbon and oxygen at the surface and in the bulk and had a sulfur-to-molybdenum ratio of ~1.7 (also, see Table 6.4). The relative concentrations of various constituents were determined from peak height sensitivity factors [6.24]. The MoS$_x$ films were exposed to air before AES analysis.

Figures 6.7(a) and (b) present atomic ratios of sulfur to molybdenum (S/Mo) for thin (~110 nm thick) and thick (~1 μm thick) MoS$_x$ films, respectively. All films were nonstoichiometric and were deposited at various laboratories in Europe and the United States by sputtering, except one of the thick films, for which a laser ablation technique was used. All the thin and all the thick sputtered MoS$_x$ films showed approximately the same S/Mo ratios, 1.6 to 1.8 and 1.3 to 1.5, respectively. The S/Mo for two thin films deposited at the same laboratory were virtually identical and agreed well with the S/Mo obtained by Rutherford backscattering spectrometry. The laser-ablated film seemed to be more like bulk molybdenite, and the sulfur seemed to be depleted by preferential sputtering to a much greater degree (x < 1.5). The concentrations of carbon and oxygen contaminants (≤5 and ~2%, respectively) in the laser-ablated MoS$_x$ film were much less than those (≤5 to 15% and ≤10 to 25%, respectively) in the two virtually identical sputtered MoS$_x$ films.

TABLE 6.4.—CONDITIONS OF BALL-ON-DISK SLIDING FRICTION EXPERIMENTS

Load, N	0.49 to 3.6
Disk rotating speed, rpm	120
Track diameter, mm	5 to 17
Sliding velocity, mm/s	31 to 107
Vacuum pressure, Pa (torr)	10^{-7} (10^{-9})
Ball material	440C stainless steel (grade 10)
Ball diameter, mm	6
Ball centerline-average surface roughness, R_a, nm	6.8 with a standard deviation of 1.8
Ball average Vickers microhardness at loads from 0.49 to 4.9 N, GPa	8.7 with a standard deviation of 0.17
Disk material	Magnetron-sputtered MoS$_x$ film on 440C stainless steel substrate
Value of x in MoS$_x$	1.7
Nominal film thickness, nm	110
Film density, g/cm^3	4.4
Film centerline-average surface roughness, R_a, nm	18.9 with a standard deviation of 5.9
Disk average Vickers microhardness at loads from 0.49 to 4.9 N, GPa	6.8 with a standard deviation of 0.14

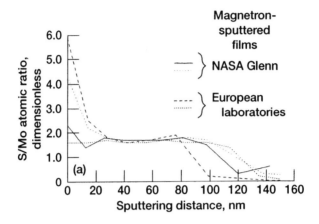

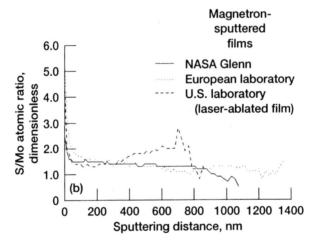

Figure 6.7.—Sulfur/molybdenum atomic ratios of
MoS$_x$ films on 440C stainless steel deposited in
Europe and the United States. (a) Thin film.
(b) Thick film.

The average Vickers microhardness values for MoS$_x$ films deposited on 440C
stainless steel disks were ~10% lower than those for uncoated 440C stainless steel
disks in the load range 0.1 to 0.25 N (Fig. 6.8). At higher loads, 1 to 5 N, however,
the microhardness values for the coated disks were ~6.8 GPa, the same as those for
the uncoated disks.

The surface morphology, surface roughness, and microstructure of the MoS$_x$
films were investigated by scanning electron microscopy (SEM), surface
profilometry, and x-ray diffraction (XRD). The film surface took on a highly dense,

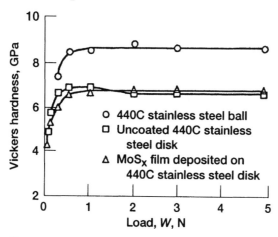

Figure 6.8.—Vickers microhardness as function of load for MoS$_x$ films on 440C stainless steel.

smooth, featureless appearance. The centerline-average surface roughness R_a was 18.9 nm with a standard deviation of 5.9 nm. An x-ray diffraction pattern showed no evidence that the film had a crystalline structure. It was either amorphous or too thin (~110 nm thick), or the crystal domain size was beyond the limits of detection using XRD.

6.3.2 Friction Behavior and Endurance Life

Figures 6.9 and 6.10 present typical coefficients of friction for MoS$_x$ films, 0.11 μm thick, deposited on sputter-cleaned 440C stainless steel disks as a function of the number of passes. Figure 6.9 extends only to 400 passes, the initial run-in period. In Fig. 6.10, the plots extend to the endurance life, the number of passes at which the coefficient of friction rose rapidly to a fixed value of ~0.15.

Quantitatively, the coefficient of friction usually started relatively high (point A) but rapidly decreased and reached its minimum value of ~0.01 (point B), sometimes decreasing to nearly 0.001 after 40 to 150 passes. Afterward, the coefficient of friction gradually increased with an increasing number of passes, as shown in Fig. 6.9. It reached its equilibrium value at point C in Fig. 6.10. From point C to point D it remained constant for a long period. At point D the coefficient of friction began to decrease and remained low from point E to point F. Finally, the sliding action caused the film to break down, whereupon the coefficient of friction rose rapidly (line F–G). The plots of Fig. 6.10 reveal the similarities in the friction behavior of MoS$_x$ films regardless of the load applied. This evidence suggests that increasing the load may not affect the wear mode and behavior of MoS$_x$ films.

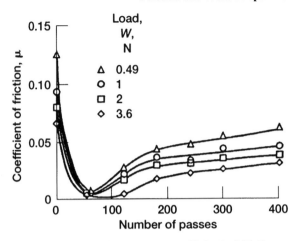

Figure 6.9.—Run-in average coefficient of friction
as function of number of passes for MoS_x films
on 440C stainless steel disks in sliding contact
with 440C stainless steel balls in ultrahigh
vacuum.

6.3.3 Effects of Load on Friction, Endurance Life, and Wear

Friction.—The friction data presented in Figs. 6.9 and 6.10 clearly indicate that
the coefficient of friction for steel balls in sliding contact with MoS_x films varies
with load. In general, the higher the load, the lower the coefficient of friction.
Therefore, the coefficients of friction as a function of load for the regions designated
in Figs. 6.9 and 6.10 were replotted in Fig. 6.11 on logarithmic coordinates. The
logarithmic plots reveal a generally strong correlation between the coefficient of
friction in the steady-state condition (C–D region) and load. The relation between
coefficient of friction μ and load W is given by $\mu = kW^{-1/3}$, which expression agrees
with the Hertzian contact model [6.25–6.29]. Similar elastic contact and friction
characteristics (i.e., load-dependent friction behavior) can also be found for bulk
materials like polymers [6.25, 6.30], diamond [6.30], and ceramics [6.31], as well
as for thin solid lubricants like sputtered molybdenum disulfide and ion-beam-
deposited boron nitride coatings [6.32–6.35].

This load-dependent (i.e., contact pressure dependent) friction behavior allows
the coefficient of friction to be deduced from the design concept (e.g., the
component design parameters). Further, a better understanding of the mechanical
factors controlling friction, such as load (contact pressure), would improve the
design of advanced bearings and the performance of solid lubricants.

Endurance life.—Figure 6.12 presents the endurance lives of the MoS_x films as
a function of load. Even in very carefully controlled conditions, repeat determina-
tions of endurance life can show considerable scatter. Although the endurance lives

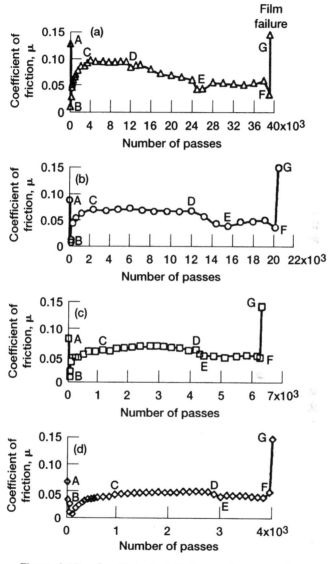

Figure 6.10.—Coefficient of friction as function of number of passes for MoS_x films on 440C stainless steel disks in sliding contact with 440C stainless steel balls in ultrahigh vacuum. (a) Load, 0.49 N. (b) Load, 1 N. (c) Load, 2 N. (d) Load, 3.6 N.

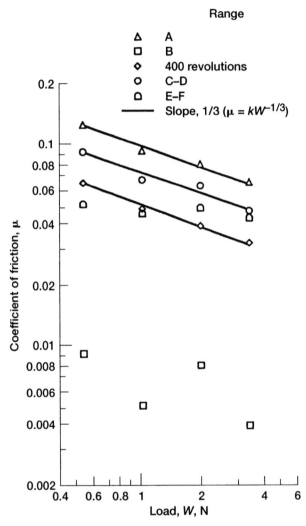

Figure 6.11.—Relation of coefficient of friction and
number of passes (from regions denoted in Figs. 6.9
and 6.10) to load for MoS_x films on 440C stainless
steel disks.

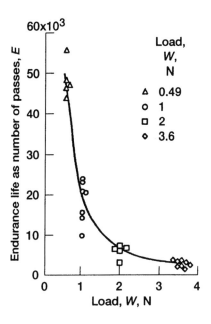

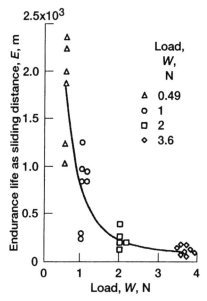

Figure 6.12.—Endurance life as function of load for MoS$_x$ films on 440C stainless steel disks in sliding contact with 440C stainless steel balls in ultrahigh vacuum; Cartesian plot.

determined by the sliding distance showed larger variation than those determined by the number of passes, the trends are similar; that is, the endurance lives of MoS_x films decreased as the load increased.

To express the relation between endurance life and load empirically, the endurance life data of Fig. 6.12 were replotted on logarithmic coordinates in Fig. 6.13. A straight line was easily placed through the data in both plots of Fig. 6.13, once again revealing the strong correlation. To a first approximation for the load range investigated, the relation between endurance life E and load W on logarithmic coordinates was expressed by $E = KW^n$, where K and n are constants for the MoS_x films under examination and where the value of n was ~1.4. The load-dependent (i.e., contact pressure dependent) endurance life allows a reduction in the time needed for wear experiments and an acceleration of life testing of MoS_x films.

Specific wear rates and wear coefficients.—An attempt to estimate average wear rates for MoS_x films was made with the primary aim of generating specific wear rates and wear coefficient data that could be compared with those for other materials in the literature. It is recognized that the contact by the ball tip is continuous and that the contact by any point on the disk track is intermittent. A fundamental parameter affecting film endurance life is the number of compression and flexure cycles (intermittent contacts) to which each element of the film is subjected. Therefore, normalizing the disk wear volume by the total sliding distance experienced by the ball is fundamentally incorrect. To account for the intermittent and fatigue aspects of this type of experiment, the volume worn away should be given by

$$\text{Wear volume} = c_1 \times \text{Normal load} \times \text{Number of passes} \qquad (6.1)$$

where the dimensional constant c_1 is an average specific wear rate expressed in cubic millimeters per newton·pass. However, a great quantity of historical ball-on-disk or pin-on-disk results have been reported using an expression of the form

$$\text{Wear volume} = c_2 \times \text{Normal load} \times \text{Sliding distance} \qquad (6.2)$$

where the dimensional constant c_2 is an average specific wear rate expressed in cubic millimeters per newton·meter. Also, a Holm-Archard relationship is the type

$$\text{Wear volume} = c_3 \times \text{Normal load} \times \text{Sliding distance} / \text{Hardness} \qquad (6.3)$$

where the nondimensional constant c_3 is the nondimensional average wear coefficient reported in the references [6.36–6.39].

Figure 6.14 presents the two specific wear rates and the wear coefficient as a function of load. That they are almost independent of load for the loads investigated suggests that increasing the load in the range 0.49 to 3.6 N may not affect the wear mode of MoS_x films.

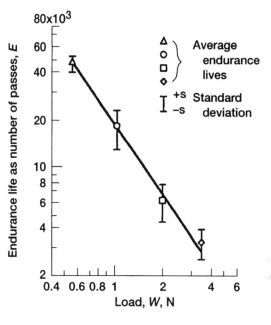

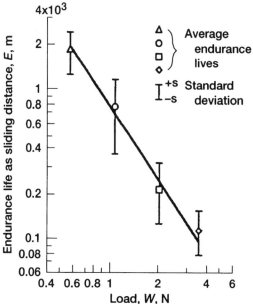

Figure 6.13.—Endurance life as function of load for MoS$_x$ films on 440C stainless steel disks in sliding contact with 440C stainless steel balls in ultrahigh vacuum; logarithmic plot.

282

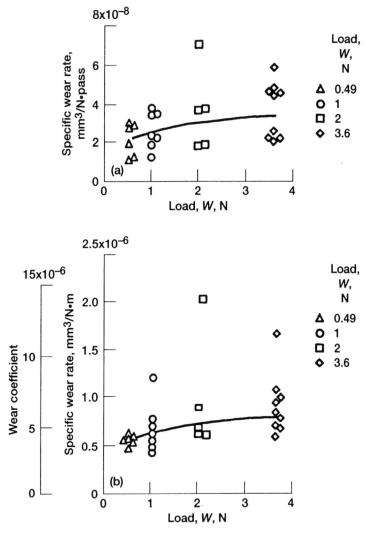

Figure 6.14.—Wear rate as function of load for MoS_x films on
440C stainless steel disks in sliding contact with 440C stainless steel balls in ultrahigh vacuum. (a) Specific wear rate,
3×10^{-8} mm^3/N·pass, and dimensional constant, c_1; (b) specific
wear rate, 8×10^{-7} mm^3/N·m, and dimensional constant, c_2;
nondimensional wear coefficient, 5×10^{-6}, and nondimensional
constant, c_3.

The worn surfaces of the MoS_x films took on a burnished appearance, and a low-wear form of adhesive wear, namely burnishing wear, was encountered. The two average specific wear rates and the nondimensional wear coefficient for the MoS_x films studied herein were $\sim3\times10^{-8}$ mm³/N·pass (c_1 in Eq. (6.1) and Fig. 6.14(a)), 8×10^{-7} mm³/N·m (c_2 in Eq. (6.2) and Fig. 6.14(b)), and 5×10^{-6} (c_3 in Eq. (6.3) and Fig. 6.14(b)).

The very concept of specific wear rates and a wear coefficient implicitly assumes a linear relation between the volume of material removed and either the number of passes for the disk (flat) or the distance slid for a ball (pin). If it were true that roughly the same amount of material is removed from a disk specimen by each pass of the disk, the relation between wear volume and number of passes would be roughly linear, and the specific wear rates and wear coefficient would be meaningful. A consequence for solid lubricating films would be that a film twice the thickness of another similar film should last twice as long. However, if material were not removed from the disk (flat) at a constant rate, measuring the wear volume after a number of passes would give only the average amount of material removed per pass. The calculated wear rate in this case, being an average, would change with the number of passes completed, and doubling a solid lubricating film's initial thickness would not double the film endurance life. Also, simple compaction of the MoS_x film under load may primarily occur during running in. Afterward, burnishing wear may dominate the overall wear rate. The specific wear rates and wear coefficient for a material such as MoS_x film, therefore, should be viewed with caution.

6.3.4 Effects of MoS_x Film Thickness on Friction and Endurance Life

The coefficient of friction and endurance life of MoS_x films depend on the film thickness, as indicated in Figs. 6.15 and 6.16. The coefficient of friction reached a minimum value with an effective or critical film thickness of 200 nm. The endurance life of the MoS_x films increased as the film thickness increased. The relation between endurance life E and film thickness h is expressed by $E = 100h^{1.21}$ at a load of 1 N and by $E = 3.36h^{1.50}$ at a load of 3 N.

6.3.5 Roles of Interface Species

The interfacial region between an MoS_x film and a substrate surface determines many characteristics of the couple. These include film adhesion or adhesion strength, wear resistance, defects, interfacial fracture, compound formation, diffusion or pseudodiffusion, or monolayer-to-monolayer change. Also, the nature of the interfacial region determines the endurance life of MoS_x films. For example, Fig. 6.17 presents the endurance lives of 110-nm-thick MoS_x films deposited on 440C stainless steel substrates with three different interfacial species: argon-sputter-cleaned steel substrate surfaces, oxidized steel surfaces, and a rhodium interlayer produced on steel surfaces.

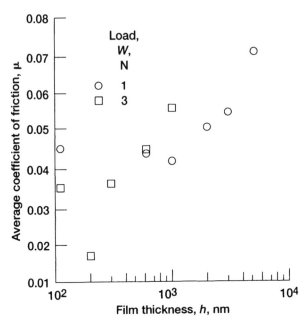

Figure 6.15.—Coefficient of friction as function of
film thickness for MoS$_x$ films on 440C stainless
steel disks in sliding contact with 440C stainless
steel balls in ultrahigh vacuum.

The argon sputter cleaning and oxidation were done in situ in the deposition
system just prior to the MoS$_x$ deposition process. Argon-sputter-cleaned 440C
stainless steel substrate disks were exposed to 1000 langmuirs (where 1 langmuir
= 130 µPa·s (1×10^{-6} torr·s)) of oxygen (O$_2$) gas. The 200-, 100-, and 30-nm-thick
rhodium interlayer films were produced on 440C stainless steel substrate disks by
sputtering using a different deposition system. Then the rhodium-coated disks were
placed in the MoS$_x$ deposition system, the system was evacuated, and the rhodium-
coated disks were argon sputter cleaned in situ just prior to the MoS$_x$ deposition
process.

Obviously, the nature and condition of the substrate surface determined MoS$_x$
film endurance life. The oxidized substrate surface clearly led to longer life. MoS$_x$
adhered well to the oxidized surfaces by forming interfacial oxide regions. Regardless of the interfacial adhesion mechanism the substrate surface must be free of
contaminants, which inhibit interaction between the surface and the atoms of the
depositing MoS$_x$ film. The sputter-cleaned 440C stainless steel surface gave more
adherent film than the rhodium surface. The rhodium metallic interlayer did not
improve MoS$_x$ adhesion. The MoS$_x$ reacted chemically with the sputter-cleaned
440C stainless steel surface more than with the rhodium interlayer surface. The

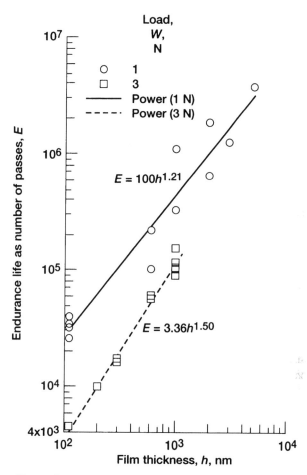

Figure 6.16.—Endurance life as function of film thickness for MoS$_x$ films on 440C stainless steel disks in sliding contact with 440C stainless steel balls in ultrahigh vacuum.

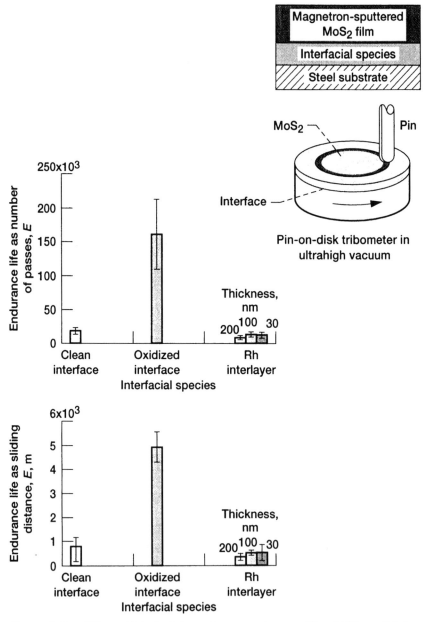

Figure 6.17.—Effect of interfacial species on endurance life of 110-nm-thick MoS$_x$ films on 440C stainless steel disks in sliding contact with 440C stainless steel balls in ultrahigh vacuum.

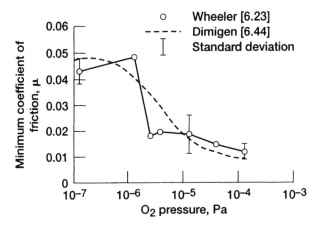

Figure 6.18.—Minimum centerline coefficient
of friction as function of oxygen pressure for
110-nm-thick MoS$_{1.7}$ films on 440C stainless
steel disks in sliding contact with 440C stainless
steel balls (diameter, 6 mm) at 1-N load [6.23]. The
solid line connects the data points as an aid to the
eye. The dashed line is the result of a similar test
reported in [6.44] (see text).

thicker the rhodium interlayer, the shorter the life. However, a thin interlayer of chemically reactive metal, such as titanium or chromium, may improve adhesion.

6.3.6 Effects of Oxygen Pressure

Molybdenum disulfide is unsuitable as a dry lubricant in air because its reacts with oxygen and/or water vapor to form corrosive products [6.40, 6.41]. The ambient atmosphere in near Earth orbit consists of atomic oxygen and outgassing products from the spacecraft at pressures in the range 10^{-7} to 10^{-4} Pa [6.42, 6.43]. However, there is no evidence that these conditions are unsuitable for the use of MoS$_x$ films. On the contrary, Dimigen et al. [6.44] observed that the coefficient of friction of magnetron-sputtered MoS$_x$ films was substantially less when run in 10^{-6} to 10^{-4} Pa of oxygen than when run in ultrahigh vacuum (see dashed line in Fig. 6.18). Dimigen et al. measured the coefficient of friction for a range of rotational speeds and oxygen gas pressures on an MoS$_{1.4}$ film by using a pin-on-disk tribometer. The dashed line in Fig. 6.18 is an attempt to represent the results of their tests at 50 rpm. They attributed these results to an adsorption-desorption phenomenon at the sliding interface rather than to a change in the chemistry or morphology of the material itself. However, the evidence for this conclusion was not clear, and the investigation of the oxygen effect was only a peripheral part of that work.

Wheeler [6.23] investigated the effect of oxygen pressure further by using 110-nm-thick, magnetron-sputtered $MoS_{1.7}$ films on 440C stainless steel disks, which were the same as the thin MoS_x films described in Section 6.3.1. The coefficient of friction of the $MoS_{1.7}$ films was substantially reduced when sliding in low partial pressures of oxygen. The coefficient of friction was 0.04 to 0.05 at oxygen pressures of 1.33×10^{-6} Pa or less. When 2.66×10^{-6} Pa or more of oxygen was present during sliding, the coefficient of friction decreased gradually to 0.01 to 0.02, independent of the pressure. As can be seen from Fig. 6.18, the data obtained by Wheeler [6.23] agree substantially with the results of Dimigen et al. [6.44] at both high and low oxygen partial pressures.

Also, Wheeler [6.23] reported that static exposure to oxygen at any pressure did not affect the subsequent friction in vacuum. The wear scars with high friction were much larger than those with low friction. Wheeler proposed that oxygen reduces the friction force by influencing material transfer to the pin in such a way as to decrease the contact area. With this hypothesis a shear strength of 4.8 ± 0.6 MPa, independent of oxygen pressure, was deduced for a representative film [6.23].

The endurance life of MoS_x films was strongly affected by gas interactions with the surface, such as the chemical reaction of the surface with a species or the adsorption of a species (physically or chemically adsorbed material). For example, Fig. 6.19 presents the endurance lives of 110-nm-thick MoS_x films in sliding contact with 440C stainless steel balls in oxygen gas at three pressures: 1×10^{-5}, 5×10^{-5}, and 1×10^{-4} Pa. The oxygen exposures at 1×10^{-5} and 5×10^{-5} Pa increased the endurance life by factors of 2 and 1.2, respectively, when compared with the average endurance life of 18 328 passes with a standard deviation of 5200 passes (the average sliding distance of 764 m with a standard deviation of 380 m) in ultrahigh vacuum at $\sim 7 \times 10^{-7}$ Pa. The oxygen exposure at 1×10^{-4} Pa provided endurance life almost the same as or slightly shorter than that in ultrahigh vacuum at $\sim 7 \times 10^{-7}$ Pa. Thus, extremely small amounts of oxygen contaminant, such as 1×10^{-5} Pa, can increase the endurance life of MoS_x films.

6.3.7 Effects of Temperature and Environment

Increasing the surface temperature of a material tends to promote surface chemical reactions. Adsorbates on a material surface from the environment affect surface chemical reactions. These chemical reactions cause products to appear on the surface that can alter adhesion, friction, and wear [6.45]. For example, Fig. 6.20 presents the steady-state coefficients of friction for 110-nm-thick, magnetron-sputtered MoS_x films in sliding contact with 440C stainless steel balls in three environments (ultrahigh vacuum, air, and nitrogen) at temperatures from 23 to 400 °C. The data presented in the figure reveal the marked differences in friction resulting from the environmental conditions. The MoS_x films had the lowest coefficient of friction in the nitrogen environment over the entire temperature range. The coefficients of friction of the MoS_x films were, in ascending order, nitrogen

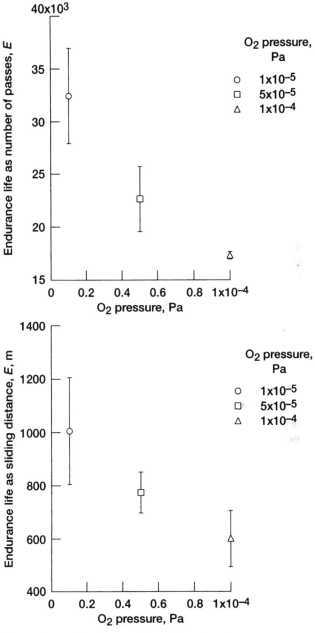

Figure 6.19.—Endurance life as function of oxygen pressure for 110-nm-thick MoS$_x$ films on 440C stainless steel disks in sliding contact with 440C stainless steel balls at 1-N load.

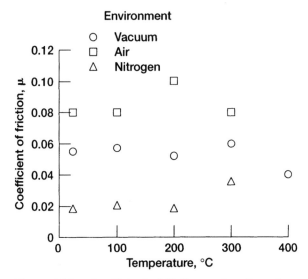

Figure 6.20.—Coefficient of friction as function of
temperature for 110-nm-thick MoS$_x$ films on 440C
stainless steel disks in sliding contact with 440C
stainless steel balls in ultrahigh vacuum, humid air,
and dry nitrogen.

< ultrahigh vacuum < air. Although the coefficient of friction varied with tempera-
ture, the variation was quite small.

References

6.1 H.P. Jost, Tribology—Origin and future, *Wear 136*: 1–17 (1990).

6.2 A.R. Lansdown, *Lubrication and Lubricant Selection—A Practical Guide*, Mechanical Engineering Publications, London, 1996.

6.3 M.E. Campbell, J.B. Loser, and E. Sneegas, *Solid Lubricants Technology Survey*, NASA
SP–5059, 1966.

6.4 M.E. Campbell, *Solid Lubricants—A Survey*, NASA SP–5059(01), 1972.

6.5 K. Miyoshi, M. Iwaki, K. Gotoh, S. Obara, and K. Imagawa, Friction and wear properties of
selected solid lubricating films, Part 1: Bonded and magnetron-sputtered molybdenum disulfide
and ion-plated silver films, NASA/TM—1999-209088/PART 1, 1999.

6.6 F.J. Clauss, *Solid Lubricants and Self-Lubricating Solids*, Academic Press, New York, 1972.

6.7 J.K. Lancaster, Solid lubricants, *CRC Handbook of Lubrication: Theory and Practice of Tribology*
(E.R. Booser, ed.), CRC Press, Boca Raton, FL, Vol. II, 1984, pp. 269–290.

6.8 K. Miyoshi, T. Spalvins, and D.H. Buckley, Tribological characteristics of gold films deposited
on metals by ion plating and vapor deposition, *Wear 108*, 2: 169–184 (1986).

6.9 M.J. Todd and R.H. Bentall, Lead film lubrication in vacuum, *International Conference on Solid
Lubrication*, SP–6, American Society of Lubrication Engineers, 1978, pp. 148–157.

6.10 J.R. Lince and P.D. Fleischauer, Solid lubrication for spacecraft mechanisms, Aerospace Corp. Report TR–97(8565)–4, 1997.

6.11 E.W. Roberts, Thin solid-lubricant films in space, *Flight-Vehicle Materials, Structures, and Dynamics* (R.L. Fusaro and J.D. Achenbach, eds.), American Society for Mechanical Engineers, New York, Vol. 4, 1993, pp. 113–132.

6.12 K. Miyoshi, M. Iwaki, K. Gotoh, S. Obara, and K. Imagawa, Friction and wear properties of selected solid lubricating films, Part 2: Ion-plated lead films, NASA/TM—2000-209088/PART 2, 2000.

6.13 H.O. Pierson, *Handbook of Carbon, Graphite, Diamond, and Fullerenes: Properties, Processing, and Applications*, Noyes Publications, Park Ridge, NJ, 1993.

6.14 J.J. Pouch and S.A. Alterovitz, eds., *Properties and Characterization of Amorphous Carbon Films*, Materials Science Forum, Trans Tech Publications, Aedermannsdorf, Switzerland, Vols. 52 & 53, 1990.

6.15 K. Miyoshi, J.J. Pouch, and S.A. Alterovitz, Plasma-deposited amorphous hydrogenated carbon films and their tribological properties, *Properties and Characterization of Amorphous Carbon Films* (J.J. Pouch and S.A. Alterovitz, eds.), Materials Science Forum, Trans Tech Publications, Aedermannsdorf, Switzerland, Vols. 52 & 53, 1990, pp. 645–656.

6.16 R.L.C. Wu, K. Miyoshi, R. Vuppuladhadium, and H.E. Jackson, Physical and tribological properties of rapid thermal annealed diamond-like carbon films, *Surf. Coat. Tech. 55, 1–3*: 576–580 (1992).

6.17 K. Miyoshi, B. Pohlchuck, K.W. Street, J.S. Zabinski, J.H. Sanders, A.A. Voevodin, and R.L.C. Wu, Sliding wear and fretting wear of diamondlike carbon-based, functionally graded nanocomposite coatings, *Wear 229*: 65–73 (1999).

6.18 A.P. Molloy and A.M. Dionne, eds., *World Markets, New Applications, and Technology for Wear and Superhard Coatings*, Gorham Advanced Materials, Inc., Gorham, ME, 1998.

6.19 K. Miyoshi, M. Murakawa, S. Watanabe, S. Takeuchi, and R.L.C. Wu, Tribological characteristics and applications of superhard coatings: CVD Diamond, DLC, and c-BN, *Proceedings of Applied Diamond Conference/Frontier Carbon Technology Joint Conference 1999*, Tsukuba, Japan, 1999, pp. 268–273. (Also NASA/TM—1999-209189, 1999.)

6.20 K. Miyoshi, Diamondlike carbon films: Tribological properties and practical applications, *New Diamond and Frontier Carbon Technology 9, 6*: 381–394 (1999).

6.21 K. Miyoshi, M. Iwaki, K. Gotoh, S. Obara, and K. Imagawa, Friction and wear properties of selected solid lubricating films, Part 3: Magnetron-sputtered and plasma-assisted, chemical-vapor-deposited diamondlike carbon films, NASA/TM—2000-209088/PART 3, 2000.

6.22 K. Miyoshi, F.S. Honecy, P.B. Abel, S.V. Pepper, T. Spalvins, and D.R. Wheeler, A vacuum (10^{-9} torr) friction apparatus for determining friction and endurance life of MoS_x films, *Tribol. Trans. 36, 3*: 351–358 (1993).

6.23 D.R. Wheeler, Effect of oxygen pressure from 10^{-9} to 10^{-6} torr on the friction of sputtered MoS_x films, *Thin Solid Films 223, 1*: 78–86 (1993).

6.24 L.E. Davis, N.C. MacDonald, P.W. Palmberg, G.E. Riach, and R.E. Weber, *Handbook of Auger Electron Spectroscopy*, Physical Electronics Division, Perkin-Elmer Corp., Eden Prairie, MN, 1979.

6.25 R.C. Bowers, Coefficient of friction of high polymers as a function of pressure, *J. Appl. Phys. 42, 12*: 4961–4970 (1971).

6.26 R.C. Bowers and W.A. Zisman, Pressure effects on the friction coefficient of thin-film solid lubricants, *J. Appl. Phys. 39, 12*: 5385–5395 (1968).

6.27 B.J. Briscoe, B. Scruton, and F.R. Willis, The shear strength of thin lubricant films, *Proc. R. Soc. London A 333*: 99–114 (1973).

6.28 B.J. Briscoe and D.C.B. Evans, The shear properties of Langmuir-Blodgett layers, *Proc. R. Soc. London A 380*: 389–407 (1982).

6.29 T.E.S. El-Shafei, R.D. Arnell, and J. Halling, An experimental study of the Hertzian contact of surfaces covered by soft metal films, *ASLE Trans. 26, 4*: 481–486 (1983).

6.30 F.P. Bowden and D. Tabor, *The Friction and Lubrication of Solids*, Pt. 2, Clarendon, Oxford, UK, 1964.

6.31 K. Miyoshi, Adhesion, friction, and micromechanical properties of ceramics, *Surf. Coat. Tech. 36, 1–2*: 487–501 (1988).

6.32 S.A. Karpe, Effects of load on friction properties of molybdenum disulfide, *ASLE Trans. 8, 2*: 164–178 (1965).

6.33 I.L. Singer, R.N. Bolster, J. Wegand, S. Fayeulle, and B.C. Strupp, Hertzian stress contribution to low friction behavior of thin MoS_2 coatings, *Appl. Phys. Lett. 57, 10*: 995–997 (1990).

6.34 L.E. Pope and J.K.G. Panitz, The effects of Hertzian stress and test atmosphere on the friction coefficients of MoS_2 coatings, *Surf. Coat. Tech. 36, 1*: 341–350 (1988).

6.35 K. Miyoshi, Fundamental tribological properties of ion-beam-deposited boron nitride films, *Materials Science Forum* (J.J. Pouch and S.A. Alterovitz, eds.), Trans Tech Publications, Aedermannsdorf, Switzerland, Vols. 54 & 55, 1990, pp. 375–398.

6.36 E. Rabinowicz, Wear coefficients, *CRC Handbook of Lubrication: Theory and Practice of Tribology* (E.R. Booser, ed.), CRC Press, Boca Raton, FL, Vol. II, 1984, pp. 201–208.

6.37 R. Holm, *Electric Contacts*, Almquist & Wiksells, Stockholm, 1946, Section 40.

6.38 J.F. Archard, Contact and rubbing of flat surfaces, *J. Appl. Phys. 24, 8*: 981–988 (1953).

6.39 E. Rabinowicz, The least wear, *Wear 100, 1*: 533–541 (1984).

6.40 A.W.J. De Gee, G. Salomon, and J.H. Zatt, On the mechanisms of MoS_2-film failure in sliding friction, *ASLE Trans. 8, 2*: 156–163 (1965).

6.41 E.W. Roberts, The tribology of sputtered molybdenum disulfide films, *Friction, Lubrication, and Wear 50 Years On*, Institute of Mechanical Engineering, London, 1987, pp. 503–510.

6.42 E.W. Roberts, Ultralow friction films of MoS_2 for space applications, *Thin Solid Films 181*: 461–473 (1989).

6.43 E.W. Roberts, Thin solid lubricant films in space, *Tribol. Int. 23, 2*: 95–104 (1990).

6.44 H. Dimigen, H. Hübsch, P. Willich, and K. Reichelt, Stoichiometry and friction properties of sputtered MoS_x layers, *Thin Solid Films 129, 1*: 79–91 (1985).

6.45 C.J. Smithells, *Metals Reference Book*, 4th ed., Plenum Press, Vol. 1, 1967.

Chapter 7
Aerospace Mechanisms and Tribology Technology: Case Studies

7.1 Introduction

In the United States space program alone a large number of spacecraft failures and anomalies have occurred (e.g., Galileo and Hubble). In addition, more demanding requirements have been causing failures or anomalies to occur during the qualification testing of future satellites and space platform mechanisms even before they are launched (e.g., GOES–NEXT, CERES, International Space Station beta joint gimbals [7.1]). In space programs throughout the world a much greater number of failures and anomalies of space mechanisms have occurred.

In the operation of space mechanisms functional reliability is, of course, vital. Even a small tribological failure can lead to catastrophic results in a spacecraft [7.2]. In this chapter an attempt is made to review aspects of real problems related to vacuum or space tribology technology and dry, solid film lubrication as case studies:

1. Case study A: Galileo spacecraft's high-gain antenna deployment anomaly
2. Case study B: Space shuttle orbiter's quad check valve failures

To understand the adhesion, friction, wear, and lubrication situation in each case study, the nature of the problem is analyzed, the tribological properties are examined, the range of potential solutions is analyzed, and recommendations are made.

7.2 Case Study A: Galileo Spacecraft's High-Gain Antenna Deployment Anomaly

7.2.1 Galileo's Partially Unfurled High-Gain Antenna: The Anomaly at 37 Million Miles

The Galileo spacecraft and its inertial upper stage booster rocket were deployed from the space shuttle *Atlantis* on October 18, 1989. Shortly thereafter the booster rocket fired and separated, sending Galileo on its six-year journey to the planet Jupiter.

Figure 7.1 shows the locations of many of Galileo's main structural and scientific components. Galileo is a spin-stabilized spacecraft and has three Earth-to-spacecraft communications antennas for commanding and returning spacecraft telemetry. Two of the antennas are low gain and the third is high gain. The umbrella-like high-gain antenna is located at the top of the spacecraft and is 4.8 m (16 ft) in diameter. It was designed to transmit data back to Earth at up to 134 000 bits per second (the equivalent of about one television picture each minute). The antenna, which is made of gold-plated metal mesh, was stowed behind a Sun shield at launch to avoid heat damage while the spacecraft flew close to the Sun.

On April 11, 1991, when the Sun-to-spacecraft distance was large enough to present no thermal danger to the antenna, the Galileo spacecraft began to deploy its high-gain antenna under computer-sequence control. Within minutes Galileo's

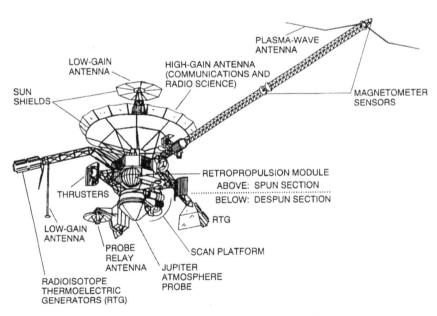

Figure 7.1.—Galileo spacecraft configuration.

flight team, watching spacecraft telemetry 37 million miles away on Earth, could see that something was wrong: The motors had stalled and something had stuck. The antenna, the 4.8-m mesh paraboloid stretched over 18 umbrella-like ribs, had opened only part way.

Galileo has been in orbit around Jupiter and its moons since 1995 and is operating as it processes and sends science data to Earth with one of the low-gain antennas. (The low-gain and high-gain antennas are part of the same assembly and face in the same direction.) Its primary mission ended in December 1997, and the spacecraft is currently in the midst of an extension known as the Galileo Europa Mission. For additional details on Galileo's high-gain antenna deployment anomaly, see references 7.3 to 7.5.

7.2.2 Anomaly Investigation

The anomaly investigation involved more than 100 people in testing, simulation, analysis, consultation, and review. After a thorough analysis of the telemetry and then ground testing and analysis, the investigation attributed the problem to the sticking of three of the antenna's 18 ribs in the stowed (or closed) position due to

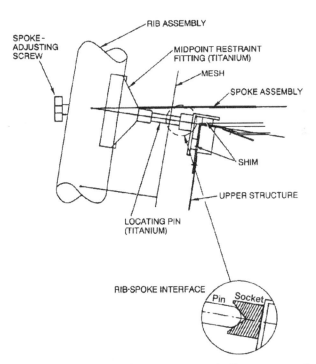

Figure 7.2.—Rib-spoke interface on Galileo spacecraft high-gain antenna.

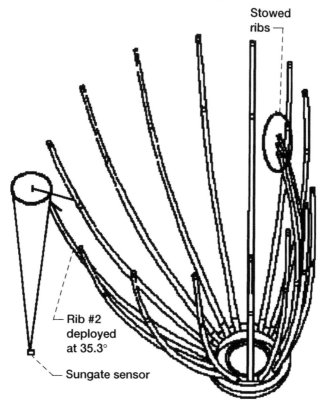

Figure 7.3.—Ribs of Galileo high-gain antenna, showing three stuck in stowed position.

high friction between their standoff pins and their sockets (Fig. 7.2). The other ribs were partially open (Fig. 7.3). The anomaly review team was confident that nothing was broken and that three adjacent antenna ribs were stuck to the central tower by friction in their paired standoff pins, which were likely misaligned enough that one pin bound the upper side of its socket and its mate the lower (Fig. 7.4).

This attribution led, in turn, to the hypothesis that repeated thermal cycling, warm-cool-warm-cool, could walk the paired pins out, freeing the ribs. The spacecraft was maneuvered to warm the high-gain antenna by solar heating and to cool it in the shade. This warming and cooling, however, did not release the stuck ribs. In addition to thermal cycling, other ideas were developed for loosening the stuck ribs, but the antenna remained stuck. After a multiyear campaign to try to free the stuck ribs, lasting until 1996, there was no longer any significant prospect of deploying the antenna.

Although freeing and using the high-gain antenna had not been ruled out, another option was taken. By using the low-gain antenna along with advanced data-compression processing techniques in the spacecraft computers and advanced

297

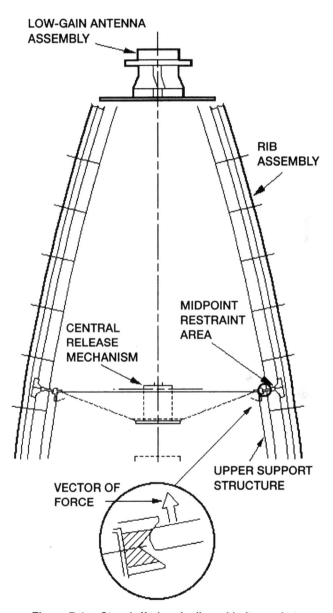

LOW-GAIN ANTENNA
ASSEMBLY

RIB
ASSEMBLY

MIDPOINT
RESTRAINT
AREA

CENTRAL
RELEASE
MECHANISM

UPPER SUPPORT
STRUCTURE

VECTOR OF
FORCE

Figure 7.4.—Standoff pin misaligned in its socket.

hardware and techniques on the ground, a significant fraction of the total planned Jupiter science data was captured.

7.2.3 Investigation of Adhesion, Friction, Wear, and Cold Welding

Experimental Conditions

To find the causes of the stuck antenna ribs in the Galileo spacecraft, sliding friction experiments were conducted using a tribometer (vacuum friction apparatus) with a pin-on-disk configuration in vacuum (space-like) environments and in air [7.6]. The materials, loads, and environments (Table 7.1) were chosen to simulate the conditions that the rib-spoke interface of the antenna's alignment pins may have experienced.

The contacting surface of the pin specimens was hemispherical with a radius of curvature of 0.5 mm. Two types of pin specimen were examined: coated titanium, 6 wt% aluminum, 4 wt% vanadium (Ti-6Al-4V) pins; and bare Ti-6Al-4V pins. The surfaces of the coated Ti-6Al-4V pins had first been coated with an electrolytically converted hard coating of titanium by using an all-alkaline bath maintained at room temperature. Then the pin surfaces were coated with an inorganic, bonded, dry film lubricant (25 mm thick) that contained a molybdenum disulfide pigment. The coatings produced antigalling and wear resistance properties on the Ti-6Al-4V pin surfaces (Fig. 7.5). Because the lubricant and the coating may have failed on the spacecraft, a bare pin-rib system was also examined as a reference.

TABLE 7.1—EXPERIMENTS FOR CASE STUDY A

(a) Materials

Contacting materials		Pretreatment		Dry film lubricant	
Pin	Disk	Pin	Disk	Pin	Disk
Ti-6Al-4V	High-nickel-content superalloy[a]	Bare	Bare	None	None
Ti-6Al-4V	High-nickel-content superalloy[a]	Electrolytically converted hard Ti coating	Bare	Bonded MoS$_2$ dry-film lubricant	None

(b) Conditions

Load, N 2.5, 4, 8.5
Mean contact pressure, GPa . . . . Approx. 1 to 1.5
Disk rotating speed, rpm . . . . 1, 10, and 40
Track diameter, mm . . . . 12 to 20
Sliding velocity, mm/s . . . . 0.5, 0.9, 8, and 36
Environment:
Air . . . . 40 percent relative humidity
Vacuum . . . . 10^{-7} Pa

[a]Composition, wt% (maximum unless shown as range):
Ni, 50–55; Cr, 17–21; Fe, 12–23; Nb + Ta, 4.75–5.5; Mo, 2.8–3.3; Co, 1; Ti, 0.65–1.15; Al, 0.2–0.8; Si, 0.35; Mn, 0.35; Cu, 0.3; C, 0.08; S, 0.015; P, 0.015; B, 0.006.

Dry film lubricant

Electrolytically converted hard coating of titanium

Ti-6Al-4V substrate

10 μm

Figure 7.5.—Tapered cross section of dry-film-lubricated Ti-6Al-4V pin at angle of 45°.

The contacting surface of the disk specimens was flat, 25 μm in diameter, and 5 mm thick. The disk specimens were uncoated, bare high-nickel-content superalloy (Inconel 718). The average surface roughness of the disks was 34 nm root mean square.

Sliding Friction in Vacuum Environment

Effect of dry film lubricant.—The coefficient of friction for the dry-film-lubricated Ti-6Al-4V in contact with the bare high-nickel-content superalloy started at 0.23 but rapidly decreased and reached an equilibrium value of about 0.04 (Fig. 7.6(a)). It remained constant for a long period of time. The friction trace fluctuated slightly with no evidence of stick-slip behavior. The sliding action finally caused the coefficient of friction to rapidly increase at 172 370 passes. Wear damage (i.e., local removal of the dry film lubricant and consequent exposure of the substrate surface), which is discussed in Section 7.2.3.4, caused the high friction at this stage.

The coefficient of friction for the unlubricated, bare Ti-6Al-4V in contact with the bare high-nickel-content superalloy started at 0.31. The friction force traces for the first few sliding passes were characterized by random fluctuation, with only occasional evidence of stick-slip behavior. The presence of oxides and contaminants on the surfaces of the bare Ti-6Al-4V pin and the bare high-nickel-content superalloy disk contributed to the low initial coefficient of friction (0.31). Stick-slip behavior became dominant after a few sliding passes. The coefficient of friction rapidly increased with an increase in the number of passes. Also, the higher the number of passes, the greater was the stick-slip behavior. At approximately 10 passes and above, the coefficient of friction reached an equilibrium value of

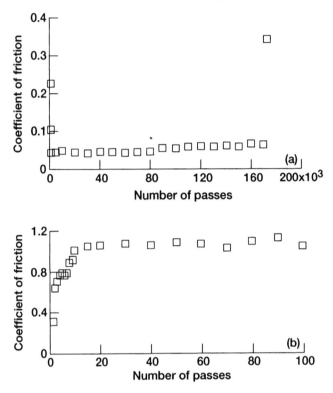

Figure 7.6.—Coefficient of friction as function of number
of passes in vacuum for dry-film-lubricated and unlubricated Ti-6Al-4V pins sliding against high-nickel-content
superalloy disks. (a) Dry-film-lubricated pin at 8.5-N
load. (b) Unlubricated pin at 4-N load.

approximately 1.1 and remained constant for a long period of time (Fig. 7.6(b)). The
traces for 10 passes and above are primarily characterized by a continuous, marked
stick-slip behavior. This type of friction is anticipated where strong metallic
interactions, particularly adhesion, occur at the interface when the oxides and
contaminants have been removed from the alloy surfaces by sliding action.

Comparing Figs. 7.6(a) and (b) shows that the adhesion and friction in the
equilibrium conditions were less for dry-film-lubricated Ti-6Al-4V than for
unlubricated, bare Ti-6Al-4V by a factor of 28. Thus, the dry film lubricant on the
Ti-6Al-4V pin surface was effective and greatly reduced adhesion and friction.

Self-healing of dry film lubricant.—At 172 370 passes of sliding contact in
vacuum (Fig. 7.6(a)) the sliding motion was stopped because of the high friction
(0.36 or greater) caused by the wear damage. After holding the conditions for
approximately 18 hr the sliding was restarted. The result is presented in Fig. 7.7(a).

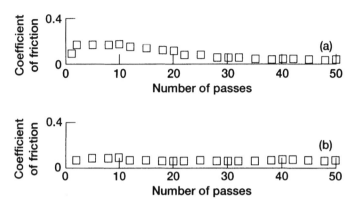

Figure 7.7.—Coefficient of friction as function of number of passes for surfaces previously worn in vacuum of dry-film-lubricated Ti-6Al-4V pin sliding against high-nickel-content superalloy disk when rerun in vacuum. (a) Load, 8.5 N; rotating speed, 40 rpm. (b) Load, 2.5 N; rotating speed, 1 rpm.

The coefficient of friction after rerun became much lower than 0.36. Further, the coefficient of friction generally decreased to about 0.05 with increasing number of passes. This result suggests that the wear damage in vacuum, which caused high friction at 172 370 passes, can self-heal when rerun in vacuum.

After 50 passes of sliding contact in the rerun process the sliding motion was stopped and the load was decreased from 8.5 N to 2.5 N. Contact was maintained for 30 s and then sliding was begun at the new load. Decreasing the load from 8.5 N to 2.5 N did not affect the coefficient of friction, as shown in Fig. 7.7(b).

Weakness of MoS₂ Dry Film Lubricant in Air

Sliding friction in air.—In air the coefficient of friction for the molybdenum disulfide (MoS_2) dry film-lubricated Ti-6Al-4V pin sliding on the high-nickel-content superalloy disk started high (approximately 0.30) in the first 23 passes but decreased and reached an equilibrium value of about 0.14 (Fig. 7.8). Continued sliding caused the coefficient of friction to increase at 270 passes. When the coefficient of friction reached an equilibrium value of about 0.32, the experiment was stopped.

The coefficients of friction obtained in air (Fig. 7.8) were much higher than those obtained in vacuum (Fig. 7.6(a)). The coefficient of friction for the dry film lubricant in vacuum was one-third of the value in air. Further, the endurance life of the dry film lubricant in vacuum (the number of passes before the onset of a marked increase in friction) was about three orders of magnitude longer than that in air.

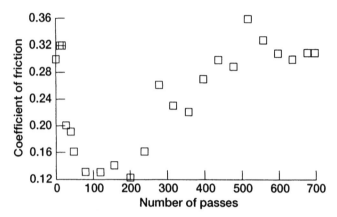

Figure 7.8.—Coefficient of friction as function of number of passes for dry-film-lubricated Ti-6Al-4V pin sliding against high-nickel-content superalloy disk at 8.5 N in humid air.

Seizure when rerun in vacuum.—After 700 passes in air (Fig. 7.8) the worn surfaces were at rest, and loose, large wear particles on the disk surface had been blown off. Then the vacuum chamber was evacuated to 10^{-7} Pa. After approximately 18 hr in contact the sliding motion was restarted. The coefficients of friction (Fig. 7.9(a)) for the pin and disk previously worn in humid air were much higher than those (Fig. 7.7(a)) for the pin and disk previously worn in vacuum. Galling increased the coefficient of friction to 0.6.

After 50 passes of sliding contact in the rerun process the sliding motion was stopped and the load was decreased from 8.5 N to 2.5 N. Contact was maintained for 30 s and then sliding was begun at the new load. The coefficient of friction continued to increase with the increasing number of passes, as shown in Fig. 7.9(b). Seizure between the pin and disk can occur when the dry film lubricant has sustained wear damage (i.e., local removal of the dry film lubricant and exposure of the substrate). Galling increased the coefficient of friction to 1.4.

Wear Damage

Contact between unlubricated Ti-6Al-4V pin and high-nickel-content superalloy disk.—Figures 7.10 to 7.13 present scanning electron microscopy (SEM) micrographs and energy dispersion x-ray analysis (EDX) spectra of the wear damage produced on the unlubricated Ti-6Al-4V pin and on the high-nickel-content superalloy disk after 100 passes at a load of 2.5 N. Substantial plastic deformation occurred on the unlubricated Ti-6Al-4V pin (Fig. 7.10). Large wear debris particles

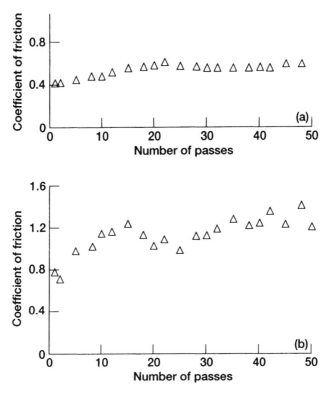

Figure 7.9.—Coefficient of friction as function of number
of passes for surfaces previously worn in humid air of
dry-film-lubricated Ti-6Al-4V pin sliding against high-
nickel-content superalloy disk when rerun in vacuum.
(a) Load, 8.5 N; rotating speed, 40 rpm. (b) Load, 2.5 N;
rotating speed, 1 rpm.

formed by plastic deformation and ductile fracture of the pin are present around the
wear scar. Also present on the wear scar are plastically deformed grooves and
clogged wear debris. Closer SEM examination and EDX analysis of the wear debris
showed it to be composed primarily of elements from the Ti-6Al-4V pin (Fig. 7.11)
and of small amounts of chromium, iron, and nickel from the disk.

Figures 7.12 and 7.13 show the Ti-6Al-4V patches on the wear track of the high-
nickel-content superalloy disk. The transfer patches occupied a large area fraction
of the overall wear track. Thus, severe damage, often called scuffing, scoring, or
galling, occurred in the unlubricated contact between the Ti-6Al-4V and the high-
nickel-content superalloy in vacuum [7.3–7.5].

Figure 7.10.—Wear scar on unlubricated Ti-6Al-4V pin after
sliding against high-nickel-content superalloy disk at
2.5 N in vacuum.

Contact between dry-film-lubricated Ti-6Al-4V pin and high-nickel-content
superalloy disk worn in vacuum and rerun in vacuum.—SEM micrographs of the
worn surfaces of the dry-film-lubricated Ti-6Al-4V pin sliding against the high-
nickel-content superalloy disk in vacuum and then rerun in vacuum show surface
smearing, tearing, and spalling of the dry film lubricant (Figs. 7.14 to 7.17). The
wear damage, which resulted from fatigue of the dry film lubricant, is often called
spalling [7.3–7.5].

Wear after initial sliding: Figure 7.14(a) shows the relatively smooth burnished
surfaces in the upper and lower regions of the wear scar and the rougher surface at
the center of the wear scar. Closer SEM examination and EDX analysis of the wear
damage at the center of wear scar showed that the flake-like wear debris
(Fig. 7.14(b)) resulted from surface smearing and tearing of the dry film lubricant.
EDX analysis of the flake-like debris indicated that it mainly contained the elements
of the MoS_2 lubricant. Figure 7.14(c) shows a crater (dark area) where fragments
of the dry film lubricant were removed and the Ti-6Al-4V was exposed. EDX
analysis of the crater indicated that it mainly contained elements of the Ti-6Al-4V.
Even with spalling the extent of removal and fragmentation of the dry film lubricant
was minimal. Nevertheless, the dry film lubricant occupied most of the overall wear
scar.

Figures 7.15 and 7.16 show a tapered cross section of the worn surface of the dry-
film-lubricated Ti-6Al-4V pin at an angle of 45° to the worn surface. The cross-
section SEM micrographs clearly indicate that most of the dry film lubricant
remained even after about 170 000 passes in vacuum. The film thickness of the

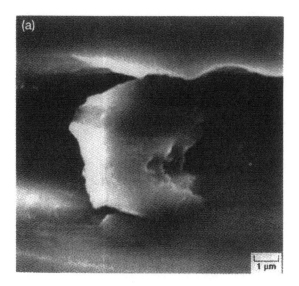

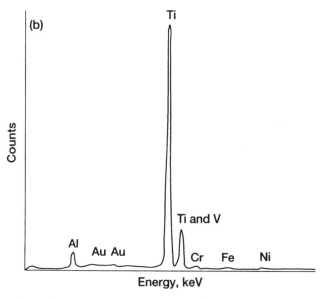

Figure 7.11.—Clogged wear debris on wear scar on unlubricated Ti-6Al-4V pin after sliding against high-nickel-content superalloy disk at 2.5 N in vacuum. (a) Secondary electron SEM image. (b) Spot EDX analysis. (Thin gold film used to reduce charging of mount is responsible for gold signal in spectrum.)

306

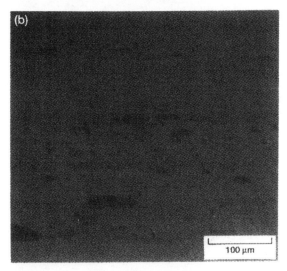

Figure 7.12.—Wear track on high-nickel-content
superalloy disk after sliding by Ti-6Al-4V pin at
2.5 N in vacuum, showing transfer patches of
Ti-6Al-4V. (a) Secondary electron SEM image.
(b) Backscatter electron SEM image.

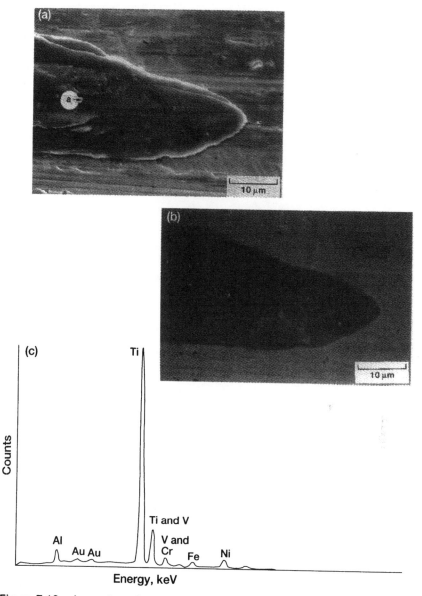

Figure 7.13.—Large transferred patch of Ti-6Al-4V on high-nickel-content superalloy disk after sliding by Ti-6Al-4V pin at 2.5 N in vacuum, showing transfer patches of Ti-6Al-4V. (a) Secondary electron SEM image. (b) Backscatter electron SEM image. (c) EDX spectrum of transferred film on disk. (Data taken at point indicated in part (a). Thin gold film used to reduce charging of mount is responsible for gold signal in spectrum.)

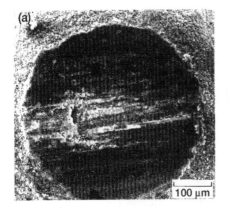

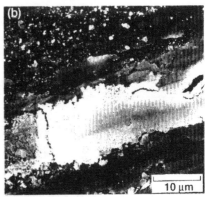

Figure 7.14.—Wear scar on dry-film-lubricated Ti-6Al-4V pin after sliding
against high-nickel-content superalloy disk in vacuum. (a) Low-magnification
overview showing relatively smooth surfaces at upper and lower areas with
spalling and tearing at center. (b) Surface smearing and tearing at center
resulting in particles separating from surface in form of flakes. (c) Spalling
at center.

remaining lubricant was about 15 μm. Close SEM examination revealed dense,
amorphous-like material in the area right underneath the worn surface (Fig. 7.16(a)).

Figure 7.17(a) shows the relatively smooth track on the high-nickel-content
superalloy disk with transfer patches of the dry film lubricant observed (even in the
low-magnification view) mainly at the center of the wear track. Closer SEM
examination of the center and upper regions of the track showed the transferred wear
particles to be in the form of flakes and powders (Figs. 7.17(b) and (c)).

Wear after sliding in air and rerun in vacuum: Figures 7.18 to 7.24 show SEM
micrographs of the worn surfaces of the dry-film-lubricated Ti-6Al-4V pin and the
high-nickel-content superalloy disk run in air for 700 passes and then rerun in

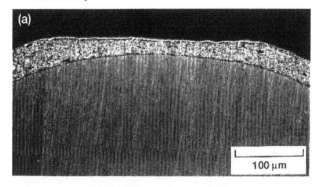

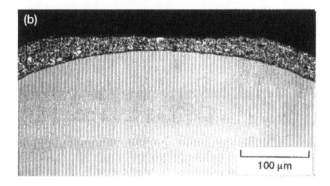

Figure 7.15.—Tapered cross section of worn surface of dry-film-lubricated Ti-6Al-4V pin at 45° angle to worn surface. (a) Secondary electron SEM image. (b) Back-scatter electron SEM image.

vacuum for 100 passes. Both plastic deformation and ductile fracture occurred in the dry-film-lubricated Ti-6Al-4V pin (Fig. 7.18(a)). The backscatter micrograph (Fig. 7.18(b)) reveals three different materials: (1) the light areas in the micrograph show where the transfer patches from the disk stayed on the Ti-6Al-4V pin, (2) the grayer areas show the Ti-6Al-4V substrate with no dry film lubricant present, and (3) the salt-and-pepper areas around the edge of the wear scar show the dry film lubricant. In addition to the major elements of the disk material, the transfer patches also contained elements such as molybdenum and sulfur from the lubricant (Fig. 7.19(a)). On the other hand, the gray areas contained only the elements of Ti-6Al-4V (Fig. 7.19(b)).

The large patches seen on the disk wear track in Fig. 7.20(a) are also apparent in the backscatter micrograph of Fig. 7.20(b), which indicated the dark areas to be

310

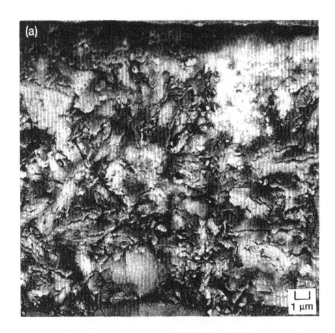

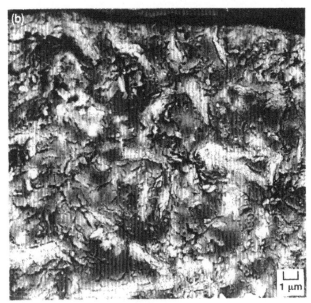

Figure 7.16.—Backscatter electron SEM image, showing comparison of microstructures. (a) Tapered cross section of worn surface. (b) Tapered cross section of as-coated area of dry film lubricant.

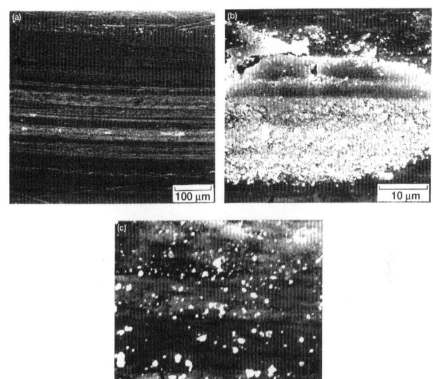

Figure 7.17.—Wear track on high-nickel-content superalloy disk after sliding by dry-film-lubricated Ti-6Al-4V pin in vacuum. (a) Low-magnification overview. (b) Detailed view showing transfer patches at center. (c) Detailed view showing powder-like wear debris of dry film lubricant.

transfer patches from the Ti-6Al-4V. Further, the EDX analysis (Fig. 7.21(a)) of a transfer patch shows that it mainly contained elements from the Ti-6Al-4V with a small amount of elements from the MoS_2 dry film lubricant. The rest of the areas in the wear track contained relatively smaller amounts of elements from the Ti-6Al-4V and the dry film lubricant (Fig. 7.21(b)). Closer SEM examination (Fig. 7.22) of the wear track revealed extensive plastic shearing of the high-nickel-content superalloy disk.

Figure 7.23, a tapered cross section of the worn surface of the dry-film-lubricated Ti-6Al-4V pin, shows clearly that the dry-film lubricant was not present on the worn surface. Further, in this overview micrograph, extrusion out of the wear scar and

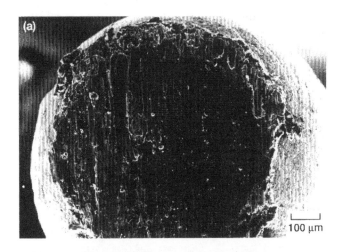

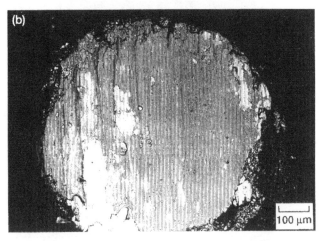

Figure 7.18.—Wear scar on dry-film-lubricated Ti-6Al-4V
pin after sliding against high-nickel-content disk in
humid air and then rerun in vacuum. (a) Secondary
electron SEM image. (b) Backscatter electron SEM
image.

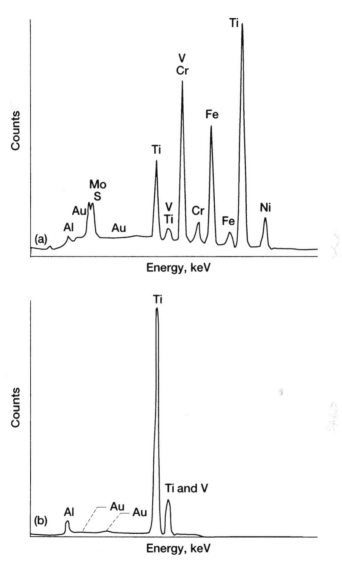

Figure 7.19.—EDX analysis of wear scar on dry-film-lubricated Ti-6Al-4V pin after sliding against high-nickel-content superalloy disk in humid air and then rerun in vacuum. (Thin gold film used to reduce charging of mount is responsible for gold signal in spectra.)

314

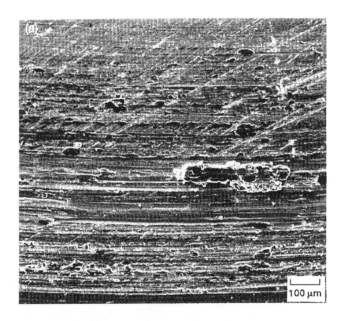

Figure 7.20.—Wear track on high-nickel-content superalloy disk after sliding by dry-film-lubricated Ti-6Al-4V pin in humid air and then rerun in vacuum. (a) Secondary electron SEM image. (b) Backscatter electron SEM image.

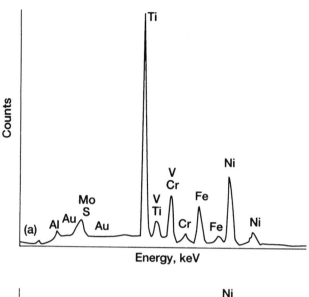

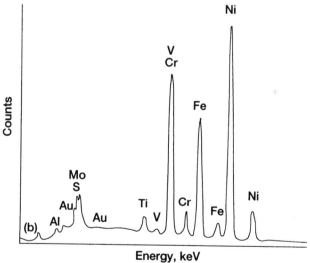

Figure 7.21.—EDX analysis of wear track on high-nickel-content superalloy disk after sliding by dry-film-lubricated Ti-6Al-4V pin in humid air and then rerun in vacuum. (Thin gold film used to reduce charging of mount is responsible for gold signal in spectra.)

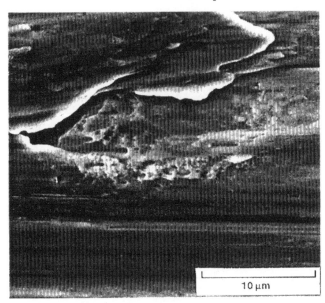

Figure 7.22.—Wear track showing extensive plastic
shearing in high-nickel-content superalloy disk after
sliding by dry-film-lubricated Ti-6Al-4V pin in humid
air and then rerun in vacuum.

transfer patches of the disk material are also well defined because of effective
atomic number contrast. Plastic deformation of the Ti-6Al-4V (extrusion out of the
wear scar) and local solid-phase welding (cold welding) between the
Ti-6Al-4V and transfer patches of the disk material (often called scuffing or
scoring) are well defined by the backscatter electron SEM images (Fig. 7.24).

Thus, the worn surfaces of the pin and disk first run in humid air and then rerun
in vacuum were completely different from the worn surfaces of the pin and disk run
only in vacuum. The surfaces worn in humid air (Figs. 7.18 and 7.20) exhibited
galling accompanied by severe surface damage and extensive transfer of the Ti-6Al-
4V to the high-nickel-content superalloy disk, or vice versa.

7.2.4 Summary of Results and Conclusions for Case Study A

In the investigation of the Galileo antenna anomaly the following results were
obtained:

1. The performance of the dry film lubricant in air was poor when compared with
   its performance in vacuum. The coefficient of friction for the dry-film-
   lubricated system in vacuum was about 0.04; the value in air was 0.13. The

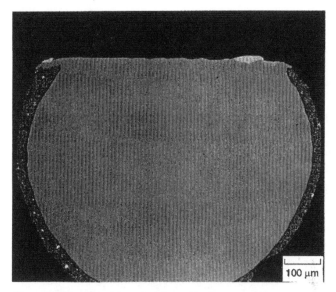

Figure 7.23.—Overview (backscatter electron SEM image) of tapered cross section at 45° angle to worn surface of dry-film-lubricated Ti-6Al-4V pin after sliding against high-nickel-content superalloy disk in humid air and then rerun in vacuum, showing extrusion out of wear scar, transfer patches of high-nickel-content superalloy, and no dry film lubricant on worn surface.

endurance life of the dry film lubricant was about three orders of magnitude longer in vacuum than in air.

2. The worn surfaces of the dry-film-lubricated Ti-6Al-4V pin and the high-nickel-content superalloy disk first run in humid air and then rerun in vacuum were completely different from the surfaces of the pin and disk run only in vacuum. Galling occurred in the former conditions, whereas spalling occurred in the latter conditions.

3. When galling occurred in the contact between the dry-film-lubricated Ti-6Al-4V pin and the high-nickel-content superalloy disk first run in humid air and then rerun in vacuum, the coefficient of friction rose to about 0.32 in humid air and to 1.4 in vacuum. The galling was accompanied by severe surface damage and extensive transfer of the Ti-6Al-4V to the high-nickel-content superalloy, or vice versa.

4. When spalling occurred in the dry-film-lubricated Ti-6Al-4V pin run against the high-nickel-content superalloy disk only in vacuum, the coefficient of friction rose to 0.36 or greater. The wear damage, however, self-healed when sliding was stopped and then rerun in vacuum, and the coefficient of friction decreased to 0.05.

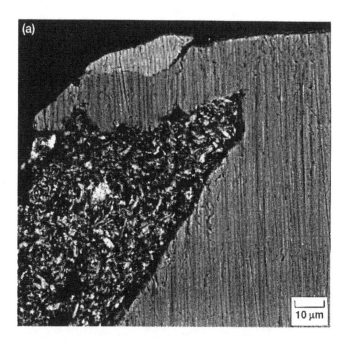

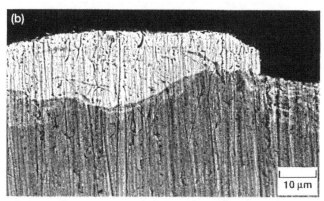

Figure 7.24.—Backscatter electron SEM images of tapered cross section of worn surface of dry-film-lubricated Ti-6Al-4V pin after sliding against high-nickel-content superalloy disk in humid air and then rerun in vacuum. (a) Showing extrusion out of wear scar. (b) Showing transfer patch of high-nickel-content superalloy on worn surface of dry-film-lubricated Ti-6Al-4V pin.

The aforementioned tribological results were also constructively reviewed in Johnson's paper [7.5]. He concluded, "The high contact stress on the V-groove pin/ socket interfaces destroyed the integrity of the lubricant film and started the chain of events that led to the deployment anomaly....The use of dry lubricant, specifically molybdenum disulfide, on a mechanism that is going to be operated in an atmosphere should be carefully evaluated. The wear rate of the MoS_2 in air is so much higher than in a vacuum that any coatings could be worn out by in-air testing and not provide the desired lubrication when needed. The pins and sockets on the high-gain antenna that received the greatest amount of relative motion due to the shipping method were the same ones that were exercised most by the vibration testing. These are also the same pins and sockets that are stuck on the spacecraft. One solution to the problem of ambient testing wearing out the lubricant coating would be to replace the lubricated components just prior to launch so there is a virgin lubricant surface for the flight operation."

As Peter Jost [7.2] stated, in the operation of space mechanisms functional reliability is, of course, vital. Even a small tribological failure can clearly lead to catastrophic results. The absence of the required knowledge of tribology can act as a severe brake on the development of new technologies. Tribological reliability of mechanical systems in the highest order will be secured in the operation of many interacting surfaces in relative motion if greater attention (such as the following examples) will be paid to tribology:

1. The effects of adhesion or cold welding on mechanical problems in deep space vehicles must be investigated.
2. Simulation experiments must be conducted on various adhesion couples in different environments and temperature ranges.
3. Particular variables contributing to the mechanism of the adhesion process must be scientifically examined.

7.3 Case Study B: Space Shuttle Orbiter's Quad Check Valve Failures

7.3.1 Quad Check Valve Failures

The space shuttle orbital maneuvering system (OMS) provides the thrust for orbit insertion, circularization, orbit transfer, rendezvous, and deorbit. The space shuttle reaction control system (RCS) provides the thrust for pitch, yaw, and roll maneuvers and for small velocity changes along the orbiter axis. The OMS/RCS left- and right-hand pods are attached to the upper aft fuselage on the left and right sides (Fig. 7.25). Each pod is divided into two compartments: the OMS housing and the RCS housing. RCS thrusters and associated propellant feed lines are also attached to the forward module.

The space shuttle quad check valve (QCV) is used on both the orbiter's OMS and RCS. The function of the QCV is to allow helium pressurant to flow downstream while precluding upstream backflow of helium and propellant vapors or liquids (Fig. 7.26). Each unit consists of four check valve module elements arranged as two parallel assemblies with two series check valve modules. A significant number of QCV failures (poppet stuck open) have resulted in the removal of QCV's from the orbiter.

7.3.2 Failure Investigation

Because of leakage and cracking pressure problems more than two dozen QCV's have been removed for repair and overhaul. Evidence not always present includes sapphire pin wear, propellant residues, and small cracks in the valve seat, which is coated with chemical-vapor-deposited tungsten and tungsten carbide (W/WC). Issues are as follows:

1. Sticking and seizing up of the W/WC poppet on the sapphire guide pin
2. Poor wear couple between the W/WC poppet and the sapphire guide pin
3. Spring force, originally designed to meet requirements for low cracking pressure, lower than necessary to close poppet

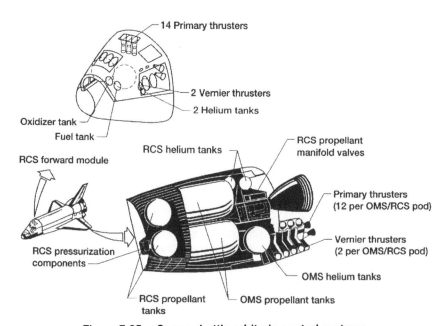

Figure 7.25.—Space shuttle orbiter's control systems.

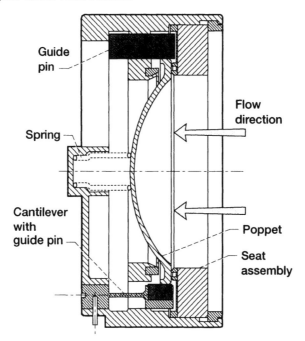

Figure 7.26.—Quad check valve design.

7.3.3 Investigation of Friction, Wear, and Lubrication

Experimental Conditions

Because QCV's in orbit are not exposed to the space vacuum environment, reciprocating sliding friction experiments were conducted in the atmosphere. The conditions and contact configurations (Table 7.2) were chosen to simulate as closely as possible the conditions that the poppet-pin interface of a QCV may have experienced. The poppet traveled back and forth, retracing its tracks on the bare surfaces of the counterpart materials: sapphire, synthetic single-crystal diamond, and natural single-crystal {111} diamond.

The poppet is made of a tungsten material with dispersed tungsten carbides (W/16 wt% WC) produced by a chemical vapor deposition (CVD) process (i.e., deposited on a commercially pure molybdenum mandrel). Further, two types of poppet were tested: one with a polytetrafluoroethylene (PTFE) top layer coated on the W/WC poppet and one without the PTFE top layer (i.e., a bare W/WC poppet). Figure 7.27 presents Vickers microhardnesses for the PTFE-coated poppet and the bare poppet as a function of indentation load. Note that the seat also uses the W/WC chemical vapor deposited on a thoriated tungsten mandrel.

TABLE 7.2.—EXPERIMENTS FOR CASE STUDY B
(a) Conditions of friction and wear experiments

Contacts	Poppet on natural single-crystal diamond {111} flat; poppet on synthetic single-crystal diamond; poppet on sapphire guide pin
Motion	Reciprocating
Load, N	3
Transverse (sliding) speed, s/track (m/s)	0.76 (1)
Track length, mm	≤0.76
Environment	Air
Temperature, °C	22
Specimen cleaning	1,1,1,-trichloroethane

(b) Contact configurations

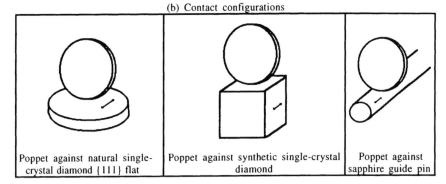

Poppet against natural single-crystal diamond {111} flat	Poppet against synthetic single-crystal diamond	Poppet against sapphire guide pin

(c) Specimen sizes

Poppet	Approximately 30 mm in diameter with rounded circular edge
Sapphire pin	3.2 mm in diameter; 9 mm long
Synthetic single-crystal diamond	1.4-mm by 1.4-mm rectangular surface; 1.1 mm thick
Natural single-crystal {111} diamond	30-mm² flat platelet; 2 mm thick

Sliding Friction in Air

PTFE-coated W/WC poppet.—Figure 7.28 shows the mean coefficients of friction for the PTFE-coated W/WC poppet in contact with bare sapphire, synthetic single-crystal diamond, and natural single-crystal {111} diamond in air. The steady-state coefficients of friction in contact with the natural and synthetic diamond were 0.06 and 0.03, respectively, and the friction was smooth with no stick-slip behavior. On the other hand, the poppet in contact with the bare sapphire had a high steady-state coefficient of friction (approximately 0.25, or greater than that of the diamonds by a factor of approximately 4 to 8). Also, the friction trace for the sapphire indicated that the friction was erratic and unstable with stick-slip behavior. In the conditions examined with the sapphire pin, PTFE was not an effective lubricating film.

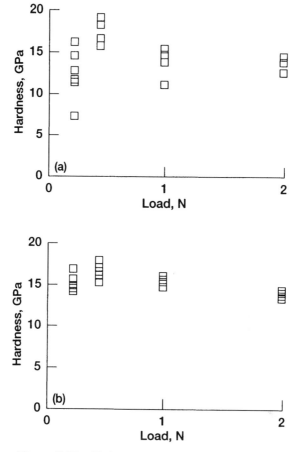

Figure 7.27.—Vickers microhardness as function of indentation load for (a) PTFE-coated W/WC poppet and (b) bare W/WC poppet.

Bare W/WC poppet.—Figure 7.29 shows the mean coefficients of friction for the bare W/WC poppet in contact with bare sapphire, synthetic single-crystal diamond, and natural single-crystal {111} diamond in air. The steady-state coefficient of friction in contact with the natural and synthetic diamond was 0.02, and the friction was smooth and calm with no stick-slip behavior. On the other hand, the bare poppet in contact with bare sapphire had a high steady-state coefficient of friction (approximately 0.7, or greater than that of the diamonds by a factor of approximately 35). Also, the friction trace for the sapphire indicated that the friction was erratic and unstable, with stick-slip behavior between 300 and 1600 passes. With the diamonds the steady-state coefficients of friction were lower when sliding against the bare poppet than when sliding against the PTFE-coated poppet.

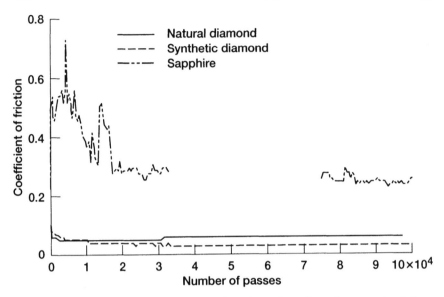

Figure 7.28.—Coefficient of friction as function of number of passes in air for PTFE-coated W/WC poppet sliding against sapphire, synthetic diamond, and natural diamond.

Wear Damage

PTFE-coated poppet.—Figure 7.30 presents SEM micrographs of the wear track on the sapphire and of the wear scar on the PTFE-coated W/WC poppet after 100 000 passes in air. The sapphire wear track contained transferred materials and wear debris. Closer SEM examination at high magnification clearly showed transfer patches of materials from the poppet. An EDX spectrum taken at the center of the sapphire wear track (Fig. 7.31(a)) indicated the presence of tungsten from the counterpart material. On the other hand, the wear scar produced on the PTFE-coated W/WC poppet had almost no transferred material from the sapphire (Figs. 7.30(b) and 7.31(b)).

The as-received surface of synthetic diamond is a cleavage face of a single crystal. This surface is flat and smooth but contains cleavage steps. The steps between smoothly cleaved planes with slightly different height levels generally have sharp edges (Fig. 7.32). Figure 7.33 presents SEM micrographs of the wear track on synthetic single-crystal diamond and of the wear scar on the PTFE-coated W/WC poppet after 100 000 passes in air. Because the sliding direction was almost perpendicular to the cleavage steps on the diamond surface in this case, the sharp edges of the cleavage steps cut and plowed the surface of the PTFE-coated W/WC poppet. Therefore, the diamond wear track contained wear debris produced by this

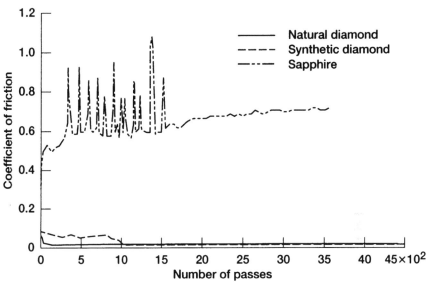

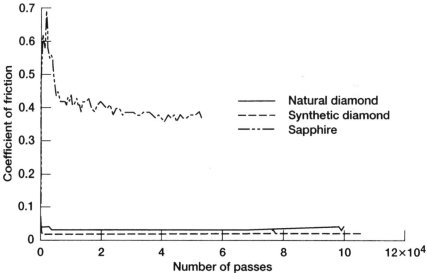

Figure 7.29.—Coefficient of friction as function of number of passes in air for bare W/WC poppet sliding against sapphire, synthetic diamond, and natural diamond.

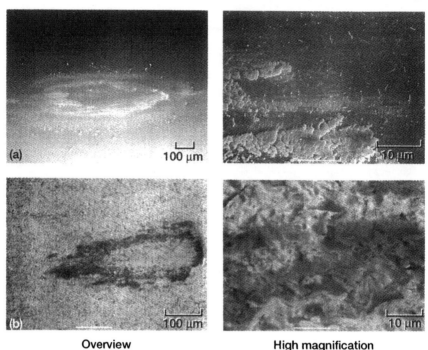

Overview   High magnification

Figure 7.30.—SEM micrographs of (a) wear track on sapphire and (b) wear scar on PTFE-coated W/WC poppet.

cutting and plowing action. Closer SEM examination at high magnification clearly showed wear debris particles and patches of materials from the poppet. An EDX spectrum taken at the patches on the diamond wear track (Fig. 7.34) indicated the presence of tungsten from the poppet materials. On the other hand, the wear scar produced on the PTFE-coated W/WC poppet had a smooth appearance, indicating complete removal of the PTFE coating in the observed area (Fig. 7.33(b)).

Because the natural single-crystal {111} diamond surface was polished, it was smooth and had no cleavage steps. Figure 7.35 presents SEM micrographs of the wear track on the natural diamond and of the wear scar on the PTFE-coated W/WC poppet after approximately 100 000 passes in air. The diamond wear track had no visible wear (Fig. 7.35(a)). Only closer SEM examination at high magnification indicated submicrometer wear debris particles from the poppet. The wear scar produced on the PTFE-coated W/WC poppet had no visible transferred material from the diamond (Fig. 7.35(b)).

Bare W/WC poppet.—Figure 7.36 presents SEM micrographs of the wear track on sapphire and of the wear scar on the bare W/WC poppet after approximately 53 000 passes in air. The sapphire wear track contained transferred materials and

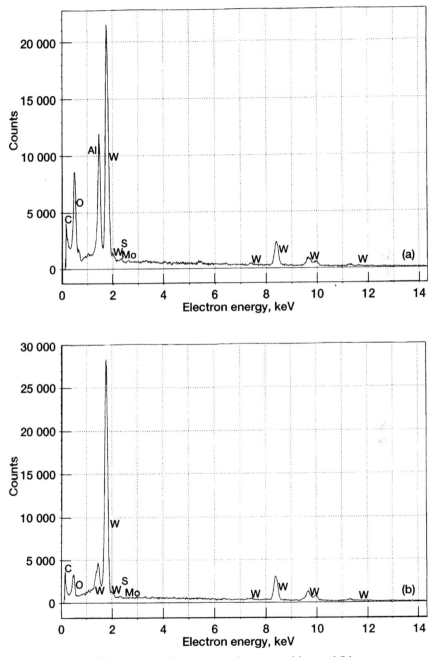

Figure 7.31.—EDX spectra of (a) wear track on sapphire and (b) wear scar on PTFE-coated W/WC poppet.

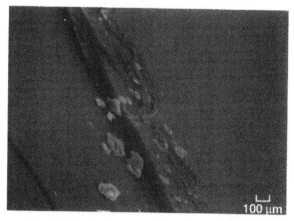

Figure 7.32.—Cleavage steps on synthetic single-
crystal diamond surface.

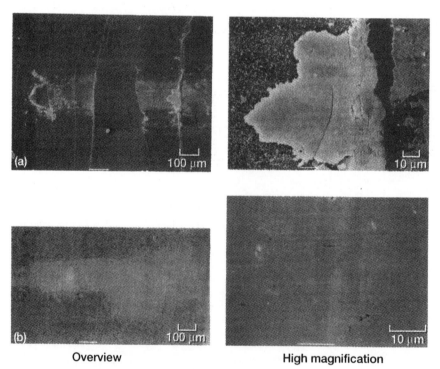

Overview   High magnification

Figure 7.33.—SEM micrographs of (a) wear track on synthetic single-crystal
diamond and (b) wear scar on PTFE-coated W/WC poppet.

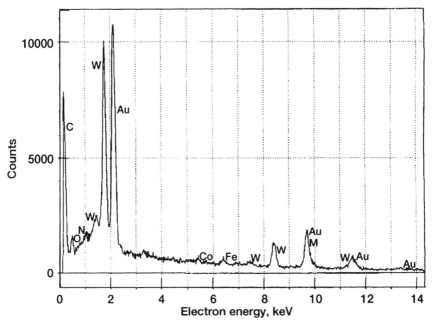

Figure 7.34.—EDX spectrum of wear track on synthetic single-crystal diamond.

wear debris. Closer SEM examination at high magnification clearly showed transferred materials from the poppet. An EDX spectrum taken at the center of the sapphire wear track indicated the presence of tungsten from the counterpart material. The wear scar produced on the bare W/WC poppet had a relatively smooth worn surface with wear debris.

Figure 7.37 presents SEM micrographs of the wear track on the synthetic single-crystal diamond and of the wear scar on the bare W/WC poppet after approximately 100 000 passes in air. In this case the sliding direction was almost parallel to the cleavage steps on the diamond surface. The diamond wear track contained wear debris particles and patches of material from the poppet (Fig. 7.37(a)). An EDX spectrum taken at the patches on the diamond wear track indicated the presence of tungsten from the poppet materials. On the other hand, the wear scar produced on the bare W/WC poppet had a smooth appearance with local fracture pits (Fig. 7.37(b)).

The polished natural single-crystal diamond had no visible wear even after approximately 100 000 sliding passes against the bare W/WC poppet. The bare W/WC poppet had a wear scar with wear debris, as shown in Fig. 7.38.

7.3.4 Summary of Results and Conclusions for Case Study B

In the investigation of failures of the space shuttle orbiter's quad check valve, the following results were obtained:

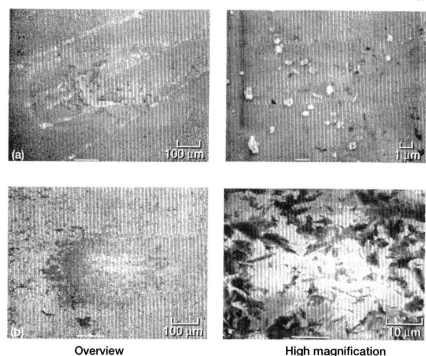

Overview   High magnification

Figure 7.35.—SEM micrographs of (a) wear track on natural single-crystal diamond and (b) wear scar on PTFE-coated W/WC poppet.

1. The coefficients of friction for a sapphire guide pin in sliding contact with both a polytetrafluoroethylene (PTFE)-coated and a bare tungsten/tungsten carbide (W/WC) poppet were high, unstable, and erratic with stick-slip behavior. Because of the wide variation in coefficient of friction resulting from high adhesion, it will be difficult to predict the functional performance of a check valve based on a W/WC poppet sliding on a sapphire guide pin.

2. The coefficients of friction of synthetic and natural diamond surfaces in contact with both a PTFE-coated and a bare W/WC poppet were generally low and did not show variable, stick-slip behavior. The steady-state coefficients of friction were lower for diamond sliding against the bare poppet than for diamond sliding against the PTFE-coated poppet. When compared with the sapphire guide pins, the synthetic and natural diamonds reduced the coefficient of friction by factors of 4 to 8 and 35 when sliding against the PTFE-coated and bare poppets, respectively.

3. More wear debris transferred to the sapphire for the PTFE-coated W/WC poppet in sliding contact with the sapphire guide pin than for the bare W/WC poppet and sapphire couple.

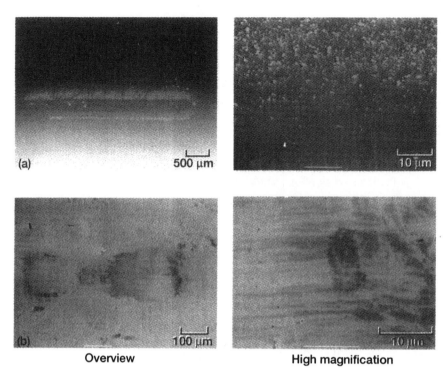

Overview   High magnification

Figure 7.36.—SEM micrographs of (a) wear track on sapphire and (b) wear
scar on PTFE-coated W/WC poppet.

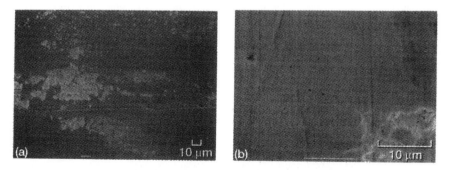

Figure 7.37.—SEM micrographs of (a) wear track on synthetic single-crystal
diamond and (b) wear scar on PTFE-coated W/WC poppet.

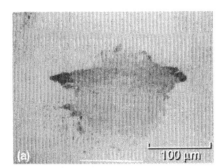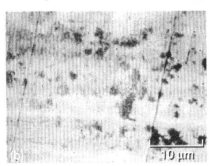

Figure 7.38.—SEM micrographs of wear scar on bare W/WC poppet after sliding against synthetic single-crystal diamond. (a) Overview. (b) High magnification.

4. Cleaved surfaces of synthetic diamond produced a great amount of wear debris from the poppet by cutting and plowing the sharp edges of the cleavage steps.
5. Polished surfaces of natural diamond had no visible wear and reduced poppet wear when compared with sapphire and cleaved synthetic diamond surfaces.

From these results the following conclusions were drawn:

1. The coefficient of friction for the sapphire guide pin is too high and dependent on surface conditions and would make the operation of a quad check valve (QCV) very difficult.
2. Diamond has a low and constant coefficient of friction with no stick-slip behavior. It can be effectively used as a guide pin material and can be a good replacement material for sapphire in space shuttle QCV's.
3. The sliding surface of diamond must be prepared with care. Surface integrity and smoothness are of paramount importance to functional reliability of the poppet-pin system.

References

7.1 W. Shapiro, et al., Space mechanisms lessons learned study, Vol. I, Summary, NASA TM–107046 (1995).

7.2 H.P. Jost, Tribology—origin and future, *Wear, 136*: 1–17 (1990) .

7.3 J. Wilson, Galileo's antenna: the anomaly at 37 million miles, JPL Universe, Jet Propulsion Laboratory, July 3, 1992.

7.4 Unfurling the HGA's enigma, Galileo Messenger, Jet Propulsion Laboratory, Aug. 1991.

7.5 M.R. Johnson, The Galileo high gain antenna deployment anomaly, *28th Aerospace Mechanisms Symposium*, NASA CP–3260, 1994, pp. 359–377.

7.6 K. Miyoshi, et al., Properties data for opening Galileo's partially unfurled main antenna, NASA TM–105355 (1992).

Chapter 8
Structures and Mechanical Properties of Natural and Synthetic Diamond

8.1 Introduction and Historical Perspective

Diamond is an allotrope of carbon, joining graphite and the fullerenes as the major pure carbon structures. Diamond has a unique combination of properties: hardness, thermal conductivity, chemical and thermal inertness, and abrasion resistance. These properties make diamond attractive for a wide range of tribological applications, including solid lubrication. Presently, modern diamonds fall into four distinct categories: natural, high-pressure synthetic, chemical vapor deposited, and diamondlike carbon.

Natural diamond is produced at high pressures and temperatures in volcanic shafts. Rough natural diamonds (i.e., uncut and unpolished) were known and prized in antiquity [8.1]. They were first reported in India 2700 years ago. From India diamond trading moved gradually westward through Persia and the Roman Empire. However, the full beauty of diamond was not uncovered until faceting and polishing techniques were developed in the 14th and 15th centuries. Natural diamond is the primary source of diamond gemstones and by far the leader in terms of monetary value. However, in the last 100 years or so the scarcity and high cost of natural diamond have been challenged by the large-scale production of synthetic diamond.

The high-pressure synthesis essentially duplicates the natural high-pressure, high-temperature process [8.1, 8.2]. The French chemist A. L. Lavoisier found in 1772 that the product of diamond combustion was singular: carbon dioxide (CO_2). In 1814 the English chemist H. Davy proved conclusively that diamond was a crystalline form of carbon. He also showed that burning diamond produced only CO_2, without the formation of aqueous vapor, indicating that it was free of hydrogen and water. Since that time many attempts have been made to synthesize diamond by trying to duplicate nature. These attempts, spread over a century, were unsuccessful. It was not until 1955 that the first unquestioned syntheses were achieved in the United States (General Electric), in Sweden (AESA), and in the Soviet Union (Institute for High-Pressure Physics). The advent of synthetic diamond and the rapid rise of industrial applications, particularly in wear, abrasive, and tool applications, have drastically altered the industry. More changes, like the recent development of chemical-vapor-deposited diamond and diamondlike carbon, will undoubtedly take place in the future. Although high-pressure synthetic diamond and natural diamond are used in the industrial market, their use in tribological applications has been limited because of their small size and high cost and their need to be bonded to a substrate in a separate operation.

Low-pressure diamond synthesis was discovered in the 1970's [8.3, 8.4]. In the 1980's synthetic diamond was produced in coating form by using a variety of low-pressure, vapor-phase synthesis techniques under relatively benign conditions. The low-pressure, vapor-phase process is based on chemical vapor deposition (CVD), a relatively inexpensive process similar to carbide and nitride processing. The material produced is often called CVD diamond, vapor-phase diamond, or diamond coating. CVD diamond coatings 1 mm or more thick are now routinely produced. After removing the substrate a free-standing shape of CVD diamond coating, normally polycrystalline, remains with good integrity and properties similar to those of single-crystal natural and high-pressure synthetic diamond.

Another new form of carbon coating was developed in the 1980's and is now available. A metastable carbon produced as a thin coating by low-pressure synthesis, it is known as diamondlike carbon (DLC). Although DLC has properties similar to those of CVD diamond, it cannot be obtained as thick monolithic shapes, at least with the present technology. A revolution in diamond technology and the diamond business is in progress as the low-pressure process becomes an industrial reality. It will soon be possible to take advantage of the outstanding properties of diamond to develop a myriad of new applications. The production of large diamond films or sheets at low cost is a distinct possibility in the not-too-distant future and may lead to drastic changes in the existing technological and business structure.

In the field of tribology synthetic diamond film technology looks promising and could edge into tribological and tooling applications. However, coating specialists and researchers still tackle such problems as enhancing adhesion to substrates (including steel), increasing bending strength, lowering deposition temperature, speeding up deposition (production) rates, and lowering costs. CVD diamond and DLC coatings offer a broader potential in the field of tribology than do natural and high-pressure synthetic diamond because size and eventually cost will be less of a limitation. For example, CVD diamond has been deposited on a 20-cm-diameter area, and DLC has been deposited on a 200-cm knife-edge. For large areas (>5 mm^2) CVD diamond and DLC are the only forms of diamond or diamondlike materials that seem economically viable. These films open the door to tribological technology and design engineering that can take full advantage of diamond's intrinsic properties to make solid-lubricating, wear-resistant, erosion- and corrosion-resistant, and protective coatings. These CVD diamond and DLC materials could offer lively new solutions to important engineering problems and exciting challenges to the tribologist or designer who seeks to cost-effectively exploit their potential. The development of safe and economic applications for diamond materials in competition with conventional tribological materials requires understanding their behavior and taking innovative approaches to manufacturing.

This chapter reviews the structures and properties of natural and synthetic diamond to gain a better understanding of the tribological behavior of these materials and related materials to be described in the following chapters.

8.2 Atomic and Crystal Structure of Diamond

Understanding diamond's behavior and properties requires a clear picture of the atomic configuration of the carbon atom and the various ways in which it bonds to other carbon atoms [8.1–8.8].

8.2.1 Tetragonal sp^3 Orbital and Trigonal sp^2 Orbital of Carbon Hybrid

The six electrons of the carbon atom in the ground state (i.e., a single atom) are configured as $1s^2 2s^2 2p^2$ (i.e., two electrons are in the K shell (1s) and four are in the L shell, two in the 2s orbital and two in the 2p orbital). The $1s^2 2s^2 2p^2$ configuration of the carbon atom does not account for the tetrahedral symmetry found in structures such as diamond or methane (CH_4). The carbon atom is bonded to four other carbon atoms in the case of diamond, or to four atoms of hydrogen in the case of methane. In both cases the four bonds are of equal strength.

To have an electron configuration that would account for this tetrahedral symmetry, the structure of the carbon atom must be altered to a state with four valence electrons instead of two, each in a separate orbital and each with its spin uncoupled from the other electrons. This alteration occurs when hybrid atomic orbitals form in which the electrons in the L shell of the ground-state atom are rearranged, one 2s electron being promoted (or lifted) to the higher orbital 2p. The new orbitals are called hybrids, since they combine the 2s and 2p orbitals. They are labeled sp^3, since they are formed from one s orbital and three p orbitals.

The hybridization accounts for the tetrahedral symmetry and the valence state of 4, with four 2sp^3 orbitals in the diamond atomic structure. The orbitals are bonded to the orbitals of four other carbon atoms with a strong covalent bond (i.e., the atoms share a pair of electrons) to form a regular tetrahedron with equal angles to each other of 109°28′, as shown in Fig. 8.1(a).

In addition to the sp^3-tetragonal hybrid orbital, two other orbitals complete the series of electronic building blocks of all carbon allotropes and compounds: the sp^2 and sp orbitals. Whereas the sp^3 orbital is the key to diamond and aliphatic compounds, the sp^2 (or trigonal) orbital is the basis of all graphitic structures and aromatic compounds: graphite, graphitic materials, amorphous carbon, and other carbon materials.

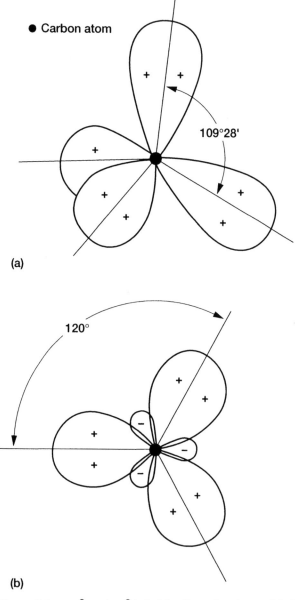

Figure 8.1.—sp³ and sp² hybridization of carbon orbitals. (a) Four sp³ orbitals of carbon atom. (Negative lobes omitted for clarity.) (b) Planar section of sp² hybrid orbitals of carbon atom.

The mechanism of the sp^2 hybridization is somewhat different from that of the sp^3 hybridization. The electrons in the L shell of the ground-state atom are rearranged, one 2s electron being promoted and combined with two of the 2p orbitals (hence, the designation sp^2) to form three sp^2 orbitals and an unhybridized free (or delocalized) p orbital electron. The valence state is now 4.

The calculated electron-density contour of the sp^2 orbitals is similar in shape to that of the sp^3 orbitals. Like the sp^3 orbital the sp^2 is directional and is called a sigma (σ) orbital; and the bond, a sigma bond. These three identical sp^2 orbitals are in the same plane, and their orientation of maximum probability forms equal angles to each other of 120° as shown in Fig. 8.1(b).

The fourth orbital (i.e., the delocalized, nonhybridized p electron) is directed perpendicular to the plane of the three sp^2 orbitals and becomes available to form the subsidiary pi (π) bond with other atoms.

8.2.2 Carbon Covalent sp^3 and sp^2 Bonds

In diamond each tetrahedron of the hybridized carbon atom combines with four other hybridized atoms (tetrahedra) to form a strongly bonded, three-dimensional, entirely covalent lattice structure, as shown schematically in Fig. 8.2. The covalent link between the carbon atoms of diamond is characterized by a short bond length (0.154 nm) and a high bond energy (711 kJ/mol).

● Carbon atom
⬭ Region of high electron probabilities (covalent bonding)

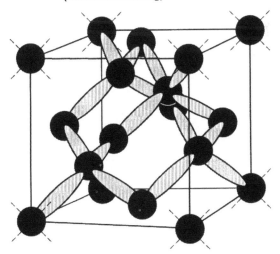

Figure 8.2.—Three-dimensional representation of sp^3 covalent bonding (lattice structure).

Diamond has two crystalline structures, one with cubic symmetry (by far the more common and stable) and one with hexagonal symmetry (found in nature as the mineral lonsdaleite). Cubic diamond will be referred to simply as "diamond" in this book. Table 8.1 summarizes the structure and properties important to tribologists and lubrication engineers. For comparison the table includes the structure and properties of graphite.

The cubic structure of diamond can be visualized as a stacking of puckered infinite layers (the {111} planes), as shown in Fig. 8.3(a). The stacking sequence of the {111} planes is ABCABC, so that every third layer is identical.

Graphite is composed of a series of stacked parallel layers (Fig. 8.3(b)) with trigonal sp^2 bonding. Within each layer the carbon atom is bonded to three others, forming a series of continuous hexagons in what can be considered as an essentially infinite two-dimensional molecule. Each sp^2-hybridized carbon atom combines with three other sp^2-hybridized atoms to form a series of hexagonal structures, all located in parallel flat layers (Fig. 8.3(b)). Like the sp^3 bond the sp^2 bond is covalent. It is also short (0.1415 nm) and strong (524 kJ/mol) because of the three sp^2 valence electrons and the small size of the atom. The fourth valency (i.e., the free delocalized electron) is oriented perpendicular to this plane. The hybridized fourth valence electron is paired with another delocalized electron from the adjacent layer by a much weaker van der Waals bond (a secondary bond arising from structural polarization) of only 7 kJ/mol (π bond). Thus, graphite's physical properties have marked layerlike anisotropy.

The most common stacking sequence of the graphite crystal is hexagonal (alpha) with a –ABABAB– stacking order (i.e., the carbon atoms in alternate layers are superimposed over each other, Fig. 8.3(b)). Carbon is the only element to have this particular layered hexagonal structure. The crystal lattice parameters (i.e., the relative positions of its carbon atoms) are $a_0 = 0.246$ nm and $c_0 = 0.6708$ nm. Hexagonal graphite is the thermodynamically stable form of graphite.

TABLE 8.1.—CRYSTAL STRUCTURE OF DIAMOND AND GRAPHITE

Structure	Diamond	Graphite
Carbon bonding (bond energy, kJ/mol)	Covalent sp^3 (711)	Covalent sp^2 (524); van der Waals (7)
Crystalline form	Cubic; hexagonal	Hexagonal; rhombohedral
Crystal lattice parameter, nm	Cubic, 0.3567	Hexagonal: $a_0 = 0.246$ $c_0 = 0.6708$
Common crystal face	{111}, {011}, {001}	{0001}, {10$\bar{1}$0}, {10$\bar{1}$1}, {$\bar{1}$012}
Cleavage and slip plane	{111}	{0001}

● Carbon atom

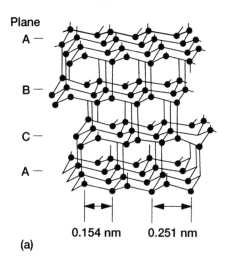

Plane
A —
B —
C —
A —

0.154 nm   0.251 nm

(a)

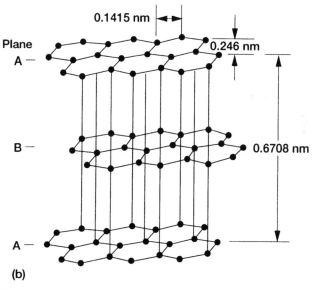

0.1415 nm

Plane
A —
0.246 nm

B —
0.6708 nm

A —

(b)

Figure 8.3.—Schematic diagrams of crystal structures of
diamond and graphite. (a) Cubic structure of diamond
(ABC sequence). (b) Hexagonal structure of graphite
(ABAB sequence).

8.3 Impurities

One universally accepted classification of diamond is based on the nature and amount of impurities contained within the structure (Table 8.2). Diamond, synthetic or natural, is never completely free of impurities. These impurities are divided into two principal types: (1) lattice impurities (type I), which consist of foreign elements incorporated into the crystalline lattice, the foreign atom replacing a carbon atom; (2) inclusions (type II), separate particles that are not part of the lattice and usually consist of aluminum, magnesium, or calcium, such as olivine and garnet. Only nitrogen and boron are known with certainty to be incorporated into the diamond lattice. These two elements have small atomic radii and fit readily within the diamond structure. Nitrogen is present in all natural diamond, with up to 2000 ppm in type Ia diamond. Boron is present in natural type IIb diamond (up to 100 ppm) and in specifically doped type IIb synthetic diamond (270 ppm). Many other impurities have been detected but are believed to be inclusions. For example, up to 10 ppm of aluminum is frequently found in natural diamond, and synthetic diamond may contain nickel or iron inclusions that occupy as much as 10% of the crystals. The properties of diamond are susceptible to these impurities, and even a minute amount of a foreign element such as nitrogen can cause drastic changes. These effects are shown in Table 8.2 and discussed in Section 8.4.

The tetrahedral bonding configuration (Fig. 8.2) and the strong carbon-carbon bond give diamond its unique properties (summarized in Table 8.3). For comparison the table also includes the properties of graphite. Determining the properties of diamond is difficult, and published data are still limited. Small specimen size, imperfections, and impurities contribute to the considerable spread in reported values often found in the literature.

TABLE 8.2.—CLASSIFICATION AND CHARACTERISTICS OF DIAMOND

Lattice impurity type	Thermal conductivity, W/m-°C	Electric resistivity, ohm-cm	Color	Impurity		Synthetic diamond	Abundance in natural diamond, percent
				Nitrogen, ppm	Others, ppm		
Ia	800	[a] 10^4–10^{16}	Clear to yellow	[b] ≈ 2000	--------	None	≈ 98
Ib	800–1700	[a] 10^4–10^{16}	Green to brown	[c] 10^2–10^3	[d] 10^4–10^5	Powder	≈ 0.1
Ib	2000	[a] 10^{16}	Yellow	[c] 1–100	--------	Single crystal	≈ 0.1
IIa	2000	[a] 10^{16}	Colorless clear	[e] ≈ 1	--------	Single crystal	1–2
IIb	------	[f] 10–10^4	Blue	≈ 1	[g] ≈ 100	Single crystal	≈ 0

[a] Insulator.
[b] Platelet form.
[c] Dispersed in lattice.
[d] Solvent metals.
[e] High purity.
[f] Semiconductor P type.
[g] Boron.

TABLE 8.3.—PROPERTIES OF DIAMOND AND GRAPHITE

Property	Diamond	Graphite
Density, g/cm^3	3.52	2.26
Young's modulus, GPa	^{a}910–1250	Single crystal: 1060 (a direction) 36.5 (c direction) Pyrolitic graphite, 28–31 Molded graphite, 5–10
Poisson's ratio	^{b}0.10–0.29	--------------------------------
Compression strength, GPa	8.68–16.53	0.065–0.089
Vickers hardness, GPa	60–100	Pyrolitic graphite, 2.4–3.6 Molded graphite, 3.9–9.8
Thermal conductivity at 25 °C, W/m-°C	600–1000 (type Ia) 2000–2100 (type II)	Pyrolitic graphite: a and b directions, 190–390 c direction, 1–3 Molded graphite, 31–159
Coefficient of thermal expansion at 25 to 200 °C, 10^{-6}/deg C	^{c}0.8–4.8	Pyrolitic graphite: a and b directions, −1 to 1 c direction, 15–25 Molded graphite, 3.2–5.7
Specific heat at 25 °C, kJ/kg-K	0.502–0.519	0.690–0.719

^{a}Typical values, 1050 and 1054.
^{b}Typical value, 0.2.
^{c}1.5 to 4.8 at 127 to 927 °C.

8.4 Mechanical Properties of Diamond

The properties described here are those of both natural and synthetic diamond. See the reference list [8.9–8.25] for further reading.

8.4.1 Strength and Hertzian Fracture Properties

Materials with high hardness are usually brittle and diamond is no exception. Diamond behaves as a brittle solid and fractures along its cleavage planes. The fracture behavior of diamond is dominated by cleavage on the {111} plane, but cleavage fracture has also been observed along the other planes, such as {211}, {110}, and {322}, since the theoretical cleavage energy differences between planes are small, as shown in Table 8.4. The cleavage energies are consistent with the order of the cleavage planes, from easiest to hardest to cleave.

Knowledge of the fracture stress and crack patterns produced in diamond is important in understanding tribological phenomena, such as wear, erosion, and abrasion. Such knowledge can be gained from crack patterns produced in natural and synthetic diamond by spherical indenters under normal loading. An indenter loaded onto the surface of a specimen generally produces a local damage zone comparable to the dimensions of the contact area (Fig. 8.4(a)). These results have interest for two reasons. First, it is important to know diamond's fracture behavior, particularly the crack patterns and fracture stress under conditions arising in its tribological applications. Second, indentation techniques are the main method for measuring the strength of materials.

TABLE 8.4.—THEORETICAL CLEAVAGE
ENERGIES FOR DIAMOND

Plane	Angle between plane and {111} plane	Cleavage energy,[a] J/m²
111	0°, 70°32′	10.6
332	10°0′	11.7
221	15°48′	12.2
331	22°0′	12.6
110	35°16′, 90°	13.0
322	11°24′	13.4
321	22°12′	14.3
211	19°28′	15.0
320	36°48′	15.3
210	39°14′	16.4
311	29°30′	16.6
100	54°44′	18.4

[a]To obtain fracture surface energy, divide by 2.

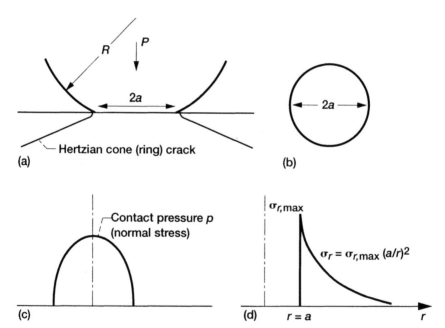

Figure 8.4.—Geometry of Hertzian cone (ring) crack and stress distributions
formed by sphere loaded normally onto plane surface of brittle material.
(a) Hertzian ring crack. (b) Contact area (radius of contact circle a).
(c) Distribution of normal stress p. (d) Distribution of tensile stress σ_r.

The basis for treating the contact problem is the elasticity analysis of Hertz [8.9]. In the Hertzian method of measuring fracture stress a hard, spherical indenter is pressed against the solid under normal loading P, and the load at fracture is measured. The radius of the contact circle a (Fig. 8.4(b)) is given by

$$a^3 = \frac{3}{4} PR\left(\frac{1-v_1^2}{E_1} + \frac{1-v_2^2}{E_2}\right) \tag{8.1}$$

where R is the radius of the spherical indenter and E_1, v_1 and E_2, v_2 are the Young's modulus and Poisson's ratio of the solid and the indenter, respectively. The radius of the contact circle is proportional to $P^{1/3}$. And then the area of contact πa^2 is proportional to $P^{2/3}$. The mean pressure (normal stress) over the contact area $P_{mean} = P/\pi a^2$ varies as $P^{1/3}$. This normal stress is not uniform over the circular area of contact but is highest at the center and falls to zero at the edge. Figure 8.4(c) shows the distribution of normal stress; the maximum normal stress, at the center of the contact circle, is 3/2 times the mean normal stress. Within the contact circle the stress field becomes largely compressive, but outside the contact circle there is a tensile stress field in the surface.

The tensile stress σ_r in the plane surface has a maximum value $\sigma_{r,max}$ at the edge of the contact area ($r = a$), given by

$$\sigma_{r,max} = (1 - 2v_1)\, P_{mean}/2 \tag{8.2}$$

$$P_{mean} = P/\pi a^2 \tag{8.3}$$

The maximum tensile stress $\sigma_{r,max}$ acts radially and parallel to the surface and gives rise to fracture.

This tensile stress falls off with increasing radial distance (Fig. 8.4(d)) according to

$$\sigma_r = \frac{(1 - 2v_1)P}{2\pi r^2} \tag{8.4}$$

$$\sigma_r = \sigma_{r,max}(a/r)^2 \tag{8.5}$$

r being the radial distance from the center of the contact. Below the surface and outside the contact area the component of tensile stress diminishes rapidly with depth and distance. A tensile "skin" layer consequently exists beyond the immediate indentation site, thereby affording highly favorable conditions for crack initiation. At some critical loading the elastic limit of the specimen will be exceeded, and irreversible deformation will occur. In a brittle specimen the critical event is marked by the sudden development of the so-called Hertzian cone (ring) crack (Fig. 8.4(a)).

With a brittle solid, such as glass, a ring crack forms in the tensile region. Such ring cracks have also been found and studied on diamond surfaces. When ring cracks develop in brittle materials, they usually start a little beyond the edge of the contact circle (i.e., not at the point predicted by the Hertzian analysis to have the maximum tensile stress). Several workers have tried to explain this result in terms of a size distribution of defects in the solid surface.

Another important factor is the interfacial shear stresses that arise when the indenter and indented materials have different elastic properties. In this situation when the solids are pressed into contact, the materials on either side of the contact interface will want to move by unequal amounts. Relative movement at the interface will be affected by frictional forces. Clearly, the Hertzian stress field, which neglects such interfacial effects, will be modified. If the indenter is the more rigid material, the interfacial stresses reduce the movement of the indented solid surface. The modified peak radial tensile stress is now less than that given by Eq. (8.4) and exists slightly farther out. In other words, the material fails at an artificially high fracture load and should do so at $r > a$. When the indenter is the more compliant material, the effect is to intensify the radial stresses, resulting in an artificially low fracture load. If this interfacial effect is small, it can be neglected.

Comparison of diamond with sapphire (alumina) and silicon.—Cleavage cracks emerging on a crystalline surface form inherent patterns on the surface. Figure 8.5, for example, shows well-developed ring cracks produced on silicon $\{100\}$, sapphire $\{0001\}$, and natural diamond $\{111\}$ during experiments in which the specimens were indented with 200-μm- or 500-μm-radius spherical diamond indenters [8.22]. Cleavage occurred mostly along the $\{111\}$ planes in both silicon and diamond and along the $\{\bar{1}012\}$ planes in sapphire. In silicon the ring cracks (following the $\{111\}$ cleavage planes) on the $\{100\}$ plane formed a square pattern in the $\langle 110 \rangle$ direction. In sapphire the ring cracks (following the $\{\bar{1}012\}$ cleavage planes) on the $\{0001\}$ plane formed a triangular or hexagonal pattern. In diamond the ring cracks (following the $\{111\}$ cleavage planes) on the $\{111\}$ plane formed a triangular or hexagonal pattern in the $\langle 110 \rangle$ direction.

Figure 8.6 presents the maximum tensile stress $\sigma_{r,\max}$, or the tensile strength, which acts normal to cleavage planes at the edge of the Hertzian contact circle and gives rise to fracture on silicon $\{100\}$, sapphire $\{0001\}$, natural diamond $\{111\}$, and synthetic diamond $\{111\}$ as a function of indenter radius. Indenter radius had a marked effect on Hertzian fracture and on the maximum tensile stress to fracture. The greater the indenter radius, the lower the maximum tensile stress to fracture. Further, the values of maximum tensile stress (tensile strength) shown in Fig. 8.6 are of the same order as the theoretical strengths for the corresponding materials. The theoretical strengths are approximately 1/10th of the Young's moduli, which are 167 GPa for silicon, 380 GPa for sapphire, and 1020 GPa for natural and synthetic diamond. The tensile strengths of the materials investigated herein were, in ascending order, silicon, sapphire, natural diamond, and synthetic diamond.

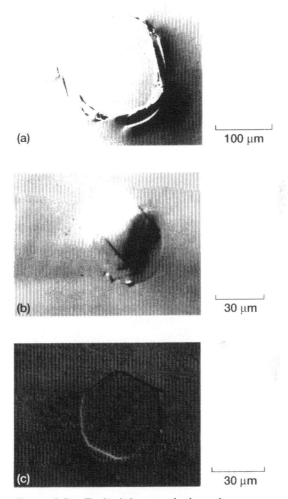

Figure 8.5.—Typical ring cracks in various specimens. (a) Silicon {100}. Indenter radius, 500 μm. (b) Sapphire {0001}. Indenter radius, 200 μm. (c) Natural diamond {111}. Indenter radius, 200 μm.

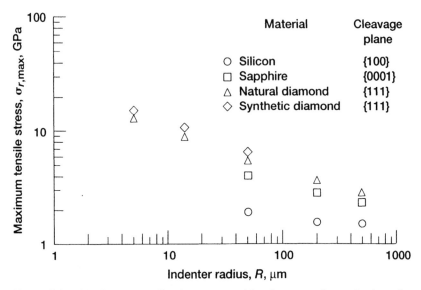

Figure 8.6.—Maximum tensile stress normal to cleavage plane at edge of
contact circle as function of indenter radius for silicon, sapphire, and
natural and synthetic diamond.

Anisotropy in tensile strength.—Figure 8.7 presents ring cracks formed on
polished {111}, {100}, and {110} surfaces of cutting-tool-grade natural diamond
by spherical diamond indenters [8.21]. All cleavage fractures occurred along the
{111} planes in diamond. On the {111} and {110} diamond surfaces ring cracks
formed a triangular or hexagonal pattern in the ⟨110⟩ direction. On the {100}
diamond surface ring cracks formed a square pattern in the ⟨110⟩ direction. As
shown in Fig. 8.8 the anisotropy in the mean maximum tensile stress to fracture on
the {111}, {100}, and {110} diamond surfaces is small, but the critical load to
fracture greatly varies with the surfaces [8.21].

Effect of impurities on tensile strength.—As shown (especially in Fig. 8.8)
determining diamond's properties is difficult, and there is considerable spread in the
values reported in the literature. Impurities, particularly nitrogen, considerably alter
the tensile strength, as shown in Fig. 8.9. When the {100} surfaces of single-crystal
synthetic diamond containing 1 ppm to nearly 300 ppm of nitrogen were indented
with a 5-μm-radius spherical diamond indenter, nitrogen altered the maximum
tensile stress to fracture (tensile strength). The tensile strength decreased with an
increase in nitrogen concentration. In other work, when the {100} surfaces of
natural diamond containing approximately 2000 ppm of nitrogen were indented
with a 5-μm-radius spherical diamond indenter, the tensile strength consistently
decreased with an increase in nitrogen concentration and ranged from 15 to 25 GPa.

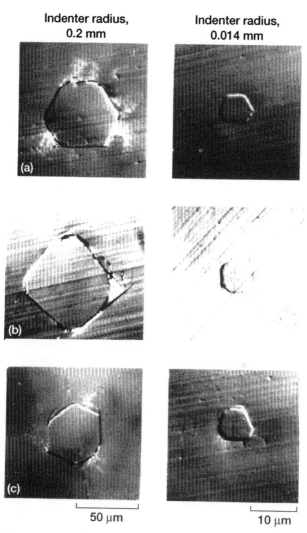

Indenter radius, 0.2 mm

Indenter radius, 0.014 mm

50 µm

10 µm

Figure 8.7.—Ring cracks on various polished surfaces of cutting-tool-grade natural diamond. (a) {111}. (b) {100}. (c) {110}.

8.4.2 Indentation Hardness

As described in the previous section, spherical indenters develop tensile stresses around the contact area that encourage brittle fracture rather than plastic flow. Pyramidal indenters (used by Knoop, Vickers, and Berkovich), however, produce rhombohedral, square, and triangular indentations, respectively, which are plastically deformed.

The property measured by an indentation hardness test is the plastic strength of the material (i.e., the amount of plastic deformation produced). All pyramidal indenters have a further advantage in that they yield values, in terms of units of pressure, that can be compared directly with other mechanical properties, such as yield stress, yield strength, and Young's modulus.

Comparison of diamond with other solids.—It has already been established that the hardness measured for crystalline solids is very much dependent on indenter shape, normal load, temperature, crystallographic orientation of the indenter with respect to the indented plane, and impurities. The diamond crystal is the hardest known material, although theoretical works suggest that the hypothetical compound β-C_3N_4 should be as hard or even harder than diamond. This fact makes it difficult to measure diamond's hardness, since only another diamond can be used as an indenter, and may help to explain the wide variations in reported values of diamond hardness. For example, although the Knoop indenter gave a hardness range of 73.5 to 88.2 GPa for diamond {111} and the Berkovich indenter a hardness range of 17.6 to 20.1 GPa for sapphire {0001}, only the median values are shown in Fig. 8.10 [8.24]. For a given crystal the Knoop hardness values are generally lowest, and the Vickers and Berkovich indenters give similar results.

Anisotropy in indentation hardness.—The indentation hardness, like the tensile strength, is intrinsically anisotropic. Figure 8.11 presents the anisotropic phenomenon in the two major types of diamond indented at room temperature with a 9.8-N load on mechanically polished surfaces. The results clearly show the type II diamonds to be significantly harder than the type I. The anisotropy in the hardness of crystals is controlled by their crystallographic structures and the relevant operative slip systems. The observed anisotropy of diamond was predicted and explained by models based on resolved shear stresses developed on the primary {111} ⟨110⟩ slip systems in the bulk of the crystal beneath the indenter. The easiest plane to indent by a Knoop indenter is the {111} in the ⟨110⟩ direction. Contrary to indentation measurements the {111} planes are generally the most difficult to abrade.

Effect of impurities on indentation hardness.—Impurities, particularly nitrogen, considerably alter the indentation hardness. Polished {100} surfaces of single-crystal synthetic diamond (type Ib) containing 1 ppm to nearly 100 ppm of nitrogen and natural diamond (type Ia) containing up to 2000 ppm of nitrogen were indented with a Knoop diamond indenter. As shown in Fig. 8.12 type Ia natural diamond invariably contained more nitrogen impurities (e.g., nitrogen platelets) and thus had greater ranges of hardness values.

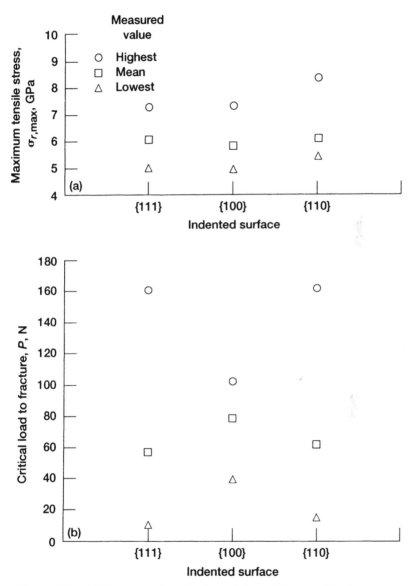

Figure 8.8.—(a) Maximum tensile stress (strength) normal to cleavage plane at edge of contact circle for and (b) critical load to fracture on {111}, {100}, and {110} surfaces of cutting-tool-grade natural diamond.

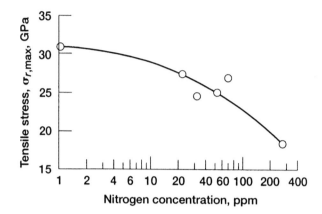

Figure 8.9.—Maximum tensile stress (strength) as function of nitrogen concentration for synthetic diamond in Hertzian contact with 5-μm-radius diamond indenter.

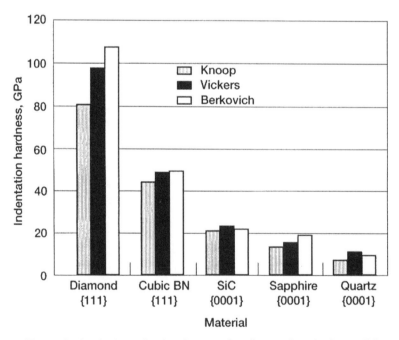

Figure 8.10.—Indentation hardnesses for diamond and other solid materials.

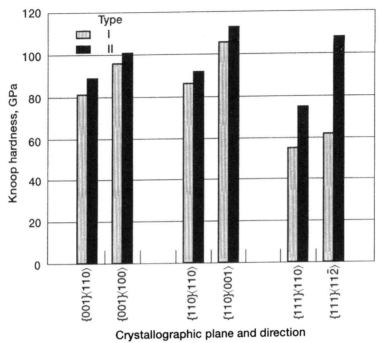

Figure 8.11.—Anisotropic Knoop hardness for two types of diamond.

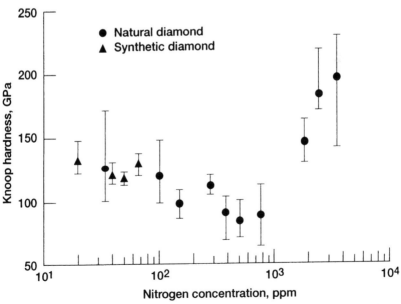

Figure 8.12.—Knoop hardness as function of nitrogen concentration for synthetic and natural diamond.

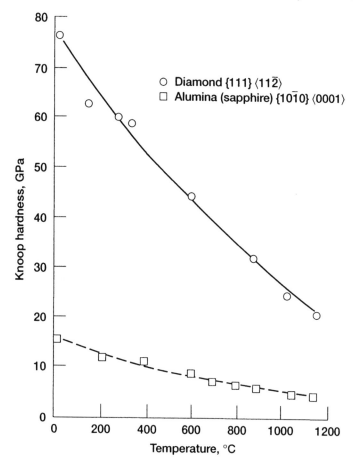

Figure 8.13.—Knoop hardness as function of temperature
for diamond {111} and sapphire (alumina) {10$\bar{1}$0 } surfaces.

Effect of temperature on indentation hardness.—Hardness values for single-crystal diamond and alumina were obtained by using a Knoop indenter oriented in $\langle 11\bar{2}\rangle$ on {111} and $\langle 0001\rangle$ on $\{10\bar{1}0\}$, respectively. The results (Fig. 8.13) confirm that, although its hardness decreased with increasing temperature and there was a corresponding increase in plastic deformation, at 1200 °C diamond was still harder than alumina at room temperature. Also, because diamond has the highest thermal conductivity of any material at room temperature, cracking due to thermal shock is not likely to be a problem, in contrast to alumina.

References

8.1 H.O. Pierson, *Handbook of Carbon, Graphite, Diamond and Fullerenes: Properties, Processing, and Applications*, Noyes Publications, Park Ridge, NJ, 1993.

8.2 J.E. Field (ed.), *The Properties of Diamond*, Academic Press, New York, 1979.

8.3 R.E. Clausing, et al. (eds.), *Diamond and Diamondlike Films and Coatings*, Plenum Press, New York, 1991.

8.4 P.D. Gigl, New synthesis techniques: properties and applications for industrial diamond, Presented at the IDA Ultrahard Materials Seminar, Toronto, Ontario, Canada, 1989.

8.5 P.G. Lurie and J.M. Wilson, The diamond surface, *Surf. Sci. 65*: 476–498 (1977).

8.6 J. Wei and J.T. Yates, Jr., Diamond surface chemistry: I-A review, *Crit. Rev. Surf. Chem. 5*: 1–71 (1995).

8.7 A.G. Guy, *Elements of Physical Metallurgy*, Addison-Wesley Publishing Co., Reading, MA, 1959.

8.8 J.P. Hirth and J. Lothe, *Theory of Dislocations*, McGraw-Hill, New York, 1968, pp. 353–363.

8.9 H. Hertz, Uber die Beruhrung fester elastischer Korper, *J. Reine und Angew. Math. 92*: 156–171 (1881).

8.10 S. Timoshenko and J.N. Goodier, *Theory of Elasticity*, McGraw-Hill, New York, 1951, p. 372.

8.11 F.C. Frank and B.R. Lawn, On the theory of Hertzian fracture, *Proc. R. Soc. London A 299*: 291–306 (1967).

8.12 B.R. Lawn, Hertzian fracture in single crystals with the diamond structure, *J. Appl. Phys. 39, 10*: 4828–4836 (1968).

8.13 B.R. Lawn and M.V. Swain, Microfracture beneath point indentations in brittle solids, *J. Mater. Sci. 10*: 113–122 (1975).

8.14 B.R. Lawn and T.R. Wilshaw, *Fracture of Brittle Solids*, Cambridge University Press, 1975.

8.15 S. Tolansky and V.R. Howes, Induction of ring cracks on diamond surfaces, *Proc. Phys. Soc. London B 70*: 521–526 (1957).

8.16 M. Seal, The abrasion of diamond, *Proc. R. Soc. London A 248*: 379–393 (1958).

8.17 V.R. Howes, The critical stress for the production of pressure crack figures on diamond faces, *Proc. Phys. Soc. 74*: 48–52 (1959) .

8.18 C.A. Brookes, Indentation hardness of diamond, *Diamond Research*, 1971, pp. 12–15.

8.19 W.R. Tyson, Theoretical strength of perfect crystals, *Phil. Mag. 14*: 925–936 (1966).

8.20 N. Ikawa, S. Shimada, and H. Tsuwa, Microfracture of diamond as fine tool material, *CIRP Ann. 31, 1*: 71–74 (1982).

8.21 N. Ikawa, S. Shimada, and T. Ono, Microstrength of diamond, *Technology Reports of Osaka University 26, 1298*: 245–254 (1976).

8.22 N. Ikawa and S. Shimada, Microstrength measurement of brittle materials, *Technology Reports of Osaka University 31, 1622*: 315–323 (1981).

8.23 S. Bhagavantam and J. Bhimasenachar, Elastic constants of diamond, *Proc. R. Soc. London A 187*: 381–384 (1946).

8.24 C.A. Brooks, Plastic deformation and anisotropy in the hardness of diamond, *Nature 228*: 660–661 (1970).

8.25 R.M. Chrenko and H.M. Strong, Physical properties of diamond, General Electric Technical Information Series No. 75CRD089, 1975, pp. 1–45.

Chapter 9
Chemical-Vapor-Deposited Diamond Film

9.1 Introduction

Chemical-vapor-deposited (CVD) diamond film technology looks promising in the field of tribology [9.1–9.5]. CVD diamond coatings offer a broader tribological potential than do natural and high-pressure synthetic diamonds because size and eventually cost will be less of a limitation. In general, however, the advantages and utility of CVD diamond as an industrial ceramic can only be realized if the price is right [9.6]. Until recently skeptics, even internationally well-known tribologists, viewed this technology merely as a rich mother lode of research papers for the academic community. According to Windischmann [9.6] that view may no longer be valid because of two advances made by a leading CVD diamond supplier in the past year:

1. Reduction of the cost of CVD diamond deposition below \$5/carat (\$8/cm^2)
2. Installation of production capacity

Thus, CVD diamond applications and business in the field of tribology, particularly cutting and forming tools, are an industrial reality [9.1].

At present CVD diamond is produced in the form of coatings or wafers. In this respect it is similar to pyrolytic graphite. Most CVD diamond is produced as films of polycrystalline diamond on ceramic or metal substrates. Coatings with a thickness of submicrometer to 1 mm or more are routinely produced. Thin CVD diamond films are used in film form; thick films are used in a free-standing shape or sheet (wafer) form after removing the substrate. The surface roughness of these coatings is typically in the range 10 nm to 1 μm R_a (centerline average) depending on the deposition conditions.

This chapter describes clean and contaminated diamond surfaces, CVD diamond film deposition technology, the characterization of CVD diamond film, and general and tribological properties of these films.

9.2 Diamond Surfaces

9.2.1 Clean Surface

Because diamond has tetrahedral, covalent bonds between each atom and its four nearest neighbors, the free surface may expose dangling bonds. The free surface has high surface energy γ, which is associated with dangling bond formation.

Crystalline diamond and silicon all have the diamond structure, where each atom in the crystal is sp^3 hybridized and tetrahedrally bonded to its four nearest neighbors. When the crystal is cut to expose a crystal plane, chemical bonds are broken along the plane, resulting in half-empty orbitals (dangling bonds) on the surface atoms [9.7, 9.8]. Because of the similarities between the bulk structures of diamond and silicon, a hypothetical surface structure of the silicon crystal, which is the ideal or bulk-terminated structure, is presented in Fig. 9.1 [9.7]. Simple counting of the

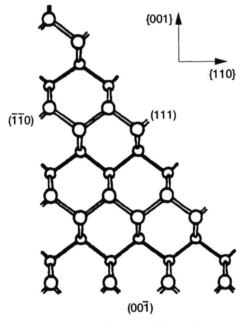

Figure 9.1.—Dangling bond formation on unreconstructed surfaces of (111), ($\bar{1}\bar{1}0$), and (00$\bar{1}$) planes of silicon. (Each silicon atom in the bulk is bonded to its four nearest neighbors in a tetrahedral configuration. The sizes of the silicon atoms are shown to decrease away from the page. From [9.7].)

broken bonds suggests that a surface atom will have two dangling bonds on the $\left(00\bar{1}\right)$ plane but only one each on the (111) and $\left(\bar{1}\,\bar{1}0\right)$ planes, which are the energetically favored ones. Because of the high energy cost associated with dangling bond formation, most diamond surfaces will become reconstructed to reduce the number of dangling bonds, thus minimizing their surface free energy. The surface valency bonds will back-bond or rehybridize.

Researchers have found that atomically clean diamond has high adhesion and friction [9.9, 9.10]. When a fresh atomically clean diamond surface comes in contact with an atomically clean surface of counterfacing material, the dangling bonds can form strong linkages with bonds on the counterfacing material surface. For example, if the surfaces of natural diamond and a metal are cleaned by argon ion bombardment, their coefficient of friction is higher than 0.4, as described in Chapter 3 and [9.10]. With argon-sputter-cleaned diamond surfaces there are probably dangling bonds ready to link up directly with those of another surface.

Because diamond has the highest atomic density of any material, with a molar density of 0.293 g-atom/cm^3, diamond is the stiffest, hardest, and least compressible of all substances. Tribologically, the real area of contact A is small. Nevertheless, the adhesion and friction of diamond are high in ultrahigh vacuum if the interface is atomically clean and dangling bonds are ready to link up with the counterfacing material surface [9.9, 9.10]. As discussed in Chapter 4, the adhesion and friction of materials are directly related to sA or γA. The dangling bonds, high shear strength s, and high surface energy γ of a clean diamond surface not only provide high friction but also produce high diamond wear due to high adhesion in ultrahigh vacuum. This subject is discussed in Section 9.6.

9.2.2 Contaminated Surface

Although diamond is generally inert to most chemical environments, the interaction of hydrogen and oxygen with diamond surfaces may play an important role in the application of diamond film technology to tribology [9.9–9.11]. A clean diamond surface exposed to atomic hydrogen can adsorb the atomic hydrogen [9.12] and form a hydrogenated diamond surface. Figures 9.2 and 9.3 present a surface structural model for an as-grown {111} diamond surface and an as-grown {100} diamond surface, respectively [9.13]. Almost all carbon atoms on the {111} surface are covered with methyl radical (CH$_3$) groups. The {100} surface is terminated by the monohydride (CH) group. The chemisorption of hydrogen on the dangling bonds preserves the sp^3 hybridization of the surface carbon atoms and stabilizes the diamond structure against the formation of graphite. When a CVD diamond film is annealed to above 1050 °C, the surface hydrogen can be desorbed, destabilizing the diamond surface structure. Consequently, a nondiamond surface layer will be formed on the diamond surface [9.13–9.15].

Significant amounts of oxygen are present on both natural surfaces and those polished in an aqueous medium [9.16]. Oxygen can be chemically bonded to

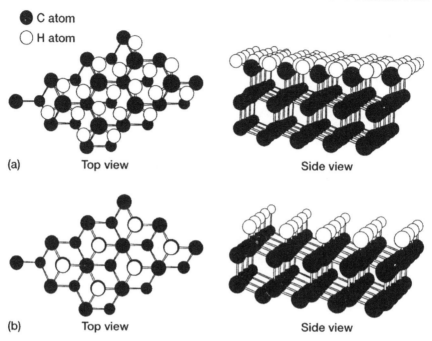

Figure 9.2.—Surface structural model for diamond {111} surface. (The sizes of the carbon and hydrogen atoms are shown to decrease away from the page.) (a) Diamond {111} 1x1 with CH₃. (b) Diamond {111} 1x1 with CH.

diamond surfaces. A hydrogenated diamond surface oxidizes above 300 °C. The oxygen-containing surface species are desorbed as carbon monoxide and carbon dioxide above 480 °C. Both oxygenation and desorption of oxygen-containing surface species simultaneously occur above 480 °C.

Both oxygen-terminated (oxidized) and hydrogen-terminated (hydrogenated) diamond surfaces provide low friction (see Section 9.6 for details). In the atmosphere diamond is known to be one of the slipperiest materials and is similar to polytetrafluoroethylene. This low friction is a surface property that apparently depends on the presence of adsorbed impurities, such as hydrocarbons from the atmosphere, on the oxidized and hydrogenated diamond surfaces.

Both oxidized and hydrogenated diamond surfaces can be fluorinated at temperatures from 300 to 1000 °C when they are treated under carbon tetrafluoride (CF_4) plasma conditions [9.16, 9.17]. The C–F bond is observed on CF_4-plasma-treated surfaces. Hydrogenated diamond surfaces fluorinate at lower temperatures than oxidized diamond surfaces. However, no fluorination occurs during thermal treatment

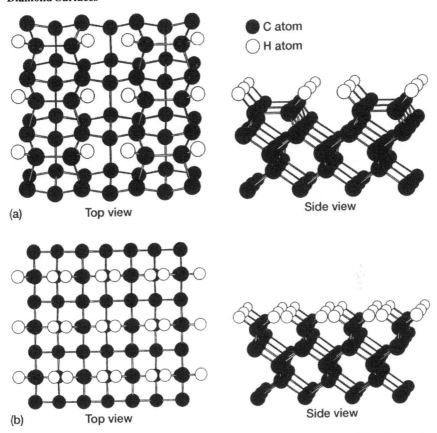

(a) Top view   Side view

(b) Top view   Side view

Figure 9.3.—Surface structural model for diamond {100} surface. (The sizes of the carbon and hydrogen atoms are shown to decrease away from the page.) (a) Diamond {100} 2x1. (b) Diamond {100} 1x1.

in a CF_4 environment. Plasma conditions, using CF_4 as a reactant precursor, are necessary for fluorination.

On the other hand, the hydrogenated diamond surface is chlorinated by thermal reaction in chlorine [9.18]. Hydrogen chemisorbed on the diamond surface is abstracted by the chlorine, resulting in chlorine chemisorption. Water vapor can react with the chlorinated diamond surface even at room temperature, producing O–H groups on the surface. A high-temperature treatment of chlorinated diamond with water further oxidizes the diamond surface, producing C=O species on the surface. Water alone, however, cannot oxidize a hydrogenated diamond surface below 500 °C.

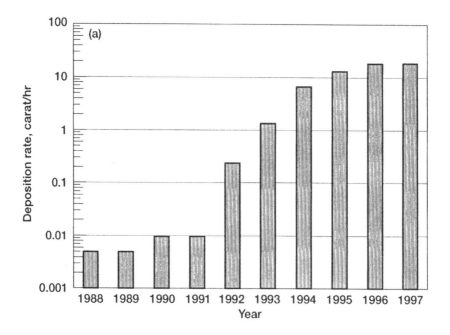

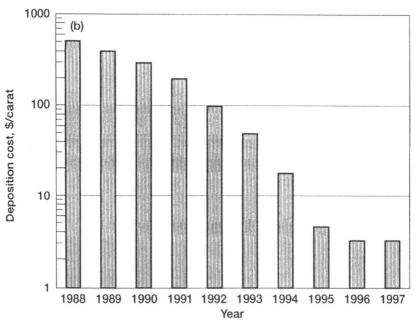

Figure 9.4.—Mass deposition rate increase and deposition cost reduction in past decade for CVD diamond produced by direct-current arc jet.

9.3 CVD Diamond Film Deposition Technology

The basic reaction in the chemical vapor deposition of diamond is simple [9.12]. It involves the decomposition of a hydrocarbon, such as methane, as follows:

$$CH_4 \rightarrow (Activation) \rightarrow C(diamond) + 2H_2$$

The carbon species must be activated, since at low pressure graphite is thermodynamically stable and, without activation, only graphite would be formed. Activation is obtained by two basic methods: high temperature and plasma, both of which require a great deal of energy.

Several CVD processes based on these two methods are presently in use. The four most important activation methods at this time are

1. High-frequency, plasma glow discharge using the microwave and radio-frequency processes
2. Plasma arcing using the direct-current arc and radiofrequency arc processes
3. Thermal CVD using the hot-filament process
4. Combustion synthesis using an oxyacetylene torch

Plasma arcing and combustion synthesis have high substrate deposition rates [9.6, 9.12]. For example, the mass deposition rate using a direct-current arc jet was 20 carats/hr in 1997 (Fig. 9.4(a)). Progress in direct-current arc jet deposition has advanced to the point that the $5/carat ($8/cm^2) barrier has been breached (Fig. 9.4(b)). In 1997 diamond made by direct-current arc jet was available at $8/cm^2 ($50/in.^{2}).

Diamond has been deposited from a large variety of precursors, including methane, aliphatic and aromatic hydrocarbons, alcohols, ketones, and solid polymers such as polyethylene, polypropylene, and polystyrene [9.12]. These substances generally decompose into two stable primary species: the methyl radicals (CH_3) and acetylene (C_2H_4). The radical is the key compound in generating the growth of CVD diamond.

9.4 Characterization of CVD Diamond

A variety of techniques can be used to characterize CVD diamond films: scanning and transmission electron microscopy (SEM and TEM), to determine surface morphology, microstructure, and grain size; surface profilometry and scanning probe microscopy such as atomic force microscopy (AFM), to measure surface roughness and to determine surface morphology; Rutherford backscattering (RBS) and hydrogen forward scattering (proton recoil detection), to determine the composition including hydrogen; Raman spectroscopy, to characterize atomic bonding state and diamond quality; and x-ray diffraction, to determine the crystal orientation of diamond growth.

It is generally accepted that, for a material to be recognized as diamond, it must have all of the following characteristics:

1. A clear, sharp diamond peak at 1332 cm^{-1} in the Raman spectrum
2. A crystalline morphology visible by SEM or TEM
3. A single-phase diamond crystalline structure detectable by x-ray or electron diffraction

Examples as case studies have been prepared focusing attention primarily on microwave-plasma-assisted CVD diamond films and are described in the following subsections.

9.4.1 Surface Morphology and Roughness

The surface morphology and roughness of CVD diamond films can be controlled by varying the deposition parameters, such as gas-phase chemistry parameters and temperatures (e.g., see Table 9.1 and [9.19]). Figure 9.5 shows SEM micrographs of fine-, medium-, and coarse-grain diamond films. Triangular crystalline facets typical of diamond are clearly evident on the surfaces of the medium- and coarse-grain diamond films, which have grain sizes estimated at 1100 and 3300 nm, respectively. The grain sizes of the fine-grain diamond were determined from bright- and dark-field electron photomicrographs to be between 20 and 100 nm. The average surface roughness of the diamond films measured by a surface profilometer increases as the grain size increases, as shown in Fig. 9.6.

Note that Fig. 2.20(b) in Chapter 2 is an atomic force microscopy image of a fine-grain CVD diamond film. The fine-grain diamond surface has a granulated or

TABLE 9.1.—DEPOSITION CONDITIONS FOR DIAMOND FILMS OF VARIOUS GRAIN SIZES

Condition	Substrate[a]						
	Si {100}	Si {100}	α-SiC	α-SiC	α-SiC	Si$_3$N$_4$	Si$_3$N$_4$
Deposition temperature, °C	1015±50	860±20	1015±50	965±50	860±20	965±50	860±20
Gaseous flow rate, cm^3/min:							
CH$_4$	3.5	4	3.5	3.5	4	3.5	4
H$_2$	500	395	500	500	395	500	395
O$_2$	0	1	0	0	1	0	1
Pressure, torr	40	5	40	40	5	40	5
Microwave power, W	1000	500	1000	1000	500	1000	500
Deposition time, hr	14	10.5	14	22	21	22	21
Thickness, nm	4200	1000	5000	8000	1000	7000	800
Grain size, nm	1100	20–100	3300	1500	22–100	1000	22–100
Surface roughness rms, nm	63	15	160	92	50	52	35

[a]Scratched with 0.5-μm diamond paste.

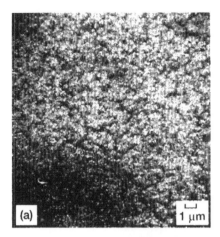

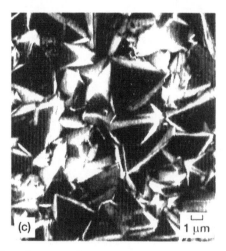

Figure 9.5.—Scanning electron micrographs of diamond films. (a) Fine-grain (20 to 100 nm) diamond film on {100} silicon substrate; rms surface roughness, 15 nm. (b) Medium-grain (1100 nm) diamond film on {100} silicon substrate; rms surface roughness, 63 nm. (c) Coarse-grain (3300 nm) diamond film on {100} α-silicon carbide substrate; rms surface roughness, 160 nm.

spherulitic morphology (i.e., the surface contains spherical asperities of different sizes).

9.4.2 Composition

Figures 9.7 and 9.8 [9.20] present a Rutherford backscattering spectrum and a hydrogen forward-scattering (HFS or proton recoil detection) spectrum, respectively, of the fine-grain CVD diamond film shown in Fig. 9.5(a). Besides carbon

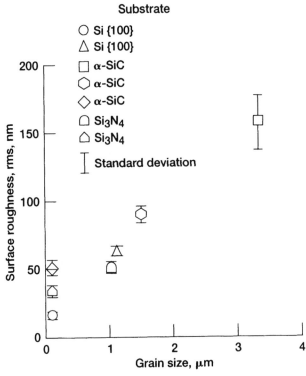

Figure 9.6.—Surface roughness as function of grain size
for diamond films.

from the diamond film and silicon from the substrate, no other elements are observed in the RBS spectrum. From both the RBS and HFS spectra it is estimated that the fine-grain diamond film consists of 97.5 at.% carbon and 2.5 at.% hydrogen. (In contrast, the medium- and coarse-grain diamond films contained less than 1 at.% hydrogen [9.19].) Both carbon and hydrogen are uniformly distributed in the fine-grain diamond film from the top of the surface to the silicon substrate.

The RBS analytical results can also be used to determine diamond film thickness. Figure 9.7 also presents a simulated RBS spectrum of the fine-grain diamond film containing a carbon-to-hydrogen ratio (C/H) of 97.5/2.5 obtained by using the RUMP computer code [9.21]. In the computer program the thickness of the diamond film is taken as a variable and is obtained from the close match between the observed and simulated RBS, as shown in Fig. 9.7. The film thickness of the fine-grain CVD diamond is 1.5 µm at the center of the substrate, and the deposition rate is estimated to be 0.14 µm/hr.

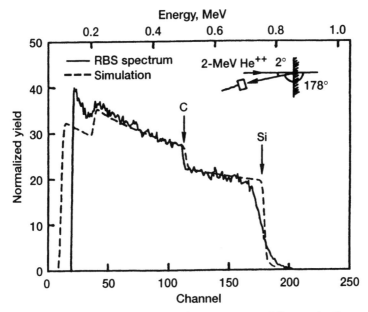

Figure 9.7.—Rutherford backscattering spectrum of fine-grain diamond film on silicon substrate. (Simulation curve was calculated by using the computer code RUMP.)

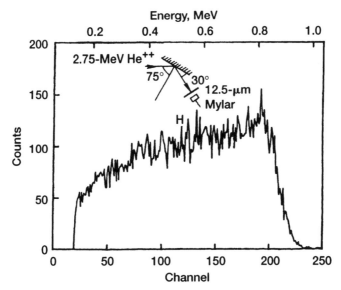

Figure 9.8.—Hydrogen forward-scattering spectrum (proton recoil analysis) of fine-grain diamond film on silicon substrate.

9.4.3 Atomic Bonding State

Figure 9.9 shows typical Raman spectra of the fine- and medium-grain diamond films. Both spectra show one Raman band centered at 1332 cm^{-1} and one at 1500 to 1530 cm^{-1}. The sharp peak at 1332 cm^{-1} is characteristic of the sp^3 bonding of the diamond form of carbon in the film. The very broad peak centered around 1530 cm^{-1} is attributed to the sp^2 bonding of the nondiamond forms of carbon (graphite and other forms of carbon) [9.22–9.24].

More diamond was produced in the medium-grain CVD diamond films (e.g., Fig. 9.9) than in the fine-grain films, as is evident from the relative intensities of the diamond and nondiamond carbon Raman bands [9.19]. However, the ratio of the intensities of the Raman responses at 1332 and 1530 cm^{-1} does not indicate the ratio of diamond to nondiamond carbon present in the film. The Raman technique is approximately 50 times more sensitive to sp^2-bonded (nondiamond) carbon than it is to sp^3-bonded (diamond) carbon [9.23]. Thus, the peak at approximately 1530 cm^{-1} for each film represents a much smaller amount of nondiamond carbon in these diamond films. The Raman spectrum of the coarse-grain diamond film was similar to that of the medium-grain diamond film.

9.4.4 Microstructure

X-ray diffraction analysis (XRD) is used to determine the structure and crystal orientation of the CVD diamond films. Figure 9.10 shows typical XRD patterns for the fine- and medium-grain diamond films [9.19]. Peaks representing only the diamond film and the silicon substrate appear in the XRD spectra. Diffraction peaks corresponding to the {111}, {220}, {311}, and {400} planes, reflective of diamond, are clearly evident. The intensity ratios I\{220\}/I\{111\} were calculated from the XRD patterns for the fine- and medium-grain diamond films and found to be 1.3 and 0.04, respectively. The powder diffraction pattern of diamond with random crystal orientation (ASTM 6–0675) gives I\{220\}/I\{111\} = 0.27.

Thus, most crystallites in the fine-grain diamond film are oriented along the ⟨110⟩ direction. Other researchers [9.22] have observed similar {110} crystal orientation texture in microcrystalline diamond film grown in a {100} silicon substrate by an activated CVD technique. Most crystallites in the medium-grain diamond films are oriented along the ⟨111⟩ direction, and the {111} planes are parallel to the surface. The well-formed triangular facets observed in SEM micrographs (Fig. 9.5) of the medium- and coarse-grain diamond films confirm the ⟨111⟩ crystal orientation.

Figure 9.11 presents the transmission electron diffraction pattern, a transmission electron bright-field micrograph, and a transmission electron dark-field micrograph of the free-standing, fine-grain CVD diamond film [9.20]. Diffraction rings and dots are observed in the selected-area diffraction (SAD) pattern (Fig. 9.11(a)). The d spacings of the diffraction rings were calculated by using an aluminum SAD

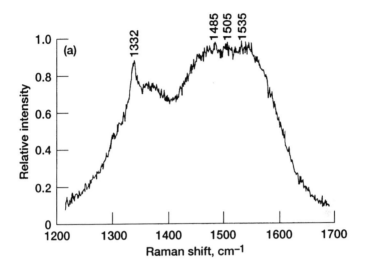

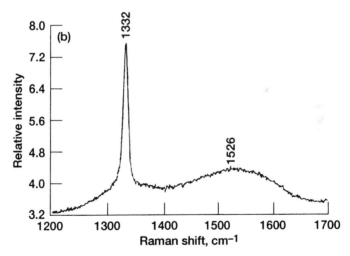

Figure 9.9.—Raman spectra of diamond films. (a) Fine-grain
(20 to 100 nm) diamond film on {100} silicon substrate.
(b) Medium-grain (1100 nm) diamond film on {100} silicon
substrate.

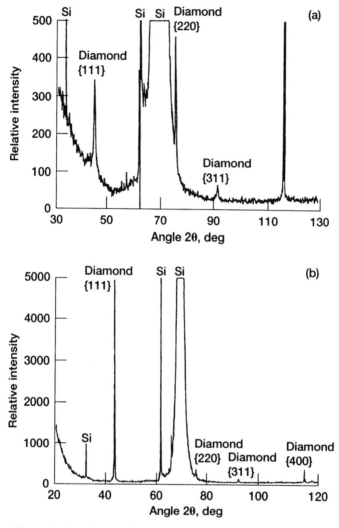

Figure 9.10.—X-ray diffraction patterns of diamond films.
(a) Fine-grain (20 to 100 nm) diamond film on {100} silicon
substrate. (b) Medium-grain (1100 nm) diamond film on
{100} silicon substrate.

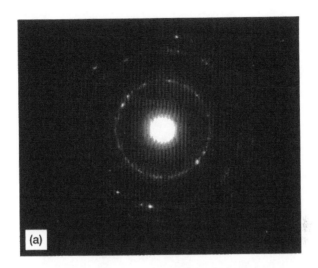

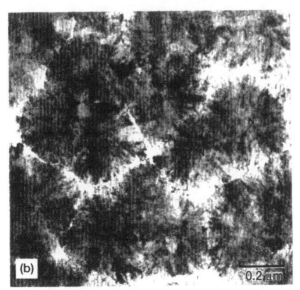

Figure 9.11.—Free-standing diamond films.
(a) Selected-area diffraction pattern.
(b) Bright-field TEM.

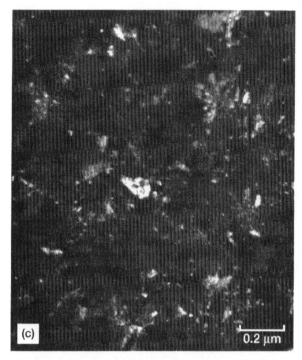

(c)

0.2 μm

Figure 9.11.—Concluded. (c) Dark-field TEM.

pattern as a calibration standard and found to match well with the known diamond
d spacings. No evidence of nondiamond carbon is found in the SAD pattern. This
observation indicates that the concentration of nondiamond component in the
diamond film is very small.

Careful observation of the bright-field micrograph (Fig. 9.11(b)) shows various
nuclei-like regions marked N. Diamond grains are distributed radially outward from
these nuclei. A grain boundary is formed where the grains from various nuclei meet.
The grain sizes were estimated from the dark-field micrograph (Fig. 9.11(c)) and
found to vary from 20 to 100 nm.

9.5 General Properties of CVD Diamond

Diamond consists of light carbon atoms held together by strong forces, and this
combination produces many extreme properties. Table 9.2 summarizes the general
properties of CVD diamond and compares them to those of single-crystal diamond,
such as type Ib. This table and the description in this section are based on a table from
Pierson [9.12] and other references [9.25–9.27]. Because of the difficulty in testing,
the effect of impurities and structural defects, and the differences between the

TABLE 9.2.—PROPERTIES OF CVD DIAMOND

Property	CVD diamond	Single-crystal diamond
Density, g/cm^3	3.51	3.515
Thermal conductivity at 25 °C, W m^{-1} °C^{-1}	2100	2200
Thermal expansion coefficient at 25 to 200 °C, °C^{-1}	2.0×10^6	$(1.5 \text{ to } 4.8) \times 10^6$
Tensile strength, GPa	1 to 5	4 to 6
Young's modulus, Pa	$(7 \text{ to } 9) \times 10^{11}$	10.5×10^{11}
Poisson's ratio, GPa	0.1 or 0.07	0.1 or 0.07
Vicker's hardness range,[a] kg/mm^3	5000 to 10 000	5700 to 10 400
Chemical properties	Both CVD sand single-crystal diamonds are resistant to all liquid organic and inorganic acids, alkalis, and solvents at room temperature. One of the most chemically resistant materials.	
Band gap, eV	5.45	5.45
Index of refraction at 10 μ m	2.34 to 2.42	2.40
Electrical resistivity, ohm·cm	10^{12} to 10^{16}	10^{16}
Dielectric constant (45MHz to 20 GHz)	5.6	5.70
Dielectric strength, V/cm	10^6	10^6
Loss tangent (45 MHz to 20 GHz)	< 0.0001	------
Saturated electron velocity	2.7	2.7
Carrier mobility:		
Electron (n)	1350 to 1500	2200
Positive hole (p)	480	1600

[a]Varies with crystal orientation (see Chapter 8).

various deposition processes, uncertainty and spread have been found in the reported property values. The values in Table 9.2 should be viewed with caution.

In diamond the light carbon atoms are densely packed, but the lattice is in many ways similar to those of other crystals, such as silicon, of the same structure. Therefore, diamond is a lightweight material, the density of both CVD diamond and single-crystal diamond being approximately 3.5 g/cm^3 (Table 9.2). Note that the closeness of carbon atoms leads to a small compressibility, as described in Chapter 8, and the specific heat is also small (6.19 J g-atom^{-1} °C^{-1}) at room temperature).

One of the many remarkable properties of diamond is its high thermal conductivity, as shown in Fig. 9.12. The data in this figure were taken from Pierson [9.12] and Windischmann [9.6]. For the highest quality single-crystal, type IIa diamonds the thermal conductivity is about 25 W cm^{-1} °C^{-1} at room temperature, or more than six times that of copper (4 W cm^{-1} °C^{-1}). Such high thermal conductivity is caused by the stiffness of the diamond bond and the diamond structure. Other types of single-crystal diamond and CVD diamond have lower thermal conductivity (Table 9.2) at room temperature because they contain larger concentrations of impurities than does type IIa diamond.

Using values of 10.5×10^{11} Pa, 5.5×10^{11} Pa, 5.3 J/m, and 0.154 nm for the Young's modulus, the shear modulus, the fracture surface energy, and the nearest neighbor distance, respectively, the theoretical strength in tension of single-crystal diamond

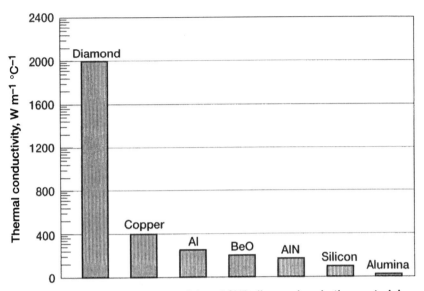

Figure 9.12.—Thermal conductivity of CVD diamond and other materials.

is 1.9×10^{11} Pa in the $\langle 111 \rangle$ direction (i.e., ~ $E/5$, where E is the modulus of elasticity [9.25]). Theoretical strength in shear equals 12.1×10^{10} Pa in the $\langle 110 \rangle \{111\}$ system (i.e., ~$G/4$, where G is the shear modulus). Actual tensile strengths for the octahedral $\{111\}$ and cubic $\{100\}$ planes of single-crystal type IIa diamond are 3.8 and 3.7 GPa, respectively. The average is ~3.75 GPa for a Poisson's ratio of 0.1 (or ~4.0 GPa for a Poisson's ratio of 0.07), roughly 1/50th of the theoretical strength. Actual tensile strengths obtained by many workers range from 4 to 6 GPa for single-crystal diamond and from 1 to 5 GPa for CVD diamond (Table 9.2). Many workers have found Young's moduli for CVD diamond in the range 7×10^{11} to 9×10^{11} Pa and Poisson's ratios of 0.1 or 0.07 [9.25–9.27].

Chapter 8 states that diamond is the hardest known material and that the hardness of single-crystal diamond varies by a factor 2 or more as a function of crystal orientation and impurities. This statement is also true for CVD diamond. Figure 9.13, based on a figure from [9.12] and other references [9.25–9.30], shows the high indentation hardness for diamond as compared with a few other hard materials.

Diamond is chemically resistant to all liquid organic and inorganic acids at room temperature. However, it can be etched by several compounds, including strong oxidizers such as sodium and potassium nitrates above 500 °C, by fluxes of sodium and potassium chlorates, and by molten hydroxides, such as sodium hydroxide (NaOH). Diamond is resistant to alkalis and solvents. At approximately 1000 °C it reacts readily with carbide-forming metals such as iron, cobalt, nickel, aluminum, tantalum, and boron. Generally speaking, diamond can be considered as one of the most chemical-resistant materials. The chemical properties of CVD diamond are similar to those of single-crystal diamond.

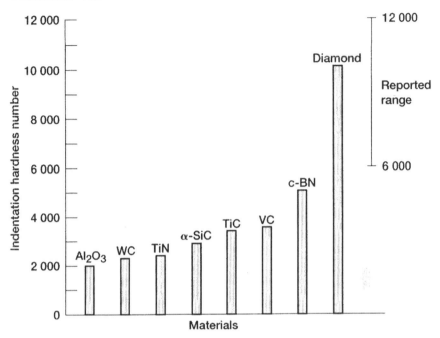

Figure 9.13.—Indentation hardness of diamond and other hard materials.

Table 9.2 also includes optical, electrical, and semiconductor properties of diamond as a reference.

9.6 Friction and Wear Properties of CVD Diamond

9.6.1 Humid Air and Dry Nitrogen Environments

The tribological properties of CVD diamond are similar to those of natural and synthetic diamond. The coefficient of friction and wear resistance of CVD diamond are generally superior in the atmosphere. However, the environment to which a CVD diamond film is exposed can markedly affect its tribological properties, such as friction and wear behavior. They vary with the environment, possessing a Jekyll-and-Hyde character [9.19].

When the fine-, medium-, and coarse-grain diamond films characterized in Section 9.4 were brought into contact with a natural diamond pin in reciprocating sliding motion in air and in nitrogen, the coefficients of friction varied as the pin traveled back and forth (reciprocating motion), retracing its tracks on the diamond films (Fig. 9.14).

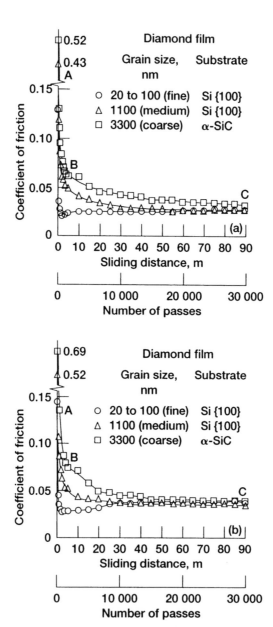

Figure 9.14.—Coefficient of friction as function of number of passes of bulk diamond pin in contact with fine-, medium-, and coarse-grain diamond films (a) in humid air (approx. 40% relative humidity) and (b) in dry nitrogen.

Both in humid air at a relative humidity of 40% and in dry nitrogen, abrasion occurred and dominated the friction and wear behavior. The bulk natural diamond pin tended to dig into the surface of diamond films during sliding and produce a wear track (groove). SEM observations of the diamond films indicated that small fragments chipped off their surfaces. When abrasive interactions between the diamond pin surface and the initially sharp tips of asperities on the diamond film surfaces were strong, the friction was high (points A in Fig. 9.14). The surface roughness of diamond films can have an appreciable influence on their initial friction (i.e., the greater the initial surface roughness, the higher the initial coefficient of friction, Fig. 9.15(a)). Similar frictional results have also been found by other workers on single-crystal diamonds [9.31] and on diamond coatings [9.32–9.34].

As sliding continued and the pin passed repeatedly over the same track, the coefficient of friction was appreciably affected by the wear on the diamond films (Fig. 9.14) (i.e., a blunting of the tips of asperities). When repeated sliding produced a smooth groove or a groove with blunted asperities on the diamond surface (Fig. 9.16), the coefficient of friction was low, and the initial surface roughness effect became negligible. Therefore, the equilibrium coefficient of friction was independent of the initial surface roughness of the diamond film (Fig. 9.15(b)).

The generally accepted wear mechanism for diamonds is that of small fragments chipping off the surface [9.9, 9.19]. This mechanism is in agreement with the wear of diamond films. The wear rate is dependent on the initial surface roughness of the diamond films (Fig. 9.17), increasing markedly with an increase in initial surface roughness. The wear rates of the diamond films in humid air and in dry nitrogen are comparable to the wear rates of single-crystal diamonds and diamond films investigated by other workers [9.32, 9.35, 9.36].

9.6.2 Ultra-High-Vacuum Environment

When the fine-, medium-, and coarse-grain diamond films were brought into contact with a natural diamond pin in unidirectional pin-on-disk sliding motion in vacuum, the coefficients of friction were high and varied with the number of passes [9.15]. In vacuum, as in humid air and in dry nitrogen, the bulk natural diamond pin dug into the surfaces of the diamond films during sliding and produced a wear track (groove, Fig. 9.18). The groove surface was generally smoother than the original surface of the diamond films. Further analysis of the grooves by scanning electron microscopy revealed that the tips of the diamond coating asperities were worn smooth and that the gaps between asperities were filled by debris.

The coefficient of friction increased with an increase in the number of passes (Fig. 9.19), just the opposite of what occurred in humid air and in dry nitrogen. Further, the initial surface roughness of the diamond film had no effect on friction. These results led us to ask the following questions: What factors determine friction behavior? Have dangling bonds been exposed during sliding and played a role in the

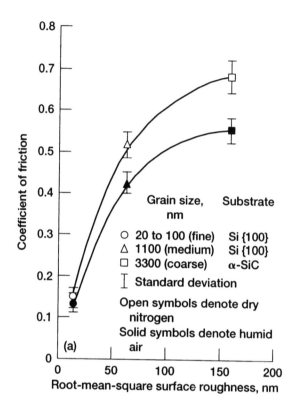

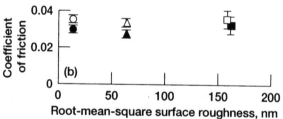

Figure 9.15.—Coefficient of friction as function of initial surface roughness of diamond films in humid air (approx. 40% relative humidity) and in dry nitrogen. (a) Initial coefficients of friction. (b) Equilibrium coefficients of friction.

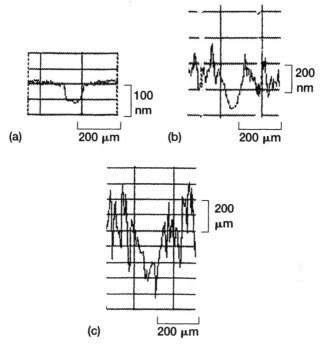

Figure 9.16.—Wear tracks (grooves) on diamond films after 30 000 passes of bulk diamond pin in dry nitrogen. (a) Fine-grain (20 to 100 nm) diamond film; rms surface roughness, 15 nm. (b) Medium-grain (1100 nm) diamond film; rms surface roughness, 63 nm. (c) Coarse-grain (3300 nm) diamond film; rms surface roughness, 160 nm.

friction behavior? Which is more important for diamond surfaces in vacuum, abrasion or adhesion?

Removing some of the contaminant surface film from the contact area of diamond films by sliding action resulted in stronger interfacial adhesion between the diamond pin and the diamond films and raised the coefficient of friction, as shown in Fig. 9.19. A contaminant surface film may be removed by repeatedly sliding the diamond pin over the same track in ultrahigh vacuum [9.36].

The friction results shown in Fig. 9.19 are in agreement with other researchers' results for single-crystal diamond rubbing against diamond and for CVD diamond sliding against CVD diamond in vacuum [9.37, 9.38]. At a pressure of 93 nPa Bowden and Hanwell [9.38] observed an initial coefficient of friction of 0.1 for diamond on diamond; within several hundred passes, however, the coefficient of friction rose rapidly to 0.9 and remained constant. Dugger, Peebles, and Pope [9.39]

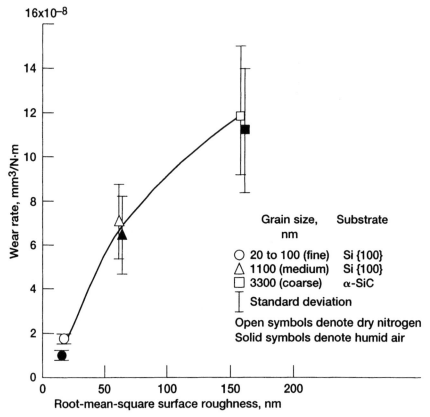

Figure 9.17.—Wear rates of diamond films as function of diamond surface roughness in humid air and in dry nitrogen.

also found that the coefficient of friction increased to 0.47 when CVD diamond slid against itself in vacuum (<0.6 μPa). In both cases the increase in friction was attributed to cleaning the adsorbed contaminants from the surface by rubbing or sliding in vacuum at room temperature.

When sliding continues, the wear dulls the tips of the diamond grains and increases the contact area in the wear track, thereby causing an increase in friction. The increase in equilibrium friction that results from cleaning off the contaminant surface film by sliding and from increasing the contact area is greater than the corresponding decrease in abrasion and friction that results from blunting the tips of surface asperities. This relationship is brought out clearly in Fig. 9.20; here the equilibrium coefficients of friction (1.5 to 1.8) are greater than the initial coefficients of friction (1.1 to 1.3) regardless of the initial surface roughness of the diamond films. In vacuum, therefore, the friction arises primarily from adhesion between the sliding surfaces of the diamond pin and the diamond films.

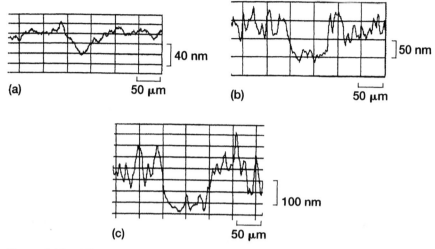

Figure 9.18.—Wear tracks (grooves) on diamond films after 100 passes of bulk diamond pin in ultrahigh vacuum. (a) Fine-grain (20 to 100 nm) diamond film on silicon substrate; rms surface roughness, 15 nm. (b) Medium-grain (1000 nm) diamond film on silicon nitride substrate; rms surface roughness, 52 nm. (c) Coarse-grain (1500 nm) diamond film on α-silicon carbide substrate; rms surface roughness, 92 nm.

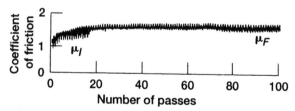

Figure 9.19.—Typical friction trace for bulk diamond pin in contact with diamond film on α-silicon carbide substrate in ultrahigh vacuum (initial coefficient of friction, μ_I; equilibrium coefficient of friction, μ_F).

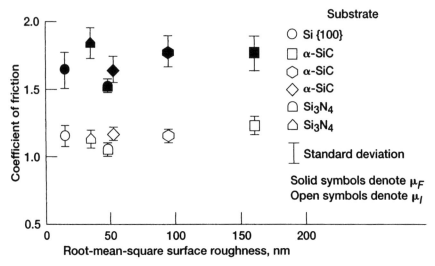

Figure 9.20.—Initial (μ_I) and equilibrium (μ_F) coefficients of friction as function of initial surface roughness of diamond films in ultrahigh vacuum.

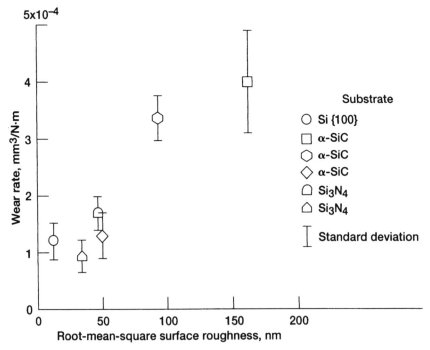

Figure 9.21.—Wear rates as function of initial surface roughness of diamond films in ultrahigh vacuum.

The wear rates of the diamond films in ultrahigh vacuum (Fig. 9.21) depended on the initial surface roughness of the diamond films, generally increasing with an increase in initial surface roughness. The wear rates of the diamond films in ultrahigh vacuum were considerably higher than those of the diamond films in humid air or in dry nitrogen (Fig. 9.17). Obviously, under these vacuum conditions adhesion between the sliding surfaces of the diamond pin and the diamond film plays an important role in the higher wear process.

Thus, under vacuum conditions it is adhesion between the sliding surfaces of the diamond pin and diamond films (due to the highly clean state) and the possible presence of dangling bonds that play a significant role in the friction and wear process. The surface roughness of the diamond films does not have much influence on the friction of diamond films in ultrahigh vacuum.

References

9.1 A.D. Molloy and A.M. Dionne, eds., *Wear and Superhard Coatings 1998*, Gorham Advanced Materials, Inc., Gorham, ME, 1998.

9.2 K. Miyoshi, ed., special issue on Tribology of Diamond and Diamondlike Films and Coatings (1), *Diamond Films and Technology 3, 3* (1994).

9.3 K. Miyoshi, ed., special issue on Tribology of Diamond and Diamondlike Films and Coatings, *Diamond Films and Technology 4, 1* (1994).

9.4 K. Miyoshi, ed., special issue on Tribology of Diamond and Diamondlike Films and Coatings, *Diamond Films and Technology 4, 2* (1994).

9.5 K. Miyoshi, ed., special issue on Tribology of Diamond and Diamondlike Films and Coatings, *Diamond Films and Technology 4, 3* (1994).

9.6 H. Windischmann, Stress issues in wafer-scale CVD diamond fabrication, *Book of Abstracts*, American Vacuum Society, 1998, p. 124; and personal communication.

9.7 D. Haneman, Surfaces of silicon, *Rep. Prog. Phys. 50, 8*: 1045–1048 (1987).

9.8 J. Wei and J.T. Yates, Jr., Diamond surface chemistry, A Review, *Critical Reviews in Surface Chemistry 5, 1–3*: 1–71 (1995).

9.9 D. Tabor, Adhesion and Friction, *The Properties of Diamond*, J.E. Field, ed. Academic Press, New York, 1979, pp. 325–350.

9.10 K. Miyoshi and D.H. Buckley, Adhesion and friction of single-crystal diamond in contact with transition metals, *Appl. Surf. Sci. 6*: 161–172 (1980).

9.11 J.A. Harrison, C.T. White, R.J. Colton, and D.W. Brenner, Molecular-dynamics simulations of atomic-scale friction of diamond surfaces, *Phys. Rev. B, 46, 15*: 9700–9708 (1992).

9.12 H.O. Pierson, *Handbook of Carbon, Graphite, Diamond and Fullerenes: Properties, Processing, and Applications*, Noyes Publications, Park Ridge, NJ, 1993.

9.13 T. Ando, T. Aizawa, K. Yamamoto, Y. Sato, and M. Kamo, The Chemisorption of hydrogen on diamond surfaces studied by high resolution electron energy-loss spectroscopy, *Diamond Rel. Mater. 3, 4–6*: 975–979 (1994).

9.14 J.E. Butler and F.G. Celri, Vapor phase diagnostics in CVD diamond deposition, *International Symposium on Diamond and Diamond-like Films* (J.P. Dismukes, ed.), The Electrochemical Society, Pennington, NJ, 1989, pp. 317–329.

9.15 T. Yamada, T.J. Chuang, H. Seki, and Y. Mitsuda, Chemisorption of fluorine, hydrogen and hydrocarbon species on the diamond C (111) surface, *Mol. Phys. 76, 4*: 887–908 (1992).

9.16   T. Ando, K. Yamamoto, M. Ishii, M. Kamo, and Y. Sato, Vapor-phase oxidation of diamond surfaces in oxygen studied by diffuse reflectance Fourier-transform infrared and temperature-programmed desorption spectroscopy, *J. Chem. Soc., Faraday Trans. 89, 19*: 3635–3640 (1993).

9.17   T. Ando, J. Tanaka, M. Ishii, M. Kamo, Y. Sato, N. Ohashi, and S.Shimosaki, Diffuse reflectance Fourier-transform infrared study of the plasma-fluorination of diamond surfaces using a microwave discharge in tetrafluoromethane, *J. Chem. Soc., Faraday Trans. 89, 16*: 3105–3109 (1993).

9.18   T. Ando, K. Yamamoto, S. Suehara, M. Kamo, Y. Sato, S. Shimosaki, and M. Nishitani-Gamo, Interaction of chlorine with hydrogenated diamond surface, *J. Chin. Chem. Soc. 42, 2*: 285–292 (1995).

9.19   K. Miyoshi, R.L.C. Wu, A. Garscadden, P.N. Barnes, and H.E. Jackson, Friction and wear of plasma-deposited diamond films, *J. Appl. Phys. 74, 7*: 4446–4454 (1993).

9.20   R.L.C. Wu, A.K. Rai, A. Garscadden, P. Kee, H.D. Desai, and K. Miyoshi, Synthesis and characterization of fine grain diamond films, *J. Appl. Phys. 72, 1*: 110–116 (1992).

9.21   L.R. Doolittle, Algorithms for the rapid simulation of Rutherford backscattering spectra, *Nucl. Instrum. Methods B9*: 344–351 (1985).

9.22   J.J. Cheng and T.D. Mautei, Effects of oxygen and pressure on diamond synthesis in a magnetoactive microwave discharge, *J. Appl. Phys. 71, 6*: 2918–2923 (1992).

9.23   K. Kobashi, K. Nishimura, V. Kawate, and T. Horiuchi, Synthesis of diamonds by use of microwave plasma chemical vapor deposition: morphology and growth of diamond film, *Phys. Rev. B 38, 6*: 4067–4084 (1988).

9.24   N. Wada and S.A. Solin, Raman efficiency measurements of graphite (and Si and Ge), *Physica B&C 105*: 353–356 (1981).

9.25   G. Davies, ed., *Properties and Growth of Diamond*, INSPEC, Institution of Electrical Engineers, London, UK, 1994.

9.26   K.E. Spear and J.P. Dismukes, eds., *Synthetic Diamond: Emerging CVD Science and Technology*, John Wiley & Sons, New York, 1994.

9.27   J.E. Field, ed., *The Properties of Diamond*, Academic Press, New York, 1979.

9.28   J. Shackelford and W. Alexander, eds., *Materials Science and Engineering Handbook*, CRC Press, Inc., Boca Raton, FL, 1992.

9.29   W.D. Kingery, *Introduction to Ceramics*, John Wiley & Sons, New York, 1960.

9.30   J.B. Wachtman, Jr., ed., *Structural Ceramics*, Academic Press, Boston, 1989.

9.31   M. Casey and J. Wilks, The friction of diamond sliding on polished cube faces of diamond, *J. Phys. D. (Appl. Phys.) 6, 15*: 1772–1781 (1973).

9.32   I.P. Hayward, Friction and wear properties of diamond and diamond coatings, *Surf. Coat. Technol. 49*: 554–559 (1991).

9.33   I.P. Hayward and I.L. Singer, Tribological behaviour of diamond coatings, *Second International Conference on New Diamond Science and Technology*, Materials Research Society, Pittsburgh, PA, 1991, pp. 785–789.

9.34   I.P. Hayward, I.L. Singer, and L.E. Seitzman, Effect of roughness on the friction of diamond on CVD diamond coatings, *Wear 157, 2*: 215–227 (1992).

9.35   D. Crompton, W. Hirst, and M.G.W. Howse, The wear of diamond, *Proc. R. Soc. London A 333-1595*: 435–454 (1973).

9.36   K. Miyoshi, R.L.C. Wu, and A. Garscadden, Friction and wear of diamond and diamondlike coatings, *Surf. Coat. Technol. 54–55, 1–3*: 428–434 (1992).

9.37   Z. Feng, Y. Tzeng, and J.E. Field, Friction of diamond on diamond in ultra-high-vacuum and low-pressure environments, *J. Phys. D. (Appl. Phys.) 25, 10*: 1418–1424 (1992).

9.38   F.P. Bowden and A.E. Hanwell, The friction of clean crystal surfaces, *Proc. R. Soc. London A 295*: 233–243 (1966).

9.39   M.T. Dugger, D.E. Peebles, and L.E. Pope, Counterface material and ambient atmosphere: role in the tribological performance of diamond films, *Surface Science Investigations in Tribology* (Y.-W. Chung, A.M. Homola, and G.B. Street, eds.), American Chemical Society, Washington, DC, 1992, p. 72.

Chapter 10
Surface Design and Engineering Toward Wear-Resistant, Self-Lubricating Diamond Films and Coatings

10.1 Introduction

High tribological reliability is of crucial importance in operating the many interacting surfaces that are in relative motion in mechanical systems [10.1]. The goals of tribological research and development are to reduce the adhesion, friction, and wear of mechanical components; to prevent their failure; and to provide long, reliable component life through the judicious selection of materials, coatings, surface modifications and treatments, operating parameters, and lubricants.

A notable amount of research effort has been put into fundamental studies of the tribological behavior of coatings. In recent years the increasing potential for the use of diamond films and diamondlike films as tribological coatings in mechanical systems has focused attention on these coating materials [10.2]. Tribological studies have been conducted with diamond and related coatings to understand better how the physical and chemical properties of these coatings will affect their behavior when in contact with themselves, ceramics, polymers, and metals [10.3–10.5].

Three surface design, surface engineering, and tribology studies have shown that the friction and wear of CVD diamond are significantly reduced in ultrahigh vacuum. This paper discusses the results of those studies: first, the friction mechanisms of clean diamond surfaces; second, the solid lubrication mechanism and the surface design of diamond surfaces; and finally, the actual tribological properties of the modified diamond surfaces and the selected materials couple. How surface modification and the selected materials couple (particularly the diamond–cubic boron nitride couple) improved the tribological functionality of coatings, giving low coefficient of friction and good wear resistance, is explained.

10.2 Friction Mechanism of Diamond Surface

10.2.1 General Friction Mechanism

The classical Bowden and Tabor model for sliding friction [10.6, 10.7], in its simple form, assumes that the friction force arises from two contributing sources. First, an adhesion force is developed at the real area of contact between the surfaces (the asperity junction). Second, a deformation force is needed to plow or cut the asperities of the harder surface through the softer. The resultant friction force is the sum of the two contributing sources: friction due to adhesion and friction due to deformation and/or fracture [10.6]. The adhesion arises from the attractive forces between the surfaces in contact. This model serves as a starting point for understanding how thin surface films can reduce friction and provide lubrication [10.8–10.10]. It should be realized, however, that one of the contributing sources acts to affect the other on many occasions. In other words, the two sources cannot be treated as strictly independent.

When a smooth diamond flat is brought into contact with a smooth spherical surface of diamond, ceramic, metal, or polymer, the plowing or cutting contribution in friction can be neglected. The friction due to adhesion is then described by the following equation [10.6]:

$$\mu = s\,A/W \tag{10.1}$$

In this equation, μ is the coefficient of friction, s is the shear strength of the real area of contact, A is the real area of contact between the surfaces, and W is the load. Also, in such a basic contact condition, if we consider the total surface energy in the real area of contact, the coefficient of friction can be expressed as a function of $\gamma\!A$

$$\mu = f(\gamma\!A) \tag{10.2}$$

Here $\gamma\!A$ is the total surface energy in the real area of contact [10.11, 10.12]. To reduce friction and to provide lubrication, therefore, the shear strength s, the real area of contact A, and the surface energy γ must be minimized.

10.2.2 Specific Friction Mechanism

Because diamond has tetrahedral, covalent bonds between each carbon atom and its four nearest neighbors, the free surface may expose dangling bonds. Such a free surface has high surface energy γ, which is associated with dangling bond formation. When an atomically clean diamond surface contacts an atomically clean surface of counterpart material, the dangling bonds can form strong linkages with bonds on the counterpart surface. Many researchers [e.g., 10.2, 10.3, 10.7, 10.13] have found that atomically clean diamond has high adhesion and friction. For example, if the surfaces of natural diamond and metal are cleaned by argon ion bombardment, the coefficient of friction is higher than 0.4 in an ultra-high-vacuum environment. The coefficient of friction increases with an increase in the total surface energy of the metal in the real area of contact γA. With the argon-sputter-cleaned diamond surface there are probably dangling bonds of carbon ready to link up directly with metal atoms on the argon-sputter-cleaned metal surface. Thus, cleaning the diamond surface provided surface defects, such as dangling bonds, and accordingly high surface energy and enhanced adhesion and shear strength at the interface. The extremely great hardness and high elastic modulus of diamond provided a small real area of contact A. Because A was small but s and γ were large, the coefficient of friction for the argon-sputter-cleaned diamond surface was high in ultrahigh vacuum (Fig. 10.1).

The situation illustrated in Fig. 10.1 applies to sliding contacts of the CVD diamond surface with itself or other materials in ultrahigh vacuum [10.2, 10.6, 10.7, 10.9–10.15]. Without sputter cleaning or heating to high temperature in a vacuum, a contaminant surface film is adsorbed on the CVD diamond surface. The contaminant surface film can be removed when it repeatedly slides over the same track of counterpart material in vacuum. Then, a fresh, clean diamond surface contacts a clean surface of counterpart material, and a strong bond forms between the two materials. As a result the coefficient of friction for the diamond film becomes

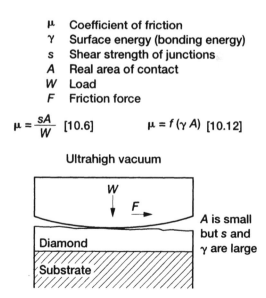

μ Coefficient of friction
γ Surface energy (bonding energy)
s Shear strength of junctions
A Real area of contact
W Load
F Friction force

$$\mu = \frac{sA}{W} \quad [10.6] \qquad \mu = f(\gamma A) \quad [10.12]$$

Ultrahigh vacuum

W
F
A is small but s and γ are large

Diamond

Substrate

Figure 10.1.—Friction mechanism of clean diamond surface in ultrahigh vacuum.

considerably high. As shown in Chapter 9 (Fig. 9.19), when a contaminant surface film was removed by repeatedly sliding a diamond pin over the same track of a diamond-coated disk in vacuum, the coefficient of friction increased from the initial value μ_I to the equilibrium value μ_F with an increasing number of passes. Figure 9.20 presented the initial and equilibrium coefficients of friction for a diamond pin sliding on various CVD diamond films in vacuum [10.13]. In all cases the equilibrium coefficients of friction (1.5 to 1.8) were greater than the initial coefficients of friction (1.1 to 1.3) regardless of the initial surface roughness of the diamond films. As shown in Fig. 9.21 the wear rate of the CVD diamond films in vacuum did depend on the initial surface roughness of the films, generally increasing with an increase in the initial surface roughness.

10.3 Solid Lubrication Mechanism and Design of Diamond Surface

According to the discussion and understanding described in the previous section, reducing the coefficient of friction requires minimizing the shear strength of the interface, the surface energy, the real area of contact, and the plowing or cutting contribution. Reducing wear generally requires minimizing these factors while

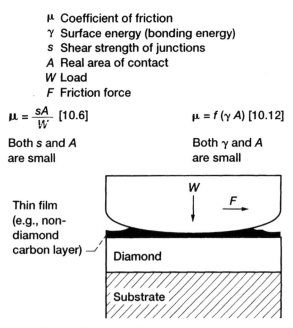

μ Coefficient of friction
γ Surface energy (bonding energy)
s Shear strength of junctions
A Real area of contact
W Load
F Friction force

$\mu = \dfrac{sA}{W}$ [10.6] $\mu = f\,(\gamma\,A)$ [10.12]

Both s and A Both γ and A
are small   are small

Thin film (e.g., non-diamond carbon layer)

W

F

Diamond

Substrate

Figure 10.2.—Lubrication mechanisms.

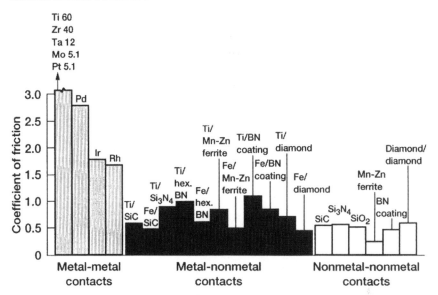

Figure 10.3.—Coefficient of friction for clean solid in sliding contact with itself or other material in ultrahigh vacuum.

maximizing the hardness, strength, and toughness of interacting materials. Toward this end, surface design and engineering can be applied to reduce the coefficient of friction and wear rate of CVD diamond.

Figure 10.2 illustrates how the minimization of the aforementioned factors can be achieved. In other words it shows how the presence of a thin film, such as nondiamond carbon on diamond, reduces the coefficient of friction. In the model presented, the thin film covers the diamond. The thin film can be any material, such as soft metal, polymer, ceramic, or a modified surface layer of the diamond, that has low shear strength or low surface energy. The underlying diamond reduces both the real area of contact and the plowing contribution of the counterpart material; a thin film or a thin surface layer reduces the shear strength and surface energy in the real area of contact. The low coefficient of friction can be attributed to the combination of the low shear strength and the low surface energy of the thin film or the thin surface layer and the small real area of contact resulting from the high elastic modulus and hardness of the underlying diamond.

The coefficient of friction for clean interacting surfaces in ultrahigh vacuum strongly depends on the materials coupled. Figure 10.3 presents examples of the coefficients of friction for clean metal-metal couples, clean metal-nonmetal couples, and clean nonmetal-nonmetal couples measured in ultrahigh vacuum. The judicious selection of counterpart materials can reduce the coefficient of friction for diamond in ultrahigh vacuum.

388

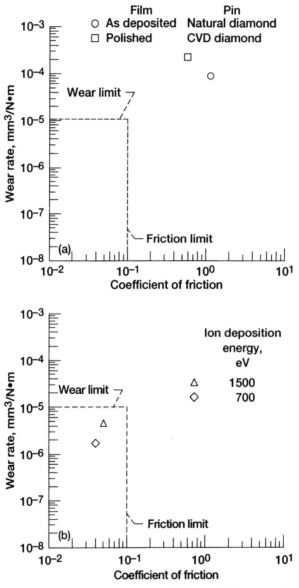

Figure 10.4.—Comparison of coefficient of friction and wear rate.
(a) As-deposited diamond and polished diamond. (b) DLC
films deposited on fine-grain diamond at 1500 and 700 eV.
(c) Carbon-ion-implanted diamond and nitrogen-ion-implanted
diamond.

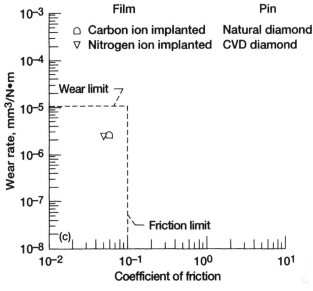

Figure 10.4.—Concluded.

10.4 Surface-Modified Diamond

10.4.1 Thin DLC Film on CVD Diamond

Figure 10.4 presents the steady-state (equilibrium) coefficients of friction and wear rates at room temperature in an ultrahigh vacuum (10^{-7} Pa). For a direct comparison the coefficients of friction and the wear rates were plotted from 10^{-2} to 10^{1} and from 10^{-8} to 10^{-3} mm^3/N·m, respectively. An effective wear-resistant, self-lubricating material must generally have a coefficient of friction less than 0.1 and a wear rate on the order of 10^{-6} mm^3/N·m.

As shown in Fig. 10.4(a) both the as-deposited, fine-grain CVD diamond film and the polished, coarse-grain CVD diamond film had high coefficients of friction (>0.4) and high wear rates (on the order of 10^{-4} mm^3/N·m), which are not acceptable for solid lubrication applications [10.13–10.15].

As shown in Fig. 10.4(b) the thin film of DLC deposited on the as-deposited, fine-grain diamond by the direct impact of an ion beam resulted in low coefficients of friction (<0.1) and low wear rates (on the order of 10^{-6} mm^3/N·m) [10.14–10.16]. The presence of a thin (<1 μm thick), amorphous, nondiamond carbon (hydrogenated carbon) film on CVD diamond greatly decreased the coefficient of friction and the wear rate. DLC on CVD diamond can be an effective wear-resistant, lubricating coating in ultrahigh vacuum.

Note that in dry nitrogen and in humid air (not shown) the coefficient of friction was less than 0.1 and the wear rate was on the order of 10^{-6} mm^3/N·m or less [10.16].

10.4.2 Thin Ion-Implanted Layer of CVD Diamond

The effect of carbon and nitrogen ion implantation on diamond's friction and wear properties was significant (Fig. 10.4(c)). Both carbon-ion-implanted diamond and nitrogen-ion-implanted diamond had low coefficients of friction (<0.1) and low wear rates (on the order of 10^{-6} mm^3/N·m), making them acceptable for solid lubrication applications [10.15, 10.16]. Bombarding diamond films with carbon ions at 60 keV or with nitrogen ions at 35 keV produced a thin, surficial layer of amorphous, nondiamond carbon (<1 μm thick). This surface layer greatly reduced the coefficient of friction and the wear rate in ultrahigh vacuum to values that are acceptable for self-lubricating, wear-resistant applications of CVD diamond films.

Note that in dry nitrogen and in humid air (not shown) the coefficient of friction was less than 0.05 and the wear rate was on the order of 10^{-6} mm^3/N·m [10.15, 10.16].

10.5 Selected Materials Couple

Boron nitride is competing with diamond and silicon carbide in most applications, including friction-reducing coatings. As for diamond a wide variety of synthesis methods are being used, and boron nitride can be grown in many phases. The cubic phase is the most desirable phase for applications [10.17]. Because cubic boron nitride (c-BN), which is chemically and thermally inert, is second only to diamond in hardness, many researchers believe that c-BN films offer great opportunities for wear parts, cutting tool inserts, rotary tools, and dies. The c-BN films are especially valuable for protective coatings on surfaces that come into contact with iron-based materials, where diamond cannot be used because of its high chemical wear due to its aggressive reaction with iron. Therefore, an

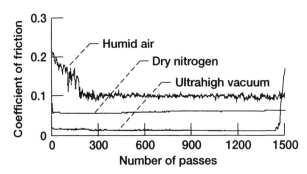

Figure 10.5.—Coefficients of friction for c-BN films in sliding contact with CVD diamond pins in humid air, dry nitrogen, and ultra-high-vacuum environments.

investigation was conducted to examine the friction of c-BN in contact with diamond in ultrahigh vacuum. Reference experiments were also conducted in dry nitrogen and in humid air. The c-BN films (approx. 0.5 μm thick) were synthesized by magnetically enhanced plasma ion plating and formed on silicon {100} wafer substrates [10.18].

Figure 10.5 shows the low coefficients of friction in ultrahigh vacuum for as-deposited c-BN films in sliding contact with CVD diamond pins as a function of the number of passes. This materials combination provided an effective self-lubricating, wear-resistant couple in ultrahigh vacuum at low numbers of passes. However, at approximately 1400 passes the sliding action caused the c-BN film to break down, whereupon the coefficient of friction rapidly increased (Fig. 10.5). The wear rate of this particular c-BN film sliding against a CVD diamond pin was on the order of 10^{-6} mm^3/N·m, but the wear rate of the CVD diamond pin was much lower.

Note that in dry nitrogen and in humid air (not shown) the coefficient of friction remained constant for a long period without breakdown even at 100 000 passes [10.16]. The endurance life of c-BN films was greater in dry nitrogen and in humid air than in ultrahigh vacuum by a factor of 60 or higher.

References

10.1   H.P. Jost, Tribology—origin and future, *Wear 136*: 1–17 (1990).

10.2   K. Miyoshi, Superhard coatings approaches to tribology: properties and applications of CVD diamond, DLC, and c-BN, *Wear and Superhard Coatings 1998*, 1998, pp. 1–15.

10.3   M.T. Dugger, D.E. Peebles, and L.E. Pope, Counterface material and ambient atmosphere: role in the tribological performance of diamond films, *Surface Science Investigations in Tribology: Experimental Approaches* (Y.-W. Chung, A.M. Homola, and G.B. Street, eds.), ACS Symposium Series 485, American Chemical Society, Washington, DC, 1992, pp. 72–102.

10.4   M.N. Gardos, Tribology and wear behavior of diamond, *Synthetic Diamond: Emerging CVD Science and Technology* (K.E. Spear and J.P. Dismukes, eds.), John Wiley & Sons, New York, 1994, pp. 419–502.

10.5   I.P. Hayward, I.L. Singer, and L.W. Seitzmann, The tribological behaviour of diamond coatings, *Proceedings of the Second International Conference on the New Diamond Science and Technology*, Materials Research Society, Pittsburgh, PA, 1991, pp. 785–789.

10.6   F.P. Bowden and D. Tabor, *The Friction and Lubrication of Solids*, Clarendon Press, Oxford, UK, 1958.

10.7   D. Tabor, Adhesion and friction, *Properties of Diamond* (J.E. Field, ed.), Academic Press, New York, 1979, pp. 325–350.

10.8   K. Holmberg and A. Matthews, *Coatings Tribology: Properties, Techniques, and Applications in Surface Engineering*, Tribology Series, 28, Elsevier, Amsterdam, The Netherlands, 1994.

10.9   I.L. Singer, Solid lubrication processes, *Fundamental of Friction: Microscopic and Macroscopic Processes*, Proceedings of NATO Advanced Study Institute (I.L. Singer and H.M. Pollock, eds.), Kluwer Academic Publishers, Dordrecht, The Netherlands, 1992, pp. 237–261.

10.10 I.M. Hutchings, *Tribology: Friction and Wear of Engineering Materials*, CRC Press, Boca Raton, FL, 1992.

10.11   K. Miyoshi, Adhesion in ceramics and magnetic media, *Proceedings of the Fifth International Congress on Tribology* (K. Holmberg, ed.), The Finnish Society for Tribology, Espoo, Finland, 1989, pp. 228–233.

10.12   K. Miyoshi, Adhesion, friction, and wear behavior of clean metal-ceramic couples, *Proceedings of the International Tribology Conference, Yokohama 1995*, Vol. III, Japanese Society of Tribologists, Tokyo, Japan, 1995, pp. 1853–1858.

10.13   K. Miyoshi, R.L.C. Wu, A. Garscadden, P.N. Barnes, and H.E. Jackson, Friction and wear of plasma-deposited diamond films, *J. Appl. Phys. 74, 7*: 4446–4454 (1993).

10.14   K. Miyoshi, Wear-resistant, self-lubricating surfaces of diamond coatings, *Proceedings of the Third International Conference on the Applications of Diamond Films and Related Materials*, NIST Special Publications Issue 885, 1995, pp. 493–500.

10.15   K. Miyoshi, Lubrication by diamond and diamondlike carbon coatings, *J. Tribol. 120*: 379–384 (1998).

10.16   K. Miyoshi, M. Murakawa, S. Watanabe, S. Takeuchi, S. Miyake, and R.L.C. Wu, CVD diamond, DLC, and c-BN coatings for solid film lubrication, *Tribol. Lett. 5*: 123–129 (1998).

10.17   J.J. Pouch and S.A. Alterovitz, eds., *Synthesis and Properties of Boron Nitride*, Materials Science Forum, Trans Tech Publications, Aedermannsdorf, Switzerland, Vols. 54 & 55, 1990.

10.18   S. Watanabe, D.R. Wheeler, P.S. Abel, K.W. Street, K. Miyoshi, M. Murakawa, and S. Miyake, Surface chemistry, microstructure, and tribological properties of cubic boron nitride films, NASA TM–113163, 1998.

Index

Printed and bound by CPI Group (UK) Ltd, Croydon, CR0 4YY

23/10/2024

01778239-0004